READINGS FOR CALCULUS

Underwood Dudley, Editor

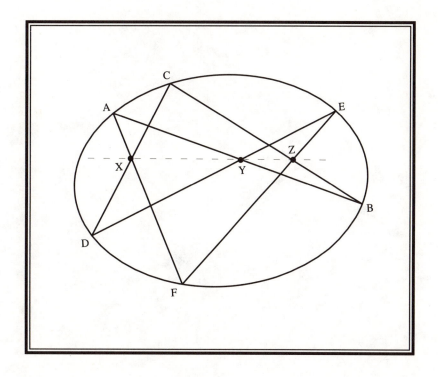

Volume 5

READINGS FOR CALCULUS

Underwood Dudley, Editor

A Project of
The Associated Colleges of the Midwest and
The Great Lakes Colleges Association

Supported by the National Science Foundation
A. Wayne Roberts, Project Director

MAA Notes Volume 31

Published and Distributed by
The Mathematical Association of America

MAA Notes and Reports Series

The MAA Notes and Reports Series, started in 1982, addresses a broad range of topics and themes of interest to all who are involved with undergraduate mathematics. The volumes in this series are readable, informative, and useful, and help the mathematical community keep up with developments of importance to mathematics.

MAA Notes

1. Problem Solving in the Mathematics Curriculum, *Committee on the Teaching of Undergraduate Mathematics,* a subcommittee of the Committee on the Undergraduate Program in Mathematics, *Alan H. Schoenfeld,* Editor

2. Recommendations on the Mathematical Preparation of Teachers, *Committee on the Undergraduate Program in Mathematics, Panel on Teacher Training.*

3. Undergraduate Mathematics Education in the People's Republic of China, *Lynn A. Steen,* Editor.

5. American Perspectives on the Fifth International Congress on Mathematical Education, *Warren Page,* Editor.

6. Toward a Lean and Lively Calculus, *Ronald G. Douglas,* Editor.

8. Calculus for a New Century, *Lynn A. Steen,* Editor.

9. Computers and Mathematics: The Use of Computers in Undergraduate Instruction, *Committee on Computers in Mathematics Education, D. A. Smith, G. J. Porter, L. C. Leinbach, and R. H. Wenger,* Editors.

10. Guidelines for the Continuing Mathematical Education of Teachers, *Committee on the Mathematical Education of Teachers.*

11. Keys to Improved Instruction by Teaching Assistants and Part-Time Instructors, *Committee on Teaching Assistants and Part-Time Instructors, Bettye Anne Case,* Editor.

13. Reshaping College Mathematics, *Committee on the Undergraduate Program in Mathematics, Lynn A. Steen,* Editor.

14. Mathematical Writing, by *Donald E. Knuth, Tracy Larrabee, and Paul M. Roberts.*

15. Discrete Mathematics in the First Two Years, *Anthony Ralston,* Editor.

16. Using Writing to Teach Mathematics, *Andrew Sterrett,* Editor.

17. Priming the Calculus Pump: Innovations and Resources, *Committee on Calculus Reform and the First Two Years,* a subcomittee of the Committee on the Undergraduate Program in Mathematics, *Thomas W. Tucker,* Editor.

18. Models for Undergraduate Research in Mathematics, *Lester Senechal,* Editor.

19. Visualization in Teaching and Learning Mathematics, *Committee on Computers in Mathematics Education, Steve Cunningham and Walter S. Zimmermann,* Editors.

20. The Laboratory Approach to Teaching Calculus, *L. Carl Leinbach et al.,* Editors.

21. Perspectives on Contemporary Statistics, *David C. Hoaglin and David S. Moore,* Editors.

22. Heeding the Call for Change: Suggestions for Curricular Action, *Lynn A. Steen,* Editor.

23. Statistical Abstract of Undergraduate Programs in the Mathematical Sciences and Computer Science in the United States: 1990–91 CBMS Survey, *Donald J. Albers, Don O. Loftsgaarden, Donald C. Rung, and Ann E. Watkins.*

24. Symbolic Computation in Undergraduate Mathematics Education, *Zaven A. Karian,* Editor.

25. The Concept of Function: Aspects of Epistemology and Pedagogy, *Guershon Harel and Ed Dubinsky,* Editors.

26. Statistics for the Twenty-First Century, *Florence and Sheldon Gordon,* Editors.

27. Resources for Calculus Collection, Volume 1: Learning by Discovery: A Lab Manual for Calculus, *Anita E. Solow,* Editor.

28. Resources for Calculus Collection, Volume 2: Calculus Problems for a New Century, *Robert Fraga,* Editor.

29. Resources for Calculus Collection, Volume 3: Applications of Calculus, *Philip Straffin,* Editor.

30. Resources for Calculus Collection, Volume 4: Problems for Student Investigation, *Michael B. Jackson and John R. Ramsay,* Editors.

31. Resources for Calculus Collection, Volume 5: Readings for Calculus, *Underwood Dudley,* Editor.

MAA Reports

1. A Curriculum in Flux: Mathematics at Two-Year Colleges, *Subcommittee on Mathematics Curriculum at Two-Year Colleges,* a joint committee of the MAA and the American Mathematical Association of Two-Year Colleges, *Ronald M. Davis,* Editor.

2. A Source Book for College Mathematics Teaching, *Committee on the Teaching of Undergraduate Mathematics, Alan H. Schoenfeld,* Editor.

3. A Call for Change: Recommendations for the Mathematical Preparation of Teachers of Mathematics, *Committee on the Mathematical Education of Teachers, James R. C. Leitzel,* Editor.

4. Library Recommendations for Undergraduate Mathematics, *CUPM ad hoc Subcommittee, Lynn A. Steen,* Editor.

5. Two-Year College Mathematics Library Recommendations, *CUPM ad hoc Subcommittee, Lynn A. Steen,* Editor.

These volumes may be ordered from the Mathematical Association of America, 1529 Eighteenth Street, NW, Washington, DC 20036. 202-387-5200   FAX 202-265-2384

Teachers may reproduce these modules for their students and they may be modified to suit particular classroom needs. However, the modules remain the property of the Mathematical Association of America and the collection may not be used for commercial gain.

Second Printing

© 1993 by the Mathematical Association of America

ISBN 0-88385-087-7

Library of Congress Catalog Number 92-62283

Printed in the United States of America

Current Printing

10 9 8 7 6 5 4 3 2 1

INTRODUCTION
RESOURCES FOR CALCULUS COLLECTION

Beginning with a conference at Tulane University in January, 1986, there developed in the mathematics community a sense that calculus was not being taught in a way befitting a subject that was at once the culmination of the secondary mathematics curriculum and the gateway to collegiate science and mathematics. Far too many of the students who started the course were failing to complete it with a grade of C or better, and perhaps worse, an embarrassing number who did complete it professed either not to understand it or not to like it, or both. For most students it was not a satisfying culmination of their secondary preparation, and it was not a gateway to future work. It was an exit.

Much of the difficulty had to do with the delivery system: classes that were too large, senior faculty who had largely deserted the course, and teaching assistants whose time and interest were focused on their own graduate work. Other difficulties came from well intentioned efforts to pack into the course all the topics demanded by the increasing number of disciplines requiring calculus of their students. It was acknowledged, however, that if the course had indeed become a blur for students, it just might be because those choosing the topics to be presented and the methods for presenting them had not kept their goals in focus.

It was to these latter concerns that we responded in designing our project. We agreed that there ought to be an opportunity for students to discover instead of always being told. We agreed that the availability of calculators and computers not only called for exercises that would not be rendered trivial by such technology, but would in fact direct attention more to ideas than to techniques. It seemed to us that there should be explanations of applications of calculus that were self-contained, and both accessible and relevant to students. We were persuaded that calculus students should, like students in any other college course, have some assignments that called for library work, some pondering, some imagination, and above all, a clearly reasoned and written conclusion. Finally, we came to believe that there should be available to students some collateral readings that would set calculus in an intellectual context.

We reasoned that the achievement of these goals called for the availability of new materials, and that the uncertainty of just what might work, coupled with the number of people trying to address the difficulties, called for a large collection of materials from which individuals could select. Our goal was to develop such materials, and to encourage people to use them in any way they saw fit. In this spirit, and with the help of the Notes editor and committee of the Mathematical Association of America, we have produced five volumes of materials that are, with the exception of volume V where we do not hold original copyrights, meant to be in the public domain.

We expect that some of these materials may be copied directly and handed to an entire class, while others may be given to a single student or group of students. Some will provide a basis from which local adaptations can be developed. We will be pleased if authors ask for permission, which we expect to be generous in granting, to incorporate our materials into texts or laboratory manuals. We hope that in all of these ways, indeed in any way short of reproducing substantial segments to

93

sell for profit, our material will be used to greatly expand ideas about how the calculus might be taught.

Though I as Project Director never entertained the idea that we could write a single text that would be acceptable to all 26 schools in the project, it was clear that some common notion of topics essential to any calculus course would be necessary to give us direction. The task of forging a common syllabus was managed by Andy Sterrett with a tact and efficiency that was instructive to us all, and the product of this work, an annotated core syllabus, appears as an appendix in Volume 1. Some of the other volumes refer to this syllabus to indicate where, in a course, certain materials might be used.

This project was situated in two consortia of liberal arts colleges, not because we intended to develop materials for this specific audience, but because our schools provide a large reservoir of classroom teachers who lavish on calculus the same attention a graduate faculty might give to its introductory analysis course. Our schools, in their totality, were equipped with most varieties of computer labs, and we included in our consortia many people who had become national leaders in the use of computer algebra systems.

We also felt that our campuses gave us the capability to test materials in the classroom. The size of our schools enables us to implement a new idea without cutting through the red tape of a larger institution, and we can just as quickly reverse ourselves when it is apparent that what we are doing is not working. We are practiced in going in both directions. Continual testing of the materials we were developing was seen as an integral part of our project, an activity that George Andrews, with the title of Project Evaluator, kept before us throughout the project.

The value of our contributions will now be judged by the larger mathematical community, but I was right in thinking that I could find in our consortia the great abundance of talent necessary for an undertaking of this magnitude. Anita Solow brought to the project a background of editorial work and quickly became not only one of the editors of our publications, but also a person to whom I turned for advice regarding the project as a whole. Phil Straffin, drawing on his association with UMAP, was an ideal person to edit a collection of applications, and was another person who brought editorial experience to our project. Woody Dudley came to the project as a writer well known for his witty and incisive commentary on mathematical literature, and was an ideal choice to assemble a collection of readings.

Our two editors least experienced in mathematical exposition, Bob Fraga and Mic Jackson, both justified the confidence we placed in them. They brought to the project an enthusiasm and freshness from which we all benefited, and they were able at all points in the project to draw upon an excellent corps of gifted and experienced writers. When, in the last months of the project, Mic Jackson took an overseas assignment on an Earlham program, it was possible to move John Ramsay into Mic's position precisely because of the excellent working relationship that had existed on these writing teams.

The entire team of five editors, project evaluator and syllabus coordinator worked together as a harmonious team over the five year duration of this project. Each member, in turn, developed a group of writers, readers, and classroom users as necessary to complete the task. I believe my chief contribution was to identify and bring these talented people together, and to see that they were supported both financially and by the human resources available in the schools that make up two remarkable consortia.

A. Wayne Roberts
Macalester College
1993

THE FIVE VOLUMES OF THE
RESOURCES FOR CALCULUS COLLECTION

1. Learning by Discovery: A Lab Manual for Calculus
Anita E. Solow, editor

The availability of electronic aids for calculating makes it possible for students, led by good questions and suggested experiments, to discover for themselves numerous ideas once accessible only on the basis of theoretical considerations. This collection provides questions and suggestions on 26 different topics. Developed to be independent of any particular hardware or software, these materials can be the basis of formal computer labs or homework assignments. Although designed to be done with the help of a computer algebra system, most of the labs can be successfully done with a graphing calculator.

2. Calculus Problems for a New Century
Robert Fraga, editor

Students still need drill problems to help them master ideas and to give them a sense of progress in their studies. A calculator can be used in many cases, however, to render trivial a list of traditional exercises. This collection, organized by topics commonly grouped in sections of a traditional text, seeks to provide exercises that will accomplish the purposes mentioned above, even for the student making intelligent use of technology.

3. Applications of Calculus
Philip Straffin, editor

Everyone agrees that there should be available some self-contained examples of applications of the calculus that are tractable, relevant, and interesting to students. Here they are, 18 in number, in a form to be consulted by a teacher wanting to enrich a course, to be handed out to a class if it is deemed appropriate to take a day or two of class time for a good application, or to be handed to an individual student with interests not being covered in class.

4. Problems for Student Investigation
Michael B. Jackson and John R. Ramsay, editors

Calculus students should be expected to work on problems that require imagination, outside reading and consultation, cooperation, and coherent writing. They should work on open-ended problems that admit several different approaches and call upon students to defend both their methodology and their conclusion. Here is a source of 30 such projects.

5. Readings for Calculus
Underwood Dudley, editor

Faculty members in most disciplines provide students in beginning courses with some history of their subject, some sense not only of what was done by whom, but also of how the discipline has contributed to intellectual history. These essays, appropriate for duplicating and handing out as collateral reading aim to provide such background, and also to develop an understanding of how mathematicians view their discipline.

ACKNOWLEDGEMENTS

Besides serving as editors of the collections with which their names are associated, Underwood Dudley, Bob Fraga, Mic Jackson, John Ramsay, Anita Solow, and Phil Straffin joined George Andrews (Project Evaluator), Andy Sterrett (Syllabus Coordinator) and Wayne Roberts (Project Director) to form a steering committee. The activities of this group, together with the writers' groups assembled by the editors, were supported by two grants from the National Science Foundation.

The NSF grants also funded two conferences at Lake Forest College that were essential to getting wide participation in the consortia colleges, and enabled member colleges to integrate our materials into their courses.

The projects benefited greatly from the counsel of an Advisory Committee that consisted of Morton Brown, Creighton Buck, Jean Callaway, John Rigden, Truman Schwartz, George Sell, and Lynn Steen.

Macalester College served as the grant institution and fiscal agent for this project on behalf of the schools of the Associated Colleges of the Midwest (ACM) and Great Lakes Colleges Association (GLCA) listed below.

ACM	GLCA
Beloit College	Albion College
Carleton College	Antioch College
Coe College	Denison University
Colorado College	DePauw University
Cornell College	Earlham College
Grinnell College	Hope College
Knox College	Kalamazoo College
Lake Forest College	Kenyon College
Lawrence University	Oberlin College
Macalester College	Ohio Wesleyan University
Monmouth College	Wabash College
Ripon College	College of Wooster
St. Olaf College	
University of Chicago	

I would also like to thank Stan Wagon of Macalester College for providing the cover image for each volume in the collection.

TABLE OF CONTENTS

About Mathematics

PREFACE

This volume contains readings meant to be useful in a calculus course. There are thirty-six selections, some old and some new, with six appearing in print for the first time.

The selections can be used in a variety of ways. Instructors may read them and use anything from them in the classroom, with or without giving credit. Students may be informed that they exist and be encouraged to browse through them. Students may be given assignments to read one or more of them. In the tradition of mathematics texts, each selection has or has had added to it a few questions and exercises. Exercises are mathematical, based on the material in the selection, while questions ask for opinions or interpretations. Either may be used to help insure that the reading assignment has been carried out. The selections can be used as bases for class discussion, not something that is typically done in calculus classes, but which could prove valuable in stimulating and maintaining student interest as well as adding variety to the classroom routine. They can be used to make writing assignments. There is no reason why students in mathematics classes cannot write papers as they do in other classes, and students should know mathematics is something that can be written about.

The content of the readings is varied. It was not the intent that this volume be a collection of supplementary topics in calculus. That purpose can be served, and served well, by the other four volumes in this series. This volume was intended to give its readers opportunities to see how calculus and mathematics fit into history and society. Some selections are not tied specifically to calculus but deal with mathematics in general. The reason for including them is that for all too many of our students, their calculus course is the last contact they will have with mathematics and it is our last chance to make them aware of the nature, and the glories, of our discipline. For most of our students, the technical details of calculus will sooner or later evaporate, but the ideas that they encounter in calculus class may stay with them longer, even for a lifetime.

Thirteen of the readings are on historical topics. This is the largest number in any category, the reason being that it is the category in which ignorance is most widespread and most in need of eradication. Newton is one of the most important persons in all of human history, and no student of calculus should be allowed to think that he lived in the nineteenth century, or the fifteenth. "Learning Calculus" contains selections that are closest to being supplementary class material, but none has been used commonly, or at all, for that purpose. "Calculus in Society" contains readings that show, in one way or another, how calculus touches the world outside the classroom. The selections in "About Mathematics" range over many topics that will almost certainly be new to calculus students.

A common objection to including readings in a calculus class is that there is no time for them. Yes, the syllabus is crowded and no matter how much time we spend on a topic our students could profit from more, but perhaps time should be made for a reading or two, or three. Consider: what is more likely to affect students for the good in the long run, one more lecture (no matter how brilliantly prepared and delivered) on a topic that will be soon forgotten by its audience, or something new and different that students have never before encountered? Perhaps we can let the $\tan(\theta/2)$ substitution go this semester, even though the class will therefore forever be ignorant of how to calculate $\int d\theta/(2 + \cos\theta)$, and consider instead who Newton was and what he did, or why mathematics is beautiful.

A common student objection to readings is that it is the business of a calculus teacher to explain how to get the answers to the problems in the textbook, and time not spent doing that is wasted. "Will this be on the test?" is one way this objection is expressed. The answer could well be, "Yes," and the next test could ask for the date of Euler's birth correct to within 75 years. It could just as well be, "No, but read it anyway. I know best what is good for you." If some students fail to do the reading, or do it half-heartedly, that is their loss. On the other hand, there is a chance that some students will find a reading interesting, or even be struck by something in it with the force of revelation and have their horizons forever widened. Their benefit may outweigh the loss to the rest of the class of one more lecture, or another five problems solved at the board.

Objections notwithstanding, the readings were selected because they were enlightening, interesting, fascinating, informative, entertaining, or all five put together. They were meant to be enjoyable, or at the least useful. I hope that you enjoy them, and that you can use some of them.

Most of the selections have some introductory remarks by the editor. These are set off by horizontal lines and are printed in one column. Questions and exercises at the end of selections are also set off by a horizontal line if they are by the editor.

There follows a list of the selections, with brief remarks on their contents, arranged in order of prerequisites as they occur in the syllabus that appears in full in

Volume I. That a selection has no prerequisites does not imply that it should be read early in a course, only that it is not necessary for students to have mastered any material in order to understand it. Also, that a selection comes late in the list does not mean that it could not be read earlier, with students being told that parts of it could be skipped over.

Introduction

Galovich, Background for Calculus
Comments on ancient Greek and medieval views of the nature of the physical world and the place of mathematics in it.

Winger, Mathematical Objectives
Why study mathematics? It is a question many students have never asked themselves, and they ought to consider it.

Ecclesiastes
Why study *anything*? The comment above applies here as well.

Gårding, The Sociology of Mathematics
An elegant essay showing the place of mathematics in society.

Davis and Hersh, Mathematics as a Social Filter
One answer to the question of why calculus is studied, namely to exclude those who fail from other things.

Spanier, Solving Equations Is Not Solving Problems
How to prepare for a career in applied mathematics.

McMillan, Applied Mathematics in Engineering
On the place of mathematics, and the applied mathematician, in industry.

Bartlett's Familiar Quotations
Quotations about mathematics, many by nonmathematicians, showing diverse views.

Moritz, Memorabilia Mathematica
Quotations about mathematics, mostly by mathematicians, showing diverse views, many of which will be new to students.

Fadiman, Anecdotes
The anecdotes, meant for a nonmathematical audience, have no mathematical content but perhaps should not be read until the names of the anecdotees have come up.

Feynman, The Relation of Mathematics to Physics
The author argues that it is impossible to understand the physical world without mathematics.

Halmos, Mathematics as a Creative Art
In this extremely readable essay, the author argues that mathematics is an art, an idea that will come as a surprise to many students.

Gardner, Nine More Problems
Examples of recreational mathematics problems, with solutions. That some people find recreation and pleasure in mathematics may not be known by all students.

Functions and Graphs

Jones, The Fabulous Fourteen of Calculus
Why the transcendental functions are different from polynomials.

The Derivative

Newton, Principia Mathematica
Newton on differentials, allowing students to see how the old differs from the new and perhaps understand the new better.

The World's First Calculus Textbook
A sketch of the contents of l'Hôpital's 1696 work, showing that calculus has changed since then.

Osen, The "Witch" of Agnesi
A sketch of the life of Maria Agnesi with essentially no mathematical content.

Muir, Leonhard Euler
A sketch of the life of Euler, also with essentially no mathematical content.

Lieber and Lieber, The Education of T. C. Mits
The authors, writing for a general audience, argue that calculus should be known by everyone. Derivatives are mentioned.

Extreme Values

Antiderivatives and Differential Equations

Gårding, the Heroic Century
A quick survey of mathematical achievements in the 17th century.

Jones and Root, Impossibility

Why cubes cannot be duplicated and why e^{x^2} has no elementary antiderivative.

Huntley, Beauty in Mathematics
Though the selection contains a dy/dx and a $\int$, it was meant for a general audience and can be read at any time. It tries to show how mathematics is beautiful.

The Definite Integral

Aaboe, Episodes from the Early History of Mathematics
A survey of the works of Archimedes, emphasizing his anticipations of calculus, which could be read at any time.

Eves, Slicing It Thin
An explanation of how Cavalieri's Principle can be used to determine areas and volumes.

Simmons, Fermat (1601-1665)
The achievements of Fermat in analytic geometry and calculus.

Kline, The Creation of the Calculus
The contributions of the other discoverer of calculus, Leibniz.

Calculus Notation
On the early notations for derivatives and integrals and their evolution. Useful for demonstrating to students that the world of mathematics is not static.

Menger, Are Variables Necessary in Calculus?
A failed attempt at reform of notation, another illustration that calculus is not unchanging. Also, seeing ideas in other notations can deepen students' understanding of them.

Cipra, Misteaks
A survey of common errors, addressed to students. It will be most appreciated after integration by partial fractions has been encountered.

Hull, Infinity: Limits and Integration
A brief discussion of the Lebesgue integral, which can be contrasted to the Riemann integral.

Bergamini, Mastering the Mysteries of Movement
The ideas of calculus, meant for a mass audience, and hence readable at any time.

Hull, Calculus and Opinions
Confidence intervals applied to opinion polls, best read by students acquainted with probability.

Sequences and Series of Numbers

Sequences and Series of Functions

Roy, Anticipations of Calculus in Medieval Indian Mathematics
A description of the Indian discovery of the series for the arctangent and thus of $\pi/4$ that has considerable mathematical content.

Eves, Transition to the Twentieth Century
Some of the mathematical figures of the eighteenth century and some of their achievements.

Fallacies
A collection of fallacies, algebraic and up. Much of it can be read earlier.

Series Solution of Differential Equations

The Integral in R_2 and R_3

The Derivative in Two and Three Variables

Bell, On the Seashore
This account of the life and works of Newton contains a partial differential equation, but it was meant for a general audience and could be read at any time after integrals have been introduced.

Underwood Dudley
DePauw University
1992

SOURCES

Aaboe, Asger. From *Episodes from the Early History of Mathematics* by Asger Aaboe. Copyright © 1964 by Yale University. Reprinted by permission of Random House, Inc.

Bartlett's *Familiar Quotations*. Fifteenth edition edited by Emily Morison Beck. Copyright © 1980 by Little, Brown and Company.

Bell, Eric Temple. From *Men of Mathematics*, pages 90-116. Copyright © 1937 by E. T. Bell renewed © 1965 by Taine T. Bell. Reprinted by permission of Simon & Schuster.

Bergamini, David. From *Life Science Library: Mathematics* by David Bergamini and the editors of Time-Life Books, copyright © 1980 by Time-Life Books Inc.

Cajori, Florian. From *A History of Mathematical Notation*, copyright © 1929 by The Open Court Publishing Co.

Cipra, Barry. From *Misteaks*, copyright © 1989 by Academic Press, Inc.

Davis, Philip J. and Rueben Hersh. From *Descartes' Dream*, copyright © 1986 by Harcourt Brace Jovanovich, Inc.

Ecclesiastes. From the King James Version of the Bible.

Eves, Howard. From *Great Moments in Mathematics (Before 1650)*, copyright © 1980 by the Mathematical Association of America. (Cavalieri's Principles.)

Eves, Howard. From *Introduction to the History of Mathematics*, fifth edition © 1983 by Saunders College Publishers, a division of Harcourt Brace Jovanovich. (Transition to the Twentieth Century.)

Fadiman, Clifton, general editor. From *The Little, Brown Book of Anecdotes*, copyright © 1985 by Little, Brown and Company.

Feynman, Richard. From *The Character of Physical Law* by Richard Feynman, copyright © 1965 by Richard Feynman. MIT Press Paperback Edition 1967.

Galovich, Steven. Published for the first time in this collection.

Gårding, Lars. From *Encounters with Mathematics*, copyright © 1977 by Springer-Verlag New York, Inc.

Gardner, Martin. From *The 2nd Scientific American Book of Mathematical Puzzles and Diversions*, copyright © 1961, 1987 by Martin Gardner. University of Chicago Press edition 1987.

Halmos, P. R. From *Selecta: Expository Writing*, copyright © 1983 by Springer-Verlag New York, Inc.

Hull, David. Published for the first time in this collection.

Huntley, H. L. From *The Divine Proportion*, copyright © 1970 by Dover Publications, Inc.

Jones, Charles A. Published for the first time in the collection.

Jones, Charles A. and Nathan Root. Published for the first time in this collection.

Kline, Morris. From *Mathematical Thought from Ancient to Modern Times* by Morris Kline. Copyright © 1972 by Morris Kline. Reprinted by permission of Oxford University Press.

L'Hôpital, Marquis de. From *Analyse des infiniment petits*, Paris, 1781.

Lieber, Lillian R. and Hugh G. Reprinted from *The Education of T. C. Mits*, drawings by Hugh Gray Lieber, words by Lillian R. Lieber, by permission of W. W. Norton & Company, Inc. Copyright © 1942, 1944 by H. G. L. R. Lieber. Copyright renewed 1972 by Lillian R. Lieber.

Maxwell, E. A. From *Fallacies in Mathematics*, copyright © 1959 by the Cambridge University Press.

McMillan, Brockway. From *The Spirit and Uses of the Mathematical Sciences*, edited by Thomas L. Saaty and F. Joachim Weyl, copyright © 1969 by McGraw-Hill, Inc.

BACKGROUND FOR CALCULUS

by Steven Galovich

History is a good thing. There are many reasons for that, and the following selection brings two of them to mind. The first is that history tells us that the world changes, sometimes slowly and sometimes violently, but that it always changes and it changes a lot. The world of one thousand years ago could almost as well have been in another galaxy for all that we have in common with it. This is a good thing to know. Animals do not know it, nor do children: they accept the world as it is with no questions asked. As far as they are concerned, it has never changed, nor will it ever. Some grown-up people, those who have not had the benefit of history, do not know it either. On the one hand, it is a sign of good mental health to have adjusted to your environment, but on the other, what may be good for one person may not be good for humanity as a whole. If you do not know how changeable the world is, you may not work to change it. This would be fine if we were surrounded by perfection, but that is not the case. There are many ways in which the world ought to be changed and the more people who are aware that change can take place the better. Mathematics has changed too, and will continue to change. Just look at history.

The second reason is that the record of the human race is not one of continued progress from lowly beginnings to what we like to think is the final culmination of evolution, namely us. It is a common view that history chronicles uninterrupted improvement, but that view is false. Mathematics provides an example. The ancient Greeks did amazing things in geometry, but after a time their progress stopped. From our point of view, they went as far up their blind alley of mathematics as they could, and before mathematics could go further it had to back up and make a fresh start in another direction. This took time, more than one thousand years. One thousand years! This shows that there is no inevitability in our mathematics and there is no necessity for it. The race survived for thousands of years without any mathematics beyond arithmetic and could have gone on for thousands of years more. We are just lucky. Progress is not inevitable, nor was it inevitable that society should be arranged exactly as it is now. The past is fixed, but the future is not. History tells us so. It's a good thing to listen to history.

The philosophers of ancient Greece sought rational explanations of natural phenomena. As far back as the fifth century B. C., they considered the question: What is the essence of things? One answer was provided by Thales, who claimed that water was the primal element of all matter. Somewhat later Anaximenes argued that air is the basic element. Each of these theories is materialistic in the sense that all matter is seen to be composed of a specific substance or specific substances. But with Pythagoras, a new viewpoint arose. Pythagoras's ideas are based on the discovery that if one plucks two taut strings whose lengths are in a 2:1 ratio, then the sounds emitted differ by one octave (the shorter string produces the higher note). If the lengths are in a 3:2 (or 4:3) ratio, then the interval between the notes is a fifth (or fourth). Thus, the sounds produced by the strings can be "explained" by the ratios of the lengths. This example illustrates the Pythagorean principle that "all things are number."

In his summary of early Greek thought, Aristotle discusses this theme (*Metaphysics* Book I, 986A):

> At the same time, however, and even earlier the so-called Pythagoreans applied themselves to mathematics ... and through studying it they came to believe that its principles are the principles of everything. And since numbers are by nature first among these principles they fancied that they could detect in numbers, to a greater extent than in fire and earth and water, many analogues of what is and comes into being—such and such a property of number being justice and such and such soul or mind, another opportunity, and similarly, more or less, with all the rest—and since they

saw further that the properties and ratios of the musical scales are based on numbers, and since it seemed clear that all other things have their whole nature modelled upon numbers, and that numbers are the ultimate things in the whole physical universe, they assumed the elements of numbers to be elements of everything, and the whole universe to be a proportion or number.

Whatever the justification, the Pythagorean doctrine has been paraphrased as follows: "All things have form, all things *are* form, and all form can be defined by number." (A. Koestler, *The Sleepwalkers*, Ch. II.)

After the Pythagoreans, the materialistic doctrine was revived by several thinkers. For example, Democritus proposed an atomic theory of matter. In addition, Empedocles put forth the notion that all matter is composed of four basic elements—earth, air, fire, and water. In the *Timaeus*, Plato combined the materialism of Empedocles with the formalism of Pythagoras. Without doubt, however, Plato's heart is with Pythagoras.

Following Empedocles, Plato asserts that there are four basic elements—earth, air, fire, and water—which, in various combinations, constitute all substances. But Plato goes beyond Empedocles. He corresponds to each element one of the five regular solids: earth-cube; air-octahedron; fire-tetrahedron; and water-icosahedron. (The fifth solid, the dodecahedron, was used by the gods "for arranging the constellations on the whole heavens." Each of the four solids can be decomposed into two "basic" types of triangles—the isosceles right triangle and the 30°–60°–90° right triangle. According to Plato, the various combinations of the four basic elements can be explained by the way the basic triangles can be combined.

> The presence in each kind of further varieties is due to the way in which the two basic triangles were put together. ... So their combination with themselves and with each other give rise to endless complexities, which anyone who is to give a likely account of reality must survey.

Thus Plato constructs what amounts to a mathematical model for the universe. Ultimately, natural phenomena are explained by mathematical laws. The elements of Empedocles become mathematical figures in Plato's

scheme. The mathematical rules governing the combining of these figures account for the ways in which the four elements can interact.

With the passing of Plato, the doctrine that the universe is subject to mathematical law diminished in importance. During the next 1600 years or so, Aristotle became the most influential natural philosopher. Aristotle deemphasized the importance of mathematics in the study of nature. He also stressed qualitative (and not quantitative) aspects of the natural world. In place of mathematical explanations of natural behavior, Aristotle substituted explanation by final cause. For example, to the question: Why does a rock fall to earth? an Aristotelian would respond: Because it seeks its natural place in the center of the universe which is the earth. Another example—question: Why does rain fall? Answer: To water man's crops. Thus one always asks Why? and the answer is given in terms of an end, a purpose, or a final cause.

According to E. A. Burtt in Aristotle's study of motions on earth, "The analysis being intended to answer the question *why* they moved rather than *how* they moved, was developed in terms of the substances concerned in any given motion, hence the prominence of such words and phrases as action, passion, efficient cause, end, natural place." (*Metaphysical Foundations of Modern Science*, p. 91)

The physical theories of Aristotle dominated medieval science through the 13th century. A few alternatives to Aristotle emerged between 1300 and 1600, but not until the appearance of Kepler and Galileo was the Aristotelian grip on science eased. Aristotle's departure was hastened by the revival of Pythagoreanism and Platonism which was spearheaded by Kepler and Galileo. The central theme in this rebirth is that nature is describable by mathematical laws.

The astronomer, Kepler, believed that through mathematics one could make sense of the chaos of the physical world. H. Zeiser said of Kepler's notion of harmony:

> Harmony is present when a multitude of phenomena is regulated by the unity of a mathematical law which expresses a cosmic idea. (Quoted by G. Holton in *Thematic Origins of Scientific Thought*, pp. 83-84)

Gerald Holton himself writes

> Kepler's unifying principle for the world of phenomena is not merely the concept of mechanical forces, but

God, expressing himself in mathematical laws. (p. 85)

But for Kepler, these underlying principles are not only mathematical but also quantitative in nature. Kepler writes

> God, who founded everything in the world according to the norm of quantity, also has endowed man with a mind which can comprehend these norms. For as the eye for color, the ear for musical sounds, so is the mind of man created for the perception not only of arbitrary entities, but rather of quantities ... (p. 84)

For example, Kepler's position is illustrated by his very own third law of planetary motion:

$$\frac{\text{square of period}}{\text{cube of mean distance from sun}} = \text{constant.}$$

With Galileo, the emphasis on a quantitative mathematical description of nature is constantly felt. There are Galileo's famous words:

> Philosophy is written in this grand book, the universe, which stands continually open to our gaze. But this book cannot be understood unless one first learns to comprehend the language and read the letters in which it is composed. It is written in the language of mathematics and its characters are triangles, circles, and other geometric figures without which it is humanly impossible to understand a single word of it; without these, one wanders about in a dark labyrinth. (*The Assayer*)

When Galileo studies nature, he does not ask *why* something occurs but *how* it occurs; i. e., how can it be described? This point of view also leads to a stress on quantity. Thus, in trying to describe how the motion of a body progresses, one is led to consider notions such as distance, time, and velocity. Other quantitative features arise naturally in other problems, for example mass, length, area. The importance of this stress on quantity for mathematics (and physics) cannot be overemphasized. For the mathematical concept of function is a direct outgrowth of quantitative physics. For example, consider Theorem II, Proposition II of the

Third Day of Galileo's *Two New Sciences*:

> The spaces described by a falling body from rest with a uniformly accelerated motion are to each other as the squares of the time-intervals employed in traversing these distances.

Let s_1 and s_2 be the distances (spaces) travelled by such a body in time intervals t_1 and t_2. Then according to this theorem $s_1/s_2 = t_1^2/t_2^2$. Suppose t_2 is a unit time interval; then $s_1 = s_2 t_1^2$ where s_2 is the distance travelled by the body from rest in a unit of time. Since t_1 can be arbitrary, we have a relationship between t_1 and s_1, i. e. for each value of t_1 we obtain a corresponding value of s_1. Thus in modern parlance, s_1 is a function of t_1.

There are scores of similar results in the *Two New Sciences*, each of which describes a functional relationship between two quantities. To summarize, the emphasis placed by Galileo on quantitative aspects of nature led to the function concept.

There are two themes in this discussion. (1) The notion that nature subscribes to a mathematical design and (2) the rise of the idea of function from the emphasis on quantity by Galileo and Kepler. In fact, the second theme is in a sense a derivative of the first. Concerning the first statement, consider the words of the Nobel Prize physicist Werner Heisenberg when comparing modern atomic theory with that of Plato:

> In modern quantum theory there can be no doubt that the elementary particles will finally also be mathematical forms, but of a much more complicated nature.

QUESTIONS

1. Children ask questions about *why* things happen or *why* they are as they are. These are the questions that Aristotle tried to answer. Children do not ask questions about *how much*, the kind that Galileo and Kepler began to answer, nearly as often, or at all. Can we then conclude that quantitative questions are superior to qualitative ones because they come further along the evolutionary track towards maturity?

2. "All things have form, all things *are* form, and all forms can be defined by number," wrote Arthur Koestler, echoing the ancient Pythagoreans. It follows

logically that all things can be defined by number. What are we to make of this? How are you going to *define* a blade of grass by number, or even several numbers? You are a thing and I am a thing; are we nothing more than a mass of numbers? Does the statement make any rational sense at all, or does it have to be understood mystically as giving an imperfect description of a nonverbal insight into the nature of things?

EPISODES FROM THE EARLY HISTORY OF MATHEMATICS

by Asger Aaboe

First, let's get straight the point of that famous anecdote. That's the one about how the king gave Archimedes the problem of determining if his new crown was pure gold or if it was part gold and part some cheaper metal. Snipping off a sample of the crown was not allowed. The story goes that the solution came to Archimedes one day while he was in the bath, and that it so excited him that he ran down the street naked yelling, "Eureka, eureka!" that is, "I have found it, I have found it!" To those of us who were not brought up in the Greek culture of the third century B. C., the surprising part of the story is Archimedes' nakedness: just think of some famous scholar running naked down the street today! But Archimedes does not live today, and when and where he did live—more than two millennia ago, on Sicily, where Greeks dominated the natives—male nakedness was common and not to be remarked on. The surprising part of the story was that Archimedes, that dignified and renowned scholar and scientist, would be running down the street yelling his head off. Greek men of the time did *not* behave in such a manner.

With that out of the way, we can go on to more important things about Archimedes. It is hard to know where to begin, since there were so many of them. Ask any mathematician who were the three greatest mathematicians who ever lived and the answer is 95% certain to be "Archimedes, Newton, and Gauss." There is no question about it. Euler—a superb technician who bubbled over with ideas, but his ideas were ... smaller. Leibniz—well, he wasn't really a *mathematician*, was he? Fermat—great, but a great might-have-been. Galois, Hilbert, Lagrange—no, there is no one else who can be mentioned in the same breath. On occasion, I have a doubt or two about Newton, but never about Gauss, and certainly never about Archimedes.

One reason for his greatness was the time in which he lived. His dates are 287-212 B. C., not too long after the time when, on the rocky peninsula of Greece, people started to ask questions that their ancestors had not. Why is the world the way it is? What is knowledge? How should a citizen behave? Why is the square on the hypotenuse of a right triangle equal to the sum of the squares on the other two sides? The Greeks were not satisfied with the answers that had served up to then: "The king says so, that's why," "That's the way it has always been," "It is the will of the gods," "Just because," or "Shut up and get back to work." They wanted *reasons*. To convince a Greek, you had to *think*. Yelling, though it helped, was not enough. The human race started to think. Parts of it, that is: at no time, including right now, has the entire race been thinking, but at all times since then at least some of it has. When Archimedes thought, he did not have all that many other thinkers to look back on, so his achievements were all the greater.

The following excerpt tells some of what Archimedes did in mathematics. He found that the value of π lies between 3 10/71 and 3 1/7. How many people today know the value of π that closely? How many think it is *equal* to 3 1/7? He found how many grains of sand it would take to fill the entire universe. How many people today would say that there is no such number, or that the number of grains would be infinite? He trisected the angle, using a compass and a straightedge that has two scratches on it. He constructed a regular heptagon. He did a lot.

He also invented calculus, almost. It is only hindsight that lets us say that, since all Archimedes thought that he was doing was solving some isolated problems about areas and volumes. But the method that he used, though based on ideas about levers, was that of dividing areas and volumes up into many small pieces, finding the areas and volumes of the pieces, and adding them up. He was able to find the surface area of a sphere and several other results that we now get by integration. If his ideas had been carried further as, almost two thousand years later, the ideas of Fermat, Barrow, Cavalieri and others would be carried further by Newton and Leibniz, then we might have had calculus two thousand years earlier and the history of the race would have been changed in ways that are hard to imagine. However, it didn't happen. Archimedes lived at the time when Greek mathematics was at its absolute never-to-be-equaled peak, in its golden age at the time when the gold was shining most brightly. After the time of Archimedes, Greek mathematics went downhill (though with occasional bumps upward, as a roller coaster does not go directly from top to bottom), until towards its end five hundred years later, all that was being written were commentaries on the works of the giants of the past. Opportunities to take the work of Archimedes further did not arise. Also, mathematicians were not as thick on the ground then as they are now, or as they were in the seventeenth century when calculus was developed. The number of people skilled in mathematics at any one time in the Greek world was very, very small. You might need more than your fingers and toes to number them all, but you and a few friends would have more than enough digits for the job. Thus, it was easy for ideas to get lost, and that is what happened to Archimedes' method of finding areas and volumes. He had no successors, and the possibility of calculus in the ancient world died with him.

---

1. Archimedes' Life

In weightiness of matter and elegance of style, no classical mathematical treatise surpasses the works of Archimedes. This was recognized already in antiquity; thus Plutarch says of Archimedes' works;

> It is not possible to find in all geome-
> try more difficult and intricate ques-
> tions, or more simple and lucid expla-
> nations. Some ascribe this to his
> genius; while others think that incredi-
> ble effort and toil produced these, to
> all appearances, easy and unlaboured
> results.

Plutarch, who lived in the second half of the first century A. D., writes this in his *Lives of the Noble Grecians and Romans*, more specifically his life of Marcellus. Marcellus was the general in charge of the Roman army that besieged, and ultimately took, the Greek colony of Syracuse on Sicily during the second Punic War (218-201 B. C.). Archimedes' ingenious war-machines played an important role in the defense of Syracuse, and for this reason Plutarch writes about him at some length.

Archimedes introduces each of his books with a dedicatory preface where he often gives some background for the problem he is about to treat. These prefaces contain precious information for the historian of mathematics, and they even throw some light on Archimedes' life. There are, furthermore, scattered references to him in the classical literature and so he becomes the Greek mathematician about whom we have the most biographical information, even though it is precious little.

Archimedes was killed in 212 B. C. during the sack of Syracuse that ended the Roman siege. Since he is said to have reached the age of 75 years, he was born

about 287 B. C. In the preface to his book *The Sand-reckoner*, he speaks of his father Pheidias, the astronomer, who is otherwise unknown. It is said that Archimedes studied in Alexandria, then the centre of learning, and it is certain that he had friends among the Alexandrian mathematicians, as we learn from his prefaces; but he spent most of his life in Syracuse where he was a friend and, as some even say, a relation of the reigning house. He spent his life pursuing interests which extended from pure mathematics and astronomy to mechanics and engineering. Indeed, it was his more practical achievements that caught the public fancy. If we may trust the stories about him, he was not above adding a dramatic touch to his demonstrations; thus Plutarch tells, in Dryden's stately translation:

> Archimedes, however, in writing to
> King Hiero, whose friend and near
> relation he was, had stated that given
> the force, any given weight might be
> moved, and even boasted, we are told,
> relying on the strength of demonstra-
> tion, that if there were another earth,
> by going into it could remove this.
> Hiero being struck with amazement at
> this, and entreating him to make good
> this problem by actual experiment,
> and show some great weight moved
> by a small engine, he fixed according-
> ly upon a ship of burden out of the
> king's arsenal, which could not be
> drawn out of the dock without great
> labour and many men; and, loading
> her with many passengers and a full
> freight, sitting himself the while far
> off, with no great endeavour, by only
> holding the head of the pulley in his
> hand and drawing the cords by de-
> grees, he drew the ship in a straight

line, as smoothly and evenly as if she had been in the sea.

The compound pulley described here was one of Archimedes' inventions. In this passage of Plutarch we also find one version of the famous saying attributed to Archimedes by Pappus: "Give me a place to stand, and I shall move the earth." As we shall see, this invention falls in well with his theoretical studies on mechanics.

Plutarch continues his story of Archimedes' demonstration by telling how Hiero, much impressed, asked him to make war-engines designed both for offense and defense. These were made and found good use under Hiero's successor and grandson Hieronymus in the defense against the Romans under Marcellus. Plutarch has a most dramatic description of the effectiveness of these machines, both for short and long ranges, and for land as well as for sea. At last the Romans became so terrified that "if they but see a little rope or a piece of wood from the wall, instantly crying out, that there it was again, Archimedes was about to let fly some engine at them, they turned their backs and fled." Marcellus laid a long siege to the city, and it was finally taken. Marcellus tried to restrain his soldiers as much as he could from pillaging and looting, and was grieved to see how little he was heeded.

> But nothing afflicted Marcellus so much as the death of Archimedes, who was then, as fate would have it, intent upon working out some problem by a diagram, and having fixed his mind alike and his eyes upon the subject of his speculation, he never noticed the incursion of the Romans, nor that the city was taken. In this transport of study and contemplation, a soldier, unexpectedly coming up to him, commanded him to follow to Marcellus; which he declining to do before he had worked out his problem to a demonstration, the soldier, enraged, drew his sword and ran him through. Others write that a Roman soldier, running upon him with a drawn sword, offered to kill him; and that Archimedes, looking back, earnestly besought him to hold his hand a little while, that he might not leave what he was then at work upon inconclusive and imperfect; but the solider, nothing moved by his entreaty, instantly killed him. Others again relate

> that, as Archimedes was carrying to Marcellus mathematical instruments, dials, spheres, and angles, by which the magnitude of the sun might be measured to the sight, some soldiers seeing him, and thinking that be carried gold in a vessel, slew him. Certain it is that his death was very afflicting to Marcellus; and that Marcellus ever after regarded him that killed him as a murderer; and that he sought for his kindred and honoured them with signal favours.

Plutarch here gives three versions of Archimedes' death, and the farther away from the event we get, the more dramatic the story becomes. In Tzetes and Zonaras we find the variant that Archimedes, drawing in the sand, said to a Roman soldier who came too close: "stand away, fellow, from my diagram" which so infuriated the soldier (who, soldier fashion, wouldn't take nothing from nobody) that he killed him. This is the origin of the modern version: "Do not disturb my circles."

This is one of the few episodes of high drama in the history of mathematics. Much later we find Galois frantically trying to write down his truly inspired ideas the night before the duel which, as he had feared, proved fatal to him. He was 21 years old. A few mathematical geniuses, for example, the Norwegian Niels Henrik Abel, died of consumption, young and poor. And Condorcet, for one, met a violent end after the French revolution. But in general mathematicians have been a pretty dull lot, compared to poets.

Archimedes became, I think, a popular image of the learned man much as Einstein did in our day, and many stories of absent-mindedness were affixed to his name. Thus we read in Plutarch that he would become so transported by his speculations that he would "neglect his person to that degree that when he was carried by absolute violence to bathe or have his body anointed, he used to trace geometrical figures in the ashes of the fire, and diagrams in the oil on his body, being in a state of entire preoccupation, and, in the truest sense, divine possession with his love and delight in science."

We also have the tale of how he, during one of his (perhaps enforced) baths, discovered the law of buoyancy still known by his name; it excited him so that he ran naked through the streets of Syracuse shouting "Huereka, huereka", which is Greek for "I have found it, I have found it". This story is found, in what I think is a slightly garbled version, in Vitruvius. This discovery enabled Archimedes to confirm Hiero's suspicion that a goldsmith, who had had Hiero's crown or golden wreath to repair, had perpetrated a fraud by substituting silver for gold. Archimedes could now, by weights, determine the crown's density, and he found it smaller than that of pure gold.

These stories of absent-mindedness appeal to our sense of the ridiculous, but it must not be forgotten that a necessary faculty for being a genius of Archimedes' order is a capacity for focusing one's entire attention on the problem at hand for a goodly time to the exclusion of everything else.

This is in essence what we know of Archimedes' life, except for his works. Some traits of personality, though, can be gleaned from his prefaces and the tales about him; thus we catch a couple of glimpses of a baroque sense of humor. We sense it in his obvious delight in the dramatic demonstration on the beach. And in the preface to his treatise *On Spirals* he tells us that it has been his habit to send some of his theorems to his friends in Alexandria, but without demonstrations, so that they themselves might have the pleasure of discovering the proofs. However, it annoyed Archimedes that some had adopted his theorems, perhaps as their own, without bothering to prove them, so he tells that he included in the last set of theorems two that were false as a warning "how those who claim to discover everything, but produce no proofs of the same, may be confuted as having actually pretended to discover the impossible".

2. Archimedes' Works

While Euclid's *Elements* was a compilation of his predecessors' results, every one of Archimedes' treatises is a fresh contribution to mathematical knowledge.

The works preserved in the Greek are (in probable chronological order):

> *On the Equilibrium of Plane Figures, I*
> *Quadrature of the Parabola*
> *On the Equilibrium of Plane Figures, II*
> *On the Sphere and the Cylinder, I, II*
> *On Spirals*
> *On Conoids and Spheroids*
> *On Floating Bodies, I, II*
> *Measurement of a Circle*
> *The Sand-reckoner*

The Greek text of these works was edited in definitive form by J. L. Heiberg. In 1906 he discovered the Greek text of yet another book, *The Method*, hitherto considered lost. It was found in the library of a monastery in Constantinople; the text was written on parch-

ment in a tenth century hand and had been washed off to make the precious parchment available for a book of prayers and ritual in the thirteenth century. Such a text, washed off and with new writing on top of it, is called a *palimpsest* (from a Greek term meaning re-scraping), and is naturally most difficult to read. Luckily, Heiberg could make out enough of this palimpsest to give us a good edition of most of this remarkable book of Archimedes as well as of other treatises of his hitherto poorly preserved or authenticated, among them *The Stomachion*, which has to do with a mathematical puzzle. *The Method* is probably the latest of his preserved works and belongs at the bottom of the above list.

Through Heiberg's sober account of his discovery there shines his joy and pride in this rare find which came as a well-earned reward to a brilliant and dedicated scholar.

T. L. Heath translated Heiberg's text into English, introducing modern mathematical notation, and this version is now readily available.

In addition to these works that have been preserved we know the titles of several treatises that are lost. Thus we are told of Archimedes' ingenious machine representing the motions of sun, moon, and celestial bodies, and that he even wrote a book on the construction of such devices called *On Sphere-making*.

In order to convey some impression of the nature and scope of Archimedes' achievements I shall describe briefly the contents of his books, though it be only briefly and incompletely.

In the books *On the Equilibrium of Plane Figures* he first proves the law of the lever from simple axioms, and later puts it to use in finding the centres of gravity of several lamina of different shapes (the notion of centre of gravity is an invention of his). This treatise and his books on floating bodies are the only non-elementary writings from antiquity on physical matters that make immediate sense to a modern reader. Book I of *On Floating Bodies* contains, as Propositions 5 and 6, Archimedes' law of buoyancy, clearly stated and beautifully justified.

But most of Archimedes' books are devoted to pure mathematics. The problems he takes up and solves are almost all of the kind which today call for a treatment involving differential and integral calculus. Thus he finds, in *On the Sphere and the Cylinder*, that the volume of a sphere is two-thirds that of its circumscribed cylinder, while its surface area is equal to the area of four great circles.

In *The Measurement of the Circle* he first proves that *the area A of a circle of radius r is equal to that of a triangle whose base is equal to the circumference C of the circle and whose height is r, or*

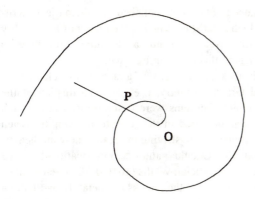

Figure 1

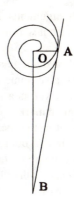

Figure 2

$$A = \frac{rC}{2}.$$

From this it follows that the ratio of the area of the circle to the square of the radius is the same as the ratio of its circumference to its diameter. This common ratio is what we call π today, and Archimedes proceeds to calculate that

$$3\frac{10}{71} < \pi < 3\frac{10}{70}$$

by computing the lengths of an inscribed and a circumscribed regular polygon of 96 sides. His upper estimate of π is, of course, the commonly used approximation 22/7.

In his book *On Spirals* he studies the curve which we appropriately call *Archimedes' spiral*; see Figure 1. If a ray from O rotates uniformly about O, like the hand of a clock, then P will trace out a spiral of this sort. Its equation in modern polar coordinates is

$$r = a\theta, \quad \theta > 0.$$

He finds many surprising properties of this curve, among them the following: Let the curve in Figure 2 from O to A be the first turn of an Archimedes' spiral (i. e., corresponding to $0 \leq \theta \leq 2\pi$); the area bounded by this curve and the line segment OA is then one-third of the circle of radius OA.

Further, if AB is tangent to the spiral at A, and OB is perpendicular to OA, then OB is equal to the circumference of the circle of radius OA. Though Archimedes does not state it explicitly, this implies that the area of triangle OAB is equal to the area of the circle of radius OA, as we can see using the above theorem from *The Measurement of the Circle*; thus Archimedes has succeeded in both *rectifying* and *squaring* the circle,

albeit with fairly complex means. [*Rectifying* a curve (in this case a circle) means to determine a straight line segment the same length as the curve; *squaring* a figure means determining a square of area equal to that of the figure.]

In *The Quadrature of the Parabola* he proves the theorem that the area of a segment of a parabola is four-thirds that of its inscribed triangle of greatest area, a theorem of which he is so fond that he gives three different proofs of it. The *Sand-reckoner*, which he addresses to Gelon, King Hiero's son, is a more popular treatise. In it he displays a number notation of his invention particularly well suited for writing very large numbers. To put this notation to a dramatic test he undertakes to write a number (10^{63}) larger than the number of grains of sand it would take to fill the entire universe, even as large a universe as the one Aristarchos assumed. Aristarchos had proposed a heliocentric planetary system, where the earth travels about a fixed sun once in one year; so in order to explain that the fixed stars apparently keep their mutual distances unchanged during the year, he was forced to maintain that the fixed star sphere was exceedingly much larger than had commonly been assumed. Here Archimedes furnishes one of our few sources of early Greek astronomy, and he even mentions his own endeavors at measuring the apparent diameter d of the sun. (His estimate is $90°/200 < d < 90°/164$; indeed, the commonly used rough approximation is $d \approx (1/2)°$. The recently discovered *Method* probably belongs at the end of a chronological list of Archimedes' works. In it he applies a certain mechanical method as he calls it—it is closely related to our integration—to a variety of problems with impressive results. The method does not carry the conviction of a proof in his eyes, but is more in the nature of plausibility arguments. He rightly emphasizes the usefulness of such arguments in surmising and formulating theorems which it will be worthwhile to try to prove rigorously.

This superficial and incomplete survey of some of Archimedes' works may give some impression of his breadth, originality, and power as a mathematician. A presentation of one of his remarkable chains of proofs in sufficient detail to do it justice lies well beyond the limits I have set myself in this book. I can only hope that a reader whose curiosity about this greatest contribution to ancient mathematics has been aroused will consult the works themselves. ...

6. Volume and Surface of a Sphere According to *The Method*

If we rotate Figure 3 about the dotted line, we generate a cone inscribed in a hemisphere which, in turn, is inscribed in a cylinder. The volumes of these three figures have the ratio 1:2:3. This beautiful theorem is a variant of Archimedes' favorite result. He was, in fact, so proud of it that he wanted a sphere with its circumscribed cylinder and their ratio (2:3) engraved on his tombstone. He got his wish, as we know from Cicero, who, when quaestor in Sicily, found Archimedes' tomb in a neglected state and restored it.

To prove this theorem in a rigorous and unexceptionable fashion is one of the chief aims of the first book of *On the Sphere and the Cylinder* (the other is to demonstrate that the surface of the sphere is four great circles). The reader of the book is forcibly impressed by the elegance of the sequence of theorems

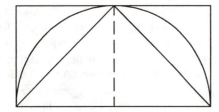

Figure 3

leading him through surprising and dramatic turns to the two final goals, but at the same time he cannot help recognizing that this sequence surely does not map the road which Archimedes first followed to discover these results.

When he writes this way, Archimedes is merely following the common practice of the Greek geome-

ters—indeed of most polished mathematical writing—which aims at convincing the reader of the validity of certain results and not at teaching him how to discover new theorems on his own.

This lack of the analytic and heuristic element in codified Greek geometry, i. e., of open display of the way in which theorems were first surmised rather than proved, was deplored in the seventeenth century when mathematicians were striving to create a new mathematical analysis (calculus and its ramifications). The English mathematician Wallis (1616-1703) even went so far as to believe that the Greeks deliberately had hidden their avenues of discovery.

Here we have one of a great many instances where lack of textual material has led modern scientists to false conclusions, for Wallis' surmise was thoroughly disproved when Heiberg found Archimedes' *Method*. Its aim is well described in the preface dedicating it to Eratosthenes. Archimedes writes here, in part (in Heath's translation):

> ... I thought fit to write out for you and explain in detail in the same book the peculiarity of a certain method, by which it will be possible for you to get a start to enable you to investigate some of the problems of mathematics by means of mechanics. The procedure is, I am persuaded, no less useful even for the proof of theorems themselves; for certain things first become clear to me by a mechanical method, although they had to be demonstrated by geometry afterwards because their investigation by the said method did not furnish an actual demonstration. But it is of course easier, when we have previously acquired, by the method, some knowledge of the questions, to supply the proof than it is to find it without any previous knowledge. This is a reason why, in the case of the theorems the proof of which Eudoxus was the first to discover, namely that the cone is the third part of the cylinder, and the pyramid of the prism, having the same base and equal height, we should give no small share of the credit to Democritus who was the first to make the assertion with regard to the said figure though he did not prove it. I am myself in the position

of having first made the discovery of the theorem now to be published by the method indicated, and I deem it necessary to expound the method partly because I have already spoken of it and I do not want to be thought to have uttered vain words, but equally because I am persuaded that it will be of no little service to mathematics; for I apprehend that some, either of my contemporaries or of my successors, will, by means of the method when once established, be able to discover other theorems in addition, which have not yet occurred to me.

QUESTIONS AND EXERCISES

1. Archimedes bounded π by 3 1/7 and 3 10/71 by inscribing and circumscribing regular polygons with 96 sides around a circle. What are the bounds on π that you get if you inscribe and circumscribe squares?

2. "The volume of a sphere is two-thirds of that of its circumscribed cylinder." Draw a sphere with radius r, circumscribe a cylinder around it, and translate the statement of Archimedes into an equation. Is it right?

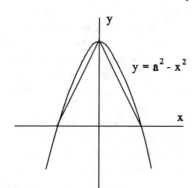

Figure 4

3. "[A sphere's] surface area is equal to the area of four great circles." Translate that into an equation.

4. Show that Archimedes was correct when he said "the area of a segment of a parabola is four-thirds that of its inscribed triangle of greatest area." Although Archimedes did not have analytic geometry, we do, so you can use Figure 4.

5. To do the same for "the area bounded by the first turn of an Archimedes' spiral is then one-third of the circle of radius OA" you need to know how to find areas of regions in polar coordinates. If you do, do so.

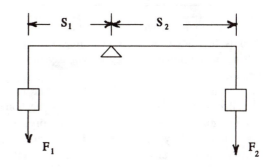

Figure 5

6. The law of the lever is that for equilibrium,
$$F_1 S_1 = F_2 S_2.$$
(see Figure 5). Let us apply it to a practical problem. The earth weighs 1.35×10^{25} pounds, more or less. If it is hung from one end of the lever and Archimedes (weight 150 pounds, say—ancient Greeks tended to be small) hangs from the other end and the fulcrum is put approximately where the moon orbits, 200,000 miles away from earth, then how far away will Archimedes have to be in order to move the earth?

ANTICIPATIONS OF CALCULUS IN MEDIEVAL INDIAN MATHEMATICS

by Ranjan Roy

Too many people are self-centered. They think that life is a drama with them as its central character, they care only about things that affect them, and they talk about themselves all the time. It is most annoying. They do not pay proper attention when I talk something *really* important, namely me.

What is true of people one at a time can remain true in a mass of them, and our western civilization can be as self-centered as the bore who goes on and on about his operation or his grandchildren. When we look at history, we tend to look at our history, ignoring what was accomplished by other cultures if it did not contribute to our civilization. It is not good to ignore things outside ourselves. We might miss out on learning something.

Indian mathematics has as long a tradition as western mathematics, and many notable achievements. Around 1150, Bhāskara was solving $61x^2 + 1 = y^2$ (the solution is not obvious: $x = 226153980$, $y = 1766319049$); such equations would not be considered in Europe until 500 years later. We do not learn about the achievements because Indian mathematics did not give rise to ours. The following selection gives one, not generally known.

The discovery of calculus is credited to Leibniz and Newton (1670s); some of their ideas were anticipated in a more classical context by a host of their European predecessors such as Cavalieri, Roberval, Fermat, Pascal, and Gregory. It is less well known that similar anticipations were made by Indian mathematicians of the fourteenth and fifteenth centuries in connection with their work in astronomy. The earliest example of such a discovery appears to be the concept of the instantaneous motion of a planet, introduced by Bhaskara II (1114-c. 1185). In applying this idea to an astronomical problem, he found that $d \sin \theta = \cos \theta$. Some later mathematicians from the southwest corner of India carried Bhaskara's ideas much further and obtained series expansions for the sine, cosine, and arctangent functions. Some details of how the series for the arctangent (and in particular a series for $\pi/4$) was obtained are given below. We shall see that, though some problems in integration and differentiation were solved, there was no general theory. It was the interest of the European mathematicians of the seventeenth century in geometrical problems (areas and tangents) which led to their discoveries in calculus. In contrast, the Indian mathematicians did not exhibit such an interest, though they used elementary geometrical methods to obtain their results.

The series for the sine, cosine, and arctangent are contained in a book by Nilakantha Somayaji (c. 1450-c. 1550) called *Tantrasangraha*. From the astronomical data given in the book, it appears to have been written around the year 1500. In the *Aryabhatiyabhasya*, a work on astronomy, Nilakantha attributes the series for the sine to Madhava, a mathematician who lived between the years 1340 and 1425. Nilakantha's teacher was Parameswara, who was a student of Madhava. It is not known whether Madhava found the other series as well or whether they were somewhat later discoveries.

The *Tantrasangraha* itself gives no proofs of the results contained in it. But proofs are given in the *Yukibhasa*, a commentary on the *Tantrasangraha*, written by Jyesthadeva. This work was composed in the sixteenth century and one conjectures that the proofs it contains are the same as those given by the earlier mathematicians.

Little is known about these mathematicians. Madhava, Parameswara, Nilakantha, and Jyesthadeva are all from Kerala on the southwest coast of India. In fact, the *Yukibhasa* was written in Malayalam, the language spoken in that region. The other books are in Sanskrit verse. Nilakantha was a man of diverse interests and also wrote on subjects outside of mathematics and astronomy. His erudite expositions on philosophy and grammar were well known and studied until recently. He attracted several gifted students, including Tuncath Ramanujan Ezuthassan, an early and important figure in Kerala literature.

We now turn to the mathematics of these people. In the *Tantrasangraha*, the series for the arctangent, sine, and cosine are given in verse which, when converted to mathematical symbols, may be written as

$$r \arctan \frac{y}{x} = \frac{1}{1} \cdot \frac{ry}{x} - \frac{1}{3} \cdot \frac{ry^3}{x^3} + \frac{1}{5} \cdot \frac{ry^5}{x^5} - \dots$$

where $y/x \leq 1$,

$$y = s - s \cdot \frac{s^2}{(2^2 + 2)r^2} + s \cdot \frac{s^2}{(2^2 + 2)r^2} \cdot \frac{s^2}{(4^2 + 4)r^2} - \dots$$

(sine)

$$r - x = r \cdot \frac{s^2}{(2^2 - 2)r^2} - r \cdot \frac{s^2}{(2^2 - 2)r^2} \cdot \frac{s^2}{(4^2 - 4)r^2} + \dots$$

(cosine).

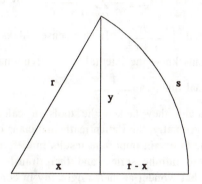

Figure 1

Here r, x, y, and s are as given in Figure 1.

The reader familiar with the usual form of the series for the arctangent, sine, and cosine functions may now easily verify that the above series can be reduced to standard form. A particular case of the first series when $x = y$ is

$$\frac{\pi}{4} = 1 - \frac{1}{3} + \frac{1}{5} - \frac{1}{7} + \dots .$$

This formula for π is generally known as the Leibniz-Gregory series: they discovered it some two hundred years after the Indian work. It was Leibniz's first important contribution to mathematics and came a few years before he began to develop the basic algorithms of calculus. Leibniz (1646-1716) was actually finding an expression for the area of a circle and was using the methods of Cavalieri, Fermat, Pascal, and other mathematicians of the mid-seventeenth century who introduced the idea of infinitesimals. Gregory (1637-1675) discovered the arctangent series as well as some others. Apparently, Gregory had found the Taylor series

formula but did not publish it because he thought that Newton (1642-1727) already knew it.

There are, however, some special features of Nilakantha's treatment of the series for $\pi/4$ which were not considered by Newton and Gregory. Nilakantha states some rational approximations for the error incurred by taking only the first n terms of the series. The expression for the approximation is then used to transform the series for $\pi/4$ into one which converges more rapidly. The error approximations are as follows:

$$\frac{\pi}{4} \equiv 1 - \frac{1}{3} + \frac{1}{5} - \dots \mp \frac{1}{n} \pm f_i(n + 1), \quad i = 1, 2, 3,$$

where

$$f_1(n) = \frac{1}{2n}, \; f_2(n) = \frac{n/2}{n^2 + 1}, \; f_3(n) = \frac{(n/2)^2 + 1}{(n^2 + 5)(n/2)}.$$

After making the transfromations, Nilakantha shows that the series become

$$\frac{\pi}{4} = \frac{3}{4} + \frac{1}{3^3 - 3} - \frac{1}{5^3 - 5} + \frac{1}{7^3 - 7} - \dots$$

and

$$\frac{\pi}{4} = \frac{4}{1^5 + 4 \cdot 1} - \frac{4}{3^5 + 4 \cdot 3} + \frac{4}{5^5 + 4 \cdot 5} - \dots .$$

Clearly, these series are more rapidly convergent than the original series for $\pi/4$. So, whereas we can be sure only of an error less than 10^{-3} for $\pi/4$ if we take the first thousand terms of $1 - 1/3 + 1/5 - \dots$, the last series guarantees this same accuracy with just its first three terms. The third approximation for error, $f_3(n)$, is also very effective in obtaining good numerical values for π without much computation. For example,

$$1 - \frac{1}{3} + \dots - \frac{1}{19} + f_3(20)$$

gives the value of $\pi/4$ correct to eight decimal places. Nilakantha himself used 104348/33215 as an approximation for π. This value is correct to nine decimal places.

We now turn to the *Yuktibhasa*'s proof of the series formula for $\pi/4$. We noted earlier that Leibniz arrived at the result by a quadrature of the circle. Jyesthadeva does it by finding the length of one-eighth of a circle. In Figure 2, the arc AC is a quarter circle with center O and radius 1. We construct the square $OABC$. The side AB is divided into n equal parts of length δ so that $n\delta = 1$ and $P_{r-1}P_r = \delta$. EF and $P_{r-1}D$ are perpendicular to OP_r. From the similarity of the triangles OEF and

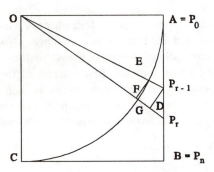

Figure 2

$OP_{r-1}D$ we get $EF = \dfrac{P_{r-1}D}{OP_{r-1}}$. Then again the similarity of the triangles $P_{r-1}P_rD$ and OAP_r gives

$$P_{r-1}D = \frac{P_{r-1}P_r}{OP_r} = \frac{\delta}{OP_r}$$

so

$$EF = \frac{\delta}{OP_{r-1}OP_r} \cong \frac{\delta}{OP_r^2} = \frac{\delta}{1 + AP_r^2} = \frac{\delta}{1 + r^2\delta^2}.$$

Since arc $EG \cong EF \cong \dfrac{\delta}{1 + r^2\delta^2}$, $\dfrac{1}{8}$ arc of circle is

$$\frac{\pi}{4} = \lim_{n \to \infty} \sum_{r=0}^{n-1} \frac{\delta}{1 + r^2\delta^2}.$$

Now we can expand $\dfrac{\delta}{1 + r^2\delta^2}$ as a geometric series

(setting $\delta = \dfrac{1}{n}$) to get

$$\frac{\pi}{4} = \lim_{n\to\infty}\left[\frac{1}{n}\sum_{r=0}^{n-1} 1 - \frac{1}{n}\sum_{r=0}^{n-1}\left(\frac{r}{n}\right)^2 \right.$$
$$\left. + \frac{1}{n}\sum_{r=0}^{n-1}\left(\frac{r}{n}\right)^4 - ... \right]$$

The limit $\lim_{n\to\infty} \dfrac{1}{n}\sum_{r=0}^{n-1}\left(\dfrac{r}{n}\right)^k$ is the same one that Wallis, Pascal, and Fermat grappled with in the 1650s. In their work, it arose as the area under $y = x^k$ over the interval $[0, 1]$. Wallis conjectured that the value of this limit is $\dfrac{1}{k+1}$ and Pascal and Fermat proved this. Since there was no analytic geometry at the time of

Nilakantha and Jyesthadeva, there is no geometrical interpretation of this limit. The *Yuktibhasa* gives an inductive proof that the limit is $\dfrac{1}{k+1}$ but the argument is not complete. Anyhow, once the limit is taken, it follows that

$$\frac{\pi}{4} = 1 - \frac{1}{3} + \frac{1}{5} - ... .$$

The more general formula for $\arctan(y/x)$ is also proved in the *Yuktibhasa* in a similar manner except that one takes an arbitrary part of the circle rather than 1/8 of it. We also do not consider here how the transformation of the series was effected by these Indian mathematicians. See problems 2 and 3 of the exercises for this.

Note that $\lim_{n\to\infty} \dfrac{1}{n}\sum_{r=0}^{n-1}\dfrac{1}{1+r^2\delta^2}$ is by definition the integral $\int_0^1 \dfrac{dx}{1+x^2}$. In this sense Nilakantha and his students knew the integral of a polynomial and of the rational function $\dfrac{1}{1+x^2}$.

Although they lacked the tools of calculus and analytic geometry, the Indian mathematicians of Kerala were able to obtain important results in differentiation, integration, infinite series, and their transformations. One can but wonder at their originality in constructing an infinite series when such a thing was unknown in their mathematical tradition. The power and intuition of these all but forgotten devotees of mathematics to work with the concepts of calculus without its framework must command our respect.

EXERCISES

1. Show that the series for the arctangent, cosine, and sine are equivalent to those given in modern textbooks.

2. This exercise shows how the transformed series are obtained. Let n be an odd integer and

$$\sigma_n = 1 - \frac{1}{3} + \frac{1}{5} - ... \mp \frac{1}{n} \pm f_i(n+1).$$

Then

$$\sigma_{n-2} = 1 - \frac{1}{3} + \frac{1}{5} - ... \mp \frac{1}{n-2} \pm f_i(n-1).$$

Subtraction gives

$$\sigma_n - \sigma_{n-2} = \mp \frac{1}{n} \pm f_i(n+1) \pm f_i(n-1).$$

Define $\mp U_n = \sigma_n - \sigma_{n-2}$. Then show that

$$\sigma_n = \sigma_{n-2} \pm U_n = \sigma_{n-4} \mp U_{n-2} \pm U_n = \dots$$

$$= \sigma_1 - U_3 + U_5 - \dots \mp U_n.$$

Deduce that (by letting $n \to \infty$ in the last equation)

$$\frac{\pi}{4} = 1 - f_i(2) - U_3 + U_5 - U_7 + \dots.$$

For $f_1(n) = \frac{1}{2n}$ show that $U_n = -\frac{1}{n^3 - n}$ and obtain the first transformed series for $\pi/4$.

3. Compute the first three terms of the second transformed series:

$$\frac{\pi}{4} = \frac{4}{1^5 + 4 \cdot 1} - \frac{4}{3^5 + 4 \cdot 3} + \frac{4}{5^5 + 4 \cdot 5} - \dots.$$

How many decimal places of accuracy for $\pi/4$ do you expect to get? Explain.

4. Show that if $g(x)$ is a continuous function for $0 \le x \le 1$ then

$$\lim_{n \to \infty} \frac{1}{n} \sum_{r=0}^{n-1} g\left(\frac{r}{n}\right) = \int_0^1 g(x)\, dx.$$

5. Use problem 4 to show that $\lim_{n \to \infty} \frac{1}{n} \sum_{r=0}^{n-1} \left(\frac{r}{n}\right)^k$

$= \frac{1}{k+1}$. Note that this limit is needed in the proof of the series formula for $\pi/4$.

6. Deduce from problem 4 that

$$\lim_{n \to \infty} \frac{1}{n} \sum_{r=0}^{n-1} \frac{1}{1 + r^2 \delta^2} = \int_0^1 \frac{dx}{1 + x^2}.$$

Here $\delta = \frac{1}{n}$. Evaluate the integral $\int_0^1 \frac{dx}{1 + x^2}$ directly.

SLICING IT THIN

by Howard Eves

There is no law of the conservation of ideas. Matter and energy can neither be created nor destroyed, but before anyone has a new idea it does not exist. It does not exist in the collective mind of the human race, that is—you can have philosophical arguments about whether it exists in some other sense. Something that does not exist cannot be thought about. It is *not there*. Then someone has a new idea, a good new idea that is useful and spreads. It now exists. People, who have a tendency to live in the present and think that the world has always been just as it is now, get used to the new idea and take it for granted. *Of course* bad money drives out good: that's why there are no silver quarters in circulation, it's *obvious*. But it wasn't always obvious, or else the principle wouldn't be called Gresham's Law. *Of course* things burn because they combine rapidly with oxygen: *everybody* knows that. But everybody didn't always know that, especially those who thought that things burned because they were rapidly giving off phlogiston. Things are so obvious, *after* they have been thought of, that the achievements of those who penetrated the unknown and brought back a new idea from the realm of nonexistence sometimes fail to get the credit they deserve. Ideas can be created, but to have a useful new idea is rare and those who have them should be greatly honored.

Bonaventura Cavalieri provides an example. Today, everyone knows that the way to find the volume of something is to divide it up into lots of little pieces and add up the volumes of the pieces using the integral. How natural and obvious! So is Cavalieri's Principle that if you have two solids, both split up into little pieces so that for each little piece of one solid there is a little piece of the other solid with exactly the same volume, then the two solids must have the same volume. Of course! Add up the volumes of the little pieces in the first solid and you get the same total as when you add up the volumes of the little pieces in the second solid. Anybody could have thought of that! No, not anybody could have. I couldn't, you couldn't, he couldn't, she couldn't, they couldn't, not one person in a million, in ten million, in one hundred million could have. Cavalieri, however, did. When we think of Cavalieri (and not many people do) we should not dismiss him with something condescending like, "Cavalieri. Oh, yes—Cavalieri's Principle. Pretty obvious, but not bad for 1635." Better that we should think, "Cavalieri! Good heavens—Cavalieri's Principle! Absolutely staggering, and in 1635 too!"

The following excerpt tells some things about Cavalieri, his Principle (Principles, actually), and what they are good for. Its author also included homework problems at the end for those people for whom no mathematics lesson is complete without exercises to do.

In the fourteenth century, the Blessed John Colombini of Siena founded a religious group known as the *Jesuats*, which was in no way related to the *Jesuits*. The order was approved by Pope Urban V in 1367. The original work of the order was the care of those stricken by the Black Death, which raged over Europe at the time, and the burial of the fatally smitten. With the passage of time the Jesuat order diminished, and in 1606 an attempt at a revival was made. But certain abuses later crept into the order, with the result that the group no longer exists. It seems that the manufacture and sale of distilled liquors, apparently in a manner unacceptable to Canon Law, along with a growing scarcity of members, led to the order's suppression by Pope Clement IX in 1668.

In 1613, only a few years after the attempted revival of the Jesuats, a young fifteen-year-old boy named Bonaventura Cavalieri was accepted as a member of the order, and then spent the rest of his life in its service. It is because of this, and because of the ultimate vanishing of the order and the natural confusion between Jesuat and Jesuit, that so many major encyclopedias, histories, and source books erroneously state that Cavalieri was a Jesuit, instead of a Jesuat, furnishing an excellent example of written histories containing a hidden perpetuated error. It is all too easy for some historian to record an erroneous and undocumented statement, and then for subsequent historians, leaning on earlier work, to repeat the falsehood. Many such erroneous statements have been widely perpetuated over considerable periods of time.

Bonaventura Cavalieri was born in Milan, Italy, in 1598, studied under Galileo, and served as a professor of mathematics at the University of Bologna from 1629 until his death in 1647 at the age of forty-nine. Cavalieri was one of the most influential mathematicians of his time, and the author of a number of works on trigonometry, optics, astronomy, and astrology. He was among the first to recognize the great value of logarithms and was largely responsible for their early introduction into Italy. But his greatest contribution to mathematics was a treatise, *Geometria indivisibilibus*, published in its first form in 1635, devoted to the pre-calculus *method of indivisibles*—a method that can, like so many things in more modern mathematics, be traced back to the early Greeks, in this case to Democritus (ca. 410 B. C.) and Archimedes (ca. 287-212 B. C.). It is quite likely that it was the attempts at integration made by Kepler that directly motivated Cavalieri. At any rate, the publishing of Cavalieri's *Geometria indivisibilibus* in 1635 marks a **great moment in mathematics**.

Cavalieri's treatise on the method of indivisibles is voluble and not clearly written, and it is not easy to learn from it precisely what Cavalieri meant by an "indivisible." It seems that an indivisible of a given planar piece is a chord of the piece, and a planar piece can be considered as made up of an infinite parallel set of such indivisibles. Similarly, it seems that an indivisible of a given solid is a planar section of that solid, and a solid can be considered as made up of an infinite parallel set of this kind of indivisible. Now, Cavalieri argued, if we slide each member of a parallel set of indivisibles of some planar piece along its own axis, so that the endpoints of the indivisibles still trace a continuous boundary, then the area of the new planar piece so formed is the same as that of the original planar piece, inasmuch as the two pieces are made up of the same indivisibles. A similar sliding of the members of a parallel set of indivisibles of a given solid will yield another solid having the same volume as the original one. (This last result can be strikingly illustrated by taking a vertical stack of cards and then pushing the sides of the stack into curved surfaces; the volume of the disarranged stack is the same as that of the original stack.) These results, slightly generalized, give the so-called *Cavalieri principles*:

1. *If two planar pieces are included between a pair of parallel lines, and if the lengths of the two segments cut by them on any line parallel to the including lines are always in a given ratio, then the areas of the two planar pieces are also in this ratio.*

2. *If two solids are included between a pair of parallel planes, and if the areas of the two sections cut by them on any plane parallel to the including planes are always in a given ratio, then the volumes of the two solids are also in this ratio.*

Cavalieri's principles constitute a valuable tool in the computation of area and volumes, and their intuitive bases can easily be made rigorous with the modern integral calculus. Accepting these principles as intuitively apparent, one can solve many problems in mensuration that normally require the more advanced techniques of the calculus.

Let us illustrate the use of Cavalieri's principles, first employing the planar case to find the area of an ellipse of semiaxes a and b, and then the solid case to find the volume of a sphere of radius r.

Consider the ellipse and the circle

$$\frac{x^2}{a^2} + \frac{y^2}{b^2} = 1, \quad a > b, \text{ and } x^2 + y^2 = a^2,$$

plotted on the same rectangular coordinate frame of reference, as shown in Figure 1.

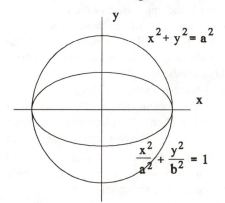

Figure 1

Solving each of the equations above for y, we find, respectively

$$y = \frac{b}{a}(a^2 - x^2)^{1/2}, \quad y = (a^2 - x^2)^{1/2}.$$

It follows that corresponding ordinates of the ellipse and the circle are in the ratio b/a. It then follows that corresponding vertical chords of the ellipse and the circle are also in this ratio, and hence, by Cavalieri's first principle, so are the areas of the ellipse and the circle. We conclude that

area of ellipse $= (b/a)$ (area of circle)
$$= (b/a)(\pi a^2) = \pi ab.$$

This is basically the procedure Kepler employed in finding the area of an ellipse of semiaxes a and b.

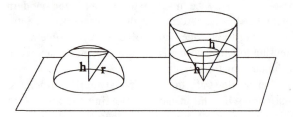

Figure 2

Now let is find the familiar formula for the volume of a sphere of radius r. In Figure 2 we have a hemisphere of radius r on the left, and on the right a circular cylinder of radius r and altitude r with a cone removed whose upper base is the upper base of the cylinder and whose vertex is the center of the lower base of the cylinder (see Figure 2). The hemisphere and the gouged-out cylinder are resting on a common plane. We now cut both solids by a plane parallel to the base plane and at a distance h above it. This plane cuts the one solid in a circular section and the other in an annular, or ring-shaped, section. By elementary geometry we easily show that each of the two sections has an area equal to $\pi(r^2 - h^2)$. It follows, by Cavalieri's second principle, that the two solids have equal volumes. Therefore the volume V of the sphere is given by

$$V = 2(\text{volume of cylinder} - \text{volume of cone})$$
$$= 2(\pi r^3 - \pi r^3/3) = 4\pi r^3/3.$$

The assumption and then consistent use of Cavalieri's second principle can greatly simplify the derivation of many of the volume formulas encountered in a beginning treatment of solid geometry. This procedure has been adopted by a number of textbook writers, and has been advocated on pedagogical grounds. For example, in deriving the familiar formula for the volume of a tetrahedron ($V = Bh/3$), the sticky part is first to show that any two tetrahedra having equivalent bases and equal altitudes on those bases have equal volumes. The inherent difficult here is reflected in all treatments

of solid geometry from Euclid's *Elements* on. With Cavalieri's second principle, however, the difficulty simply melts away.

Cavalieri's hazy conception of indivisibles, as sort of atomic parts of a figure, led to much discussion and serious criticism by some students of the subject, particularly by the Swiss goldsmith and mathematician Paul Guldin (1577-1642). Cavalieri recast his treatment in the vain hope of meeting these objections. The French geometer and physicist Gilles Persone de Roberval (1602-1675) ably employed the method and claimed to be an independent inventor of it. The method, or some process very like it, was effectively used by Evangelista Torricelli (1608-1647), Pierre de Fermat (1601?-1665), Blaise Pascal (1623-1662), Grégoire de Saint-Vincent (1584-1667), Isaac Barrow (1630-1677), and others. In the course of the work of these men, results were reached which are equivalent to performing such integrations as

$$\int x^n \, dx, \quad \int \sin x \, dx, \quad \int \sin^2 x \, dx, \quad \int x \sin x \, dx.$$

Two planar pieces that can be placed so that they cut off equal segments in each member of a family of parallel lines, or two solids that can be placed so that they intercept equiareal sections on each member of a family of parallel planes, are said to be *Cavalieri congruent*. Two figures that are Cavalieri congruent have, of course, equal areas (in the one case) or equal volumes (in the other case). Among the curiosities concerning this type of congruence are the following:

1. Though there cannot exist a polygon to which a given circle is Cavalieri congruent, there does exist a polyhedron (actually a tetrahedron) to which a given sphere is Cavalieri congruent.

2. Though there exist tetrahedra of the same volume that are not Cavalieri congruent, any pair of triangles of the same area are Cavalieri congruent.

Exercises

19.1 Establish Cavalieri's principles by modern integration.

19.2 Find, by Cavalieri's first principle, the area enclosed by the curve

$$b^2 y^2 = (b + x)^2(a^2 - x^2),$$

where $b \geq a > 0$.

19.3 An oblique plane through the center of the base of a right circular cylinder cuts off from the cylinder a cylindrical wedge, called a *hoof*. Find, by Cavalieri's second principle, the volume of a hoof in terms of the radius r of the associated cylinder and the altitude h of the hoof.

19.4   Show that there cannot exist a polygon to which a given circle is Cavalieri congruent.

19.5   Find a polyhedron to which a given sphere is Cavalieri congruent.

EXERCISES

1.   "By elementary geometry we easily show that each of the two sections has an area equal to $\pi(r^2 - h^2)$." Use elementary geometry, trigonometry, or both to show that this is correct.

2.   Do exercise 19.2 by integrating.   Using the symmetry of the curve with about the x-axis, the area is

$$2 \int_0^a \frac{(b + x)\sqrt{a^2 - x^2}}{b} \, dx \, ;$$

splitting the integral into two pieces and evaluating each piece (either by using integral tables or by hand) will give the result. A computer program that does integration would be even faster. *My* program told me that the answer was $a^3/3b + \pi a^2/4$. (The antiderivative had an arcsin in it.)

3.   Do exercise 19.1.

THE HEROIC CENTURY

by Lars Gårding

How does a person learn about the history of mathematics? One way is to take a course in the subject, if you can find one. That may be a good idea, but taking courses can have disadvantages. For one thing, teachers of courses sometimes have the idea that their students have the aim of becoming professionals in their disciplines, and teach accordingly. For another, a course may tell you more than you want to know: if you want to learn only what an educated person should know about the history of mathematics it is not really necessary to know the exact date of the publication of *Principia Mathematica* (1687), precisely where Euler did most of his work (Russia and Germany), or how many people it took to prove the theorems of Bolzano-Weierstrass and Mittag-Leffler (three), but there are other things that it *is* necessary to know, so students tend to spend too much time memorizing. Courses can work sometimes, but there is no guarantee of success.

The best way to learn about the history of mathematics is to read about it. That does not mean to *study* it, overlining in yellow all the names and dates and the first sentence of every paragraph. The best way is to read a lot, in a lot of different books. If you read four different histories of mathematics and find that Newton gets a chapter to himself in each of them, you tend to get the idea that Newton is important, and you may even remember why he is important without having to go to the effort of consciously memorizing. Similarly, if the name of Condorcet comes up only once, you can conclude without even having to think about it that Condorcet is not a major figure in the history of mathematics and you do not have to know that he was born on September 17, 1743, nor that he ended his life a suicide, in prison, on April 8, 1794. Also, reading several different accounts of the same events gives you several different perspectives on them and you can understand them better than if you read the same story four times. Yet further, your skills of critical reading can be developed. For a trivial example, in the excerpt that follows the author says that Fermat died in 1663. In another excerpt another author says that he died in 1665. Someone is making a mistake. For a less trivial example, different authors have different opinions on the purpose and value of Newton's writings on alchemy and religion. Reading the divergent views, considering the evidence, and making up your own mind is a valuable exercise. What is education for, if not to teach people how to think for themselves? Reading widely and reflecting on what you read is a good way to reach that goal.

The following excerpt deals with all of mathematics in the seventeenth century and thus it overlaps other excerpts that deal with narrower topics. But it does no harm to see again the names of Fermat, Descartes, Leibniz, and so on. The more often you are exposed to something, the more likely you are to catch it. The more mud you throw at a wall, the more is likely to stick. That last image may not be quite appropriate, but you get the idea.

The theory of differentiation and integration is called *infinitesimal calculus*, which means computation with infinitely small entities. It dates from the seventeenth century, which saw the birth of modern mathematics. For some time the strict proofs of the Greeks had been abandoned in favor of heuristic reasoning. Audaciously exploring new approaches, mathematicians surpassed everything that had been done before. The century had long religious wars, severe crop failures, and serious outbreaks of the plague, but for science and mathematics it was a time of unprecedented discoveries. The development was rapid. The works of Galilei (from 1604) on accelerated motion are almost childish compared to what Leibniz and the brothers Bernoulli did at the end of the century, solving many different problems of infinitesimal calculus and variational calculus with modern methods and notation.

Infinitesimal calculus was born out of efforts to compute areas of plane figures bounded by curved lines and the volumes of bodies bounded by curved surfaces.

Certain of these computations, which are easy calculus exercises, had given the Greeks a great deal of trouble. In fact, almost everything that Archimedes wrote had to do with such problems, and his results mark the high tide of Greek mathematics. Before him Eudoxus from Chnidos had computed the volume of the cone and the pyramid, but Archimedes computed the volume and the area of the sphere, the area of the parabolic segment, the center of gravity of the triangle, and the area enclosed by a certain spiral now called the spiral of Archimedes. Of these, the volume of the cone and the pyramid, the area of the parabolic segment, the center of gravity of the triangle, and the area of the segment of the spiral all depend on the same integral, namely $\int x^2\, dx$, but there are no signs that Archimedes saw the connection between these problems. Every one of them he solves with different and often very ingenious methods. In his work on the spiral he even says that the problem treated there has no connection with certain other problems, e. g., the volume of the paraboloid segment which he has just mentioned and which leads to the integral $\int x\, dx$. Archimedes' proofs are very strict. He uses the "method of exhaustion" of Eudoxus, in which figures with known areas are circumscribed and inscribed into the figure whose area one wants to compute. There is a similar method for arc lengths, and Archimedes used it to prove that the number π lies between 3 1/7 and 3 10/71. The proof is perfectly rigorous and seventeenth century mathematicians often used Archimedes as a model when they wanted to give absolutely convincing proofs. But the rigor also made the proofs cumbersome and difficult. It is reasonable to assume that in most cases Archimedes had known the solutions of his problems before he worked out the proofs. Through a letter from Archimedes to Eratosthenes of Alexandria, discovered in 1906, we even know his method for this. He called it the mechanical method. The gist of it was to consider, e. g., a plane figure as something with a weight composed of straight lines each one with a weight proportional to its length. The actual work is then done by balancing various geometric figures against each other and looking for centers of gravity. About this Archimedes wrote:

> This procedure is, I am persuaded, no less useful even for the proofs of the theorems themselves; for certain things became clear to me by the mechanical method, although they had to be demonstrated by geometry afterwards because their investigation by the said method did not furnish an actual demonstration ... and I deem it

necessary to expound the method partly because I have spoken of it and I do not want to be thought to have uttered vain words, but equally because I am persuaded that it will be of no little service to mathematics; for I apprehend that some, either of my contemporaries or of my successors will, by means of the method when once established, be able to discover other theorems in addition, which have not yet occurred to me.

But it was a long time before these prophetic remarks came true. There are several reasons why Archimedes' work was not continued by others, e. g., Archimedes' superior gifts and the sterilizing effect that the Roman conquest had on Greek science in general. But the main cause is probably the Greek geometric method itself. It is impeccable but it does not reveal the right connections and therefore makes new discoveries difficult. The progress predicted by Archimedes did not occur until the seventeenth century.

The main works of Greek literature and philosophy were printed in Italy before 1520, and the first edition of Archimedes' work in Basel in 1544. Contemporary mathematics, represented by Cardano, was then rather algebraic, and Archimedes' influence came later with Galilei and Kepler, both of whom were astronomers and physicists more than they were mathematicians. But from their time, the early seventeenth century, until about 1670, mathematicians are constantly quoting Archimedes. He is translated and commented upon, and everybody declares him a paragon and a source of inspiration.

In the beginning of the seventeenth century, scientists had to work under very primitive conditions, but later these improved enormously through the consolidation of the universities and the foundation of scientific societies and periodicals. Mathematicians had no periodical before 1665, when the newly founded Royal Society started publishing *Philosophical Transactions*. Before this time they had to write letters to each other or print books, often at their own expense when no patron of science was available. Publishers and printers who could do such a job were scarce, and sometimes not very honest. After the printing a new ordeal awaited the author. The general uncertainty about the foundations of infinitesimal calculus made it very easy for the rivals to find the weak points and criticize them. There were many bitter controversies of this kind, fought in bad faith by both parties. It is not surprising that many preferred to work in peace and quiet, just telling good friends about new results. Certain mathematical dilettantes, e. g., Mersenne in Paris and Collins in London, conducted a large correspondence supplying the mathematical news service with excerpts from letters. Students traveled a lot and in this way new ideas spread efficiently but perhaps in not a very orderly fashion. The random contacts and the fact that all mathematicians were working on essentially the same problems made priority fights very common. The quarrel between Newton and Leibniz about who invented infinitesimal calculus was well-known at the time, even to nonmathematicians.

Roughly speaking, there are two major periods of the seventeenth century, the time before and that after 1670. The most important names of the first period are the Italians Galilei (1564-1642) and Cavalieri (1598-1647), the German astronomer Kepler (1571-1630), the Frenchmen Fermat (1601-1663), Descartes (1596-1650), and Blaise Pascal (1623-1662), the Dutchman Huygens (1629-1695), and the Englishmen Wallis (1616-1703) and Barrow (1630-1677). All these did work preparatory to infinitesimal calculus proper which then, in a brief period after 1670, was created by Newton (1642-1727), the German Leibniz (1646-1716), and the Scotsman James Gregory (1638-1675). Before going into details we should perhaps say a few words about the actors in the drama. Galilei, professor in Pisa and Padua, broke with Aristotelian physics and discovered the laws of falling bodies. He constructed telescopes and made some fundamental astronomical discoveries. He believed in the theory of Copernicus that the earth and the planets move around the sun, but the Church forced him to deny this heresy in 1633. Cavalieri was professor of mathematics in Bologna and a friend of Galilei's. Kepler succeeded the Danish astronomer Tycho Brahe as imperial mathematician in Prague; he deduced, from Brahe's observations, that the planets move in elliptic orbits around the sun. Fermat, a lawyer in Toulouse, worked with number theory and analysis, and he corresponded with Pascal and Descartes. The second of these last two was a nobleman, a soldier, a philosopher and a teacher of royalty. He died in Stockholm at the court of Queen Christina. Descartes' great mathematical discovery was analytic geometry. At the age of 16, Pascal discovered a fundamental theorem about conic sections. Later, he wrote about probability theory and computed areas and centers of gravity. He had several religious crises, and his contemporaries knew him more as a philosopher and religious writer than as a mathematician. Huygens studied law and thought of becoming a diplomat but soon made a name for himself as a scientist. His analysis of progressive waves and the refraction of light is valid even today. Huygens was

elected member of the French Academy of Sciences in 1666 and lived in Paris for a long time. There he met Leibniz and got him interested in the new mathematics. Wallis started as a theologian and became professor of mathematics in Oxford in 1649. Barrow, Newton's teacher, was professor of mathematics in Cambridge and later retired to a parsonage. He left his chair to Newton, and it is likely that Newton had some influence on Barrow's work. Gregory was professor of mathematics at St. Andrews in Scotland. With this background, we are ready now for the two main characters, Newton and Leibniz.

Newton came to Cambridge in 1660, at the age of 17. Nine years later he succeeded Barrow and planned to publish a treatise on derivatives and series containing the fundamental theorems of infinitesimal calculus. It remained in manuscript, but was printed after his death and became known as the theory of fluxions. Newton considered the derivative as a velocity, and called it a *fluxion*. In his main work, *Philosophie Naturalis Principia Mathematica*, known as *Principia* and printed in 1687, Newton proved that the movements of celestial bodies can be deduced from the law of motion (the force equals the time derivative of momentum) and the law of gravitation. *Principia* was the first big success for the combination of physics and mathematics, and it has been followed by many others for almost 300 years. The firmly rooted prestige now enjoyed by this couple started with Newton's work. Its first unparalleled success has led to sometimes exaggerated hopes that mathematics in combination with, e. g., biology or economics will yield the same brilliant results.

Most of Newton's contemporaries thought that comets were the work of God or the Devil, and were portentous signs of coming events. After *Principia* educated people could no longer have this faith, but philosophy and religion soon adjusted to the fact that the movements of celestial bodies are as predictable as those of the wheels of a clockwork. According as new planets were discovered it was difficult to maintain that God intended them to be five in order to join the sun and the moon in a sacred seven, but *Principia* did not shake God's position as the Creator. On the contrary, Creation appeared as an even greater miracle than before. Politically, and in religious matters, Newton was a conservative and he had a firm belief in God. Among his unpublished manuscripts there are long investigations of religious chronology and the topography of hell. The spirit of the times was such that they are consistent with *Principia*. After 1690 Newton served for some time as director of the Royal Mint and an M. P. for Cambridge University.

Leibniz started his career as a precocious student in

Leipzig, and after 1676 earned his living as diplomat, genealogist, and librarian to the house of Hanover. One of its members became king of England in 1714, under the name of George I. He was supported by the Whig party. Since Newton was a fanatical Tory, he and Leibniz were political adversaries, which is assumed to explain some of the animosity between them. The contact with Huygens in Paris in 1673 was the beginning of Leibniz's career as a mathematician. He visited London many times and exchanged letters with Newton, Collins, Huygens, and many others. Leibniz founded the academies of science of Leipzig and Berlin, and published most of his mathematical papers in *Acta Eruditorum*, the journal of the Leipzig Academy. He was a pioneer in symbolic logic, but was also a philosopher in the classical tradition, who occupied himself with explaining the universe and proving God's existence. His most important philosophical work remained in manuscript. What he published was more or less tailored to the taste of ruling princes. In any case they had no difficulty in accepting his famous dictum that we live in the best of all possible worlds.

Here we leave the personalities and return to mathematics. We shall follow the development through an analysis of three themes: mathematical rigor versus heuristic reasoning, connections between problems, and the balance between geometry and algebra.

Among the Greeks philosophical and logical aspects and mathematical rigor were dominant, and there was almost nothing left for heuristic arguments. There was a large gap between the initial idea and the artfully executed, polished proof, and this must have had a frustrating effect on ingenious mathematicians. To a large extent, progress in the seventeenth century is due to the fact that mathematical rigor was neglected in favor of heuristic reasoning. Archimedes' mechanical method was not known, but mathematicians started arguing as he had done. They talked about "infinitely small quantities" and "sums of infinitely small quantities." A plane figure, for instance, was considered to be composed of parallel line intervals and—now comes the meaningless but useful point of view—its area as the sum of the areas of corresponding infinitely narrow rectangles. Aided by such arguments it is easy to convince oneself that if the figure is doubled in every direction, its area gets four times as large. In the same way, if a body is doubled in every direction, its volume is multiplied by eight. This observation in general form is due to Cavalieri and was called *Cavalieri's principle*. Before coming to believe it Cavalieri checked it against all the areas and volumes computed by Archimedes.

Similar arguments were used to treat the second main problem of infinitesimal calculus, the determina-tion of tangents and the calculation of arclength. The existence of tangents was postulated and the curve itself was thought of as composed of infinitely small line segments sometimes considered as parts of tangents, sometimes as chords. The arclength was supposed to be the sum of the lengths of these infinitely small parts.

Not everybody was content with these arguments and Fermat, for instance, generally took great care to provide strict proofs in every special case. Others did not, but all had the feeling of being on safe ground. "It would be easy," Fermat says somewhere, "to give a proof in the manner of Archimedes, but I content myself by saying this once in order to avoid endless repeti-tions." Pascal assures his readers that the two meth-ods—Archimedes and the principle of infinitesimals-—only differ in the manner of speaking, and Barrow remarks nonchalantly that "it would be easy to add a long proof in the style of Archimedes, but to what purpose?" As time passes the references to Archimedes tend to become mere formalities, often used to give an air of respectability to methods which Archimedes himself would certainly not have endorsed.

Some one hundred years later, all these consid-erations were taken care of by the concept of a limit. It is also not difficult to find quotations from, e. g., Pascal or Newton, where this concept occurs more or less explicitly, but we have only to read them in context to realize that the time was not yet ripe for a systematic theory. Instead, the really significant steps forward were made when Newton and Leibniz turned away from the past, justifying infinitesimal calculus by its fertility and coherence rather than by rigorous proofs.

With this we come to the second theme, the connec-tion between problems. Nowadays, we compute vol-umes, areas, and arclengths using a single operation, *integration*; and we treat problems about tangents, maxima, and minima using another one, *differentiation*. These two, on the other hand, are connected via the main theorem of integral calculus. The Greeks had different methods for all these geometric problems, and not even Archimedes saw that they are connected when seen from a higher and more abstract point of view. But the seventeenth century continued where Archime-des had left off. Within a period of 50 years, Cavalieri, Fermat, Huygens, Barrow, and Wallis succeeded in reducing many computations of volumes, areas, and arclengths to the problem of integrating (or, using their own term, *finding the quadrature* of) certain simple functions, e. g., entire or even fractional powers of x. Some of these quadratures were found and others were guessed at. Using special methods they had also found the tangents of certain curves. The decisive steps were then taken by Newton and Leibniz, both of whom introduced a special notation for the derivative of a function, and by Leibniz, who did the same for the integral and gave the algebraic formulas governing the use of these notations. This made all previous work obsolete. The new calculus had lucid formulas and simple procedures for computing volumes, areas, arclengths, and tangents using only the two basic operations, integration and differentiation. Leibniz's notations became universally accepted. The first calculus textbook, published shortly before 1700 by the Marquess de l'Hospital, was an adaptation of a manuscript by Johann Bernoulli, a pupil of Leibniz. Together with infinitesimal calculus, a new tool of analysis was invented, the power series. In 1668 Mercator had made the sensational discovery that the logarithm could be developed into a power series, and Newton, Gregory, and Leibniz competed to find power series for the basic functions.

Our third theme is the balance between geometry and algebra. The mathematics of the Greeks was geometric. In the works of Archimedes there is not a single formula; everything is expressed in words and figures. The seventeenth century found this geometric method to be a straitjacket and finally got rid of it. The beginnings had already been made. The Arabic numerals, based on the positional system, had proved themselves superior in practice to the Roman numerals, computations with letters had been gradually accepted, and algebra and the theory of equations as we know them now had been studied intensively in Italy in the sixteenth century. The usefulness of algebra had become obvious. Using simple school algebra it is child's play to prove, for instance, the basic lemma of Archimedes' treatise on spirals, whereas without algebra it is a feat. Galilei sticks to geometry but Fermat already uses algebra rather freely, and analysis gets less and less geometric until Leibniz creates the new calculus giving analysis an algebraic form. But the geometric tradition put up a stubborn resistance. The most striking example of this is *Principia*, where the terminology and the proofs are geometric, although Newton himself had achieved his results using calculus. Only afterwards did he give them a geometric form.

After its breakthrough around 1700, infinitesimal calculus was to consolidate, grow, and find a host of new applications. But the heroic time was over—the rapid development, the great discoveries, and the hard fights.

EXERCISES AND QUESTIONS

1. Harvard College was founded in the seventeenth century—in 1636, to be exact—and by the end of the century there were several institutions of higher learning in the American colonies. Why is there no mention of American mathematics in the excerpt?

2. "Barrow remarks nonchalantly that 'it would be easy to add a long proof in the style of Archimedes, but to what purpose?'" To what purpose indeed? Why should mathematicians bother to give formal proofs of things that they know are true?

3. "In 1668 Mercator had made the sensational discovery that the logarithm could be developed into a power series."

(a) (an exercise) Derive the power series for $\ln(1 - x)$ by writing the geometric series that adds up to $1/(1 - x)$ and integrating it.

(b) (a question) How did Mercator find his series? To find the answer, you will have to do some library research.

FERMAT (1601-1665)

by George F. Simmons

When Fermat's name is mentioned (which does not happen very much in casual conversation) people think of Fermat's Last Theorem if they think of anything. This is yet one more example of the injustices of history. Immortality is often given to the wrong people, and for the wrong reasons. Fermat deserves immortality, but not for his wrong statement about proving that $x^n + y^n = z^n$ has no non-zero integer solutions when $n \geq 3$. Fermat did analytic geometry. He was the first to write down equations for straight lines, parabolas, circles, and ellipses. Even though his coordinate system had no negative coordinates and his equations were things like "B quad. – A quad. aequetur E quad." (that is, $B^2 - A^2 = E^2$), he knew that curves had equations and equations represented curves. Fermat did calculus. He found tangent lines, he found maximums and minimums, and he did integrals, all before Newton or Leibniz were born. Fermat did probability. Pascal asked him a question about dice (that Pascal couldn't answer), Fermat answered it, and the exchange of letters marked the beginning of a huge and useful part of mathematics. As if that wasn't enough, Fermat did optics and number theory too. Fermat had the greatest mathematical mind of the seventeenth century, except maybe for Newton's, and that's why he deserves immortality.

All the while Fermat was doing these amazing things, he was earning his living as a lawyer and judge. His mathematical work was for his leisure time, more or less as a hobby. That is why he did not write up and publish his discoveries, and that is one reason why his work is not better appreciated. It is too bad that the seventeenth century was not a time when a mathematician could be paid for doing mathematics. If Fermat had been able to devote all of his time and all of his astonishing mind to the subject, there is no telling how the course of mathematical history would have been changed. His time would have been much better spent on mathematics than on the legal wranglings of Frenchmen, especially since there is some evidence that he was not a very good judge. Perhaps thoughts of mathematics were distracting him from the dreary disputes that he no doubt had to spend his days listening to.

We will never know what would have been, but we can learn about what was. The following selection gives a short sketch of the essential Fermat.

... a master of masters.
    E. T. Bell

Pierre de Fermat was perhaps the greatest mathematician of the seventeenth century, but his influence was limited by his lack of interest in publishing his discoveries, which are known mainly from letters to friends and marginal notes in his copy of the *Arithmetica* of Diophantus. By profession he was a lawyer and a member of the provincial supreme court in Toulouse, in southwestern France. However, his hobby and private passion was mathematics, and his casual creativity was one of the wonders of the age to the few who knew about it.

His letters suggest that he was a shy and retiring man, courteous and affable, but slightly remote. His outward life was as quiet and orderly as one would expect of a provincial judge with a sense of responsibility toward his work. Fortunately, this work was not too demanding, and left ample leisure for the extraordinary inner life that flourished by lamplight in the silence of his study at night. He was a lover of classical learning, and his own mathematical ideas grew in part out of his intimate familiarity with the works of Archimedes, Apollonius, Diophantus, and Pappus. Though he was a genius of the first magnitude, he seems to have thought of himself as at best a rather clever fellow with a few good ideas, and not at all in the same class with the masters of Greek antiquity.

Father Mersenne in Paris heard about some of Fermat's researches from a mutual friend, and wrote to him in 1636 inviting him to share his discoveries with the Parisian mathematicians. If Fermat was surprised to receive this letter, Mersenne was even more surprised at the reply, and at the cascade of letters that followed over the years, to him and also to other members of his circle. Fermat's letters were packed with ideas and discoveries, and were sometimes accompanied by short expository essays in which he briefly described a few of his methods. These essays were handwritten in Latin and were excitedly passed from one person to another in the Mersenne group. To the mathematicians in Paris, who never met him personally, he sometimes seemed to be a looming, faceless shadow dominating all their efforts, a mysterious magician buried in the country who invariably solved the problems they proposed and in return proposed problems they could not solve—and then genially furnished the solutions on request. He enjoyed challenges himself, and naively took it for granted that his correspondents did too. For instance, Mersenne once wrote to him asking whether the very large number 100,895,598,169 is prime or not. Such questions often take years to answer, but Fermat replied without hesitation that this number is the product of 112,303 and 898,423, and that each of these factors is prime—and to this day no one knows how he did it. The unfortunate Descartes locked horns with him several times, on issues that he considered crucial both to his reputation as a mathematician and to the success of his philosophy. As an outsider Fermat knew nothing about Descartes's monumental egotism and touchy disposition, and with calm courtesy demolished him on each occasion. Wonder, exasperation, and chagrin were apparently common emotions among those who came into contact with Fermat's mind.

He invented analytic geometry in 1629 and described his ideas in a short work entitled *Introduction to Plane and Solid Loci*, which circulated in manuscript from early 1637 on but was not published during his lifetime. The credit for this achievement has usually been given to Descartes on the basis of his *Geometry*, which was published late in 1637 as an appendix to his famous *Discourse on Method*. However, nothing that we would recognize as analytic geometry can be found in Descartes' essay, except perhaps the idea of using algebra as a language for discussing geometric problems. Fermat had the same idea, but did something important with it: he introduced perpendicular axes and found the general equations of straight lines and circles and the simplest equations of parabolas, ellipses, and hyperbolas; and he further showed in a fairly complete and systematic way that every first- or second-degree equation can be reduced to one of these types. None of this is in Descartes' essay; but to give him his due, he did introduce several notational conventions that are still with us—which gives his work a modern appearance—while Fermat used an older and now archaic

algebraic symbolism. The result is that superficially Descartes' essay looks as if it might be analytic geometry, but isn't; while Fermat's doesn't look it, but is. Descartes certainly knew some analytic geometry by the late 1630s; but since he had possession of the original manuscript of the *Introduction* several months before the publication of his own *Geometry*, it may be surmised that much of what he knew he learned from Fermat.

The invention of calculus is usually credited to Newton and Leibniz, whose ideas and methods were not published until about 20 years after Fermat's death. However, if differential calculus is considered to be the mathematics of finding maxima and minima of functions and drawing tangents to curves, then Fermat was the true creator of this subject as early as 1629, more than a decade before either Newton or Leibniz was born. With his usual honesty on such matters, Newton stated—in a letter that was discovered only in 1934—that his own early ideas about calculus came directly "from Fermat's way of drawing tangents."

So few curves were known before Fermat's time that no one had felt any need to improve upon the old and comparatively useless idea that a tangent is a line that touches a curve at one and only one point. However, with the aid of his new analytic geometry, Fermat was able not only to find the equations of familiar classic curves, but also to construct a multitude of new curves by simply writing down various equations and considering the corresponding graphs. This great increase in the variety of curves that were available for study aroused his interest in what came to be called "the problem of tangents."

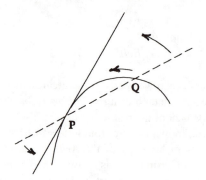

Figure 1

What Newton acknowledged in the remark quoted above is that Fermat was the first to arrive at the modern concept of the tangent line to a given curve at a given point *P*. (Figure 1.) In essence, he took a second nearby point Q on the curve, drew the secant line PQ, and considered the tangent line at P to be the limiting position of the secant as Q slides along the curve toward P. Even more important, this qualitative idea served him as a stepping-stone to quantitative methods for calculating the exact slope of the tangent.

Fermat's methods were of such critical significance for the future of mathematics and science that we pause briefly to consider how they arose.

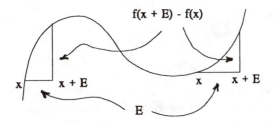

Figure 2

While sketching the graphs of certain polynomial functions $y = f(x)$, he hit upon a very ingenious idea for locating points at which such a function assumes a maximum or minimum value. He compared the value $f(x)$ at a point x with the value $f(x + E)$ at a nearby point $x + E$ (see Figure 2). For most x's the difference between these values, $f(x + E) - f(x)$, is not small compared with E, but he noticed that at the top or bottom of a curve this difference is much smaller than E and diminishes faster than E does. This idea gave him the approximate equation

$$\frac{f(x + E) - f(x)}{E} = 0,$$

which becomes more and more nearly correct as the interval E is taken smaller and smaller. With this in mind, he next put $E = 0$ to obtain the equation

$$\left[\frac{f(x + E) - f(x)}{E} \right]_{E=0} = 0,$$

According to Fermat, this equation is exactly correct at the maximum and minimum points on the curve, and solving it yields the values of x that correspond to these points. The legitimacy of this procedure was a subject of acute controversy for many years. However, students of calculus will recognize that Fermat's method amounts to calculating the derivative

$$f'(x) = \lim_{E \to 0} \frac{f(x + E) - f(x)}{E}$$

and setting this equal to zero, which is just what we do in calculus today, except that we customarily use the symbol Δx in place of his E.

In one of the first tests of his procedure, he gave the following proof of Euclid's theorem that the largest rectangle with a given perimeter is a square. If B is half the perimeter and one side is x, then $B - x$ is the adjacent side, and the area is $f(x) = x(B - x)$. To maximize this area by the process described above, compute

$$f(x + E) - f(x) = (x + E)[(B - (x + E)] - x(B - x)$$

$$= EB - 2Ex - E^2,$$

$$\frac{f(x + E) - f(x)}{E} = B - 2x - E,$$

and

$$\left[\frac{f(x + E) - f(x)}{E} \right]_{E=0} = B - 2x.$$

Fermat's equation is therefore $B - 2x = 0$, so

$$x = \frac{B}{2}, \quad B - x = \frac{B}{2},$$

and the largest rectangle is a square. When he reached this conclusion, he remarked with justifiable pride, "We can hardly expect to find a more general method." He also found the shape of the largest cylinder that can be inscribed in a given sphere (ratio of height to diameter of base = $\sqrt{2}/2$) and solved many similar problems that are familiar in calculus courses today.

Fermat's most memorable application of his method of maxima and minima was his analysis of the refraction of light. The qualitative phenomenon had of course been known for a very long time: that when a ray of light passes from a less dense medium into a denser medium—for instance, from air into water—it is refracted toward the perpendicular (Figure 3). The quantitative description of refraction was apparently discovered experimentally by the Dutch scientist Snell in 1621. He found that when the direction of the incident ray is altered, the ratio of the sines of the two indicated angles remains constant,

$$\frac{\sin \alpha}{\sin \beta} = a \text{ constant.}$$

This sine law was first published by Descartes in 1637

(without any mention of Snell), and he purported to prove it in a form equivalent to

$$\frac{\sin \alpha}{\sin \beta} = \frac{v_w}{v_a},$$

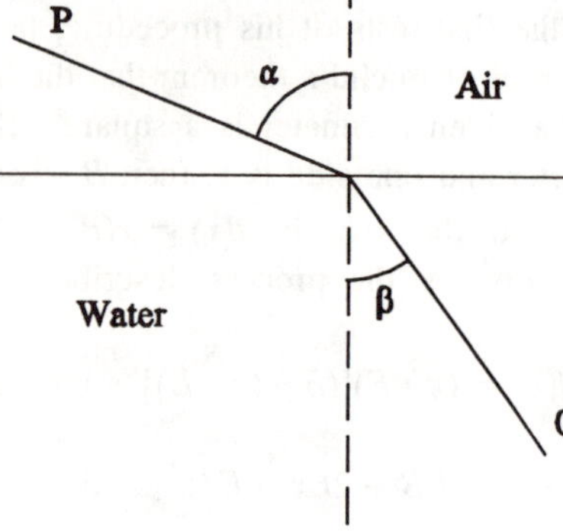

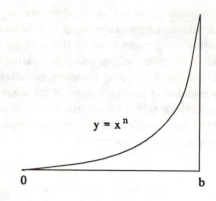

Figure 4

Figure 3

where v_a and v_w are the velocities of light in air and water. Descartes based his argument on a fanciful model and on the metaphysically inspired opinion that light travels faster in a denser medium. Fermat rejected both the opinion ("shocking to common sense") and the argument ("demonstrations which do not force belief cannot bear this name"). After many years of passive skepticism, he actively confronted the problem in 1657 and proved the correct law himself,

$$\frac{\sin \alpha}{\sin \beta} = \frac{v_a}{v_w}.$$

The foundation of his reasoning was the hypothesis that the actual path along which the ray of light travels from P to Q is that which minimizes the total time of travel—now known as *Fermat's principle of least time*. The principle of least time led to the calculus of variations created by Euler and Lagrange in the next century, and on from this discipline to Hamilton's principle of least action, which has been one of the most important unifying ideas in modern physical science.

Fermat's method for finding tangents developed out of his approach to problems of maxima and minima, and was the occasion of yet another clash with Descartes. When the famous philosopher was informed of Fermat's method by Mersenne, he attacked its generality, challenging Fermat to find the tangent to the curve $x^3 - y^3 = 3axy$, and foolishly predicted that he would fail. Descartes was unable to cope with this problem himself, and was intensely irritated when Fermat solved it easily.

These successes in the early stage of differential

calculus were matched by comparable achievements in integral calculus. We mention only one: his calculation of the area under the curve $y = x^n$ from $x = 0$ to $x = b$ for any positive integer n (see Figure 4). In modern notation, this amounts to the evaluation of the integral

$$\int_0^b x^n \, dx = \frac{b^{n+1}}{n+1}.$$

The Italian mathematician Cavalieri had proved this formula by increasingly laborious methods for $n = 1, 2, ..., 9$, but bogged down at $n = 10$. Fermat devised a beautiful new approach that worked with equal ease for all n's.

In the light of all these accomplishments, it may reasonably be asked why Newton and Leibniz are commonly regarded as the inventors of calculus, and not Fermat. The answer is that Fermat's activities came a little too early, before the essential features of the subject had fully emerged. He had pregnant ideas and solved many individual calculus problems; but he did not isolate the explicit calculation of derivatives as a formal process, he had no notion of indefinite integrals, he apparently never noticed the Fundamental Theorem of Calculus that binds together the two parts of the subject, and he didn't even begin to develop the rich structure of computational machinery on which the more advanced applications depend. Newton and Leibniz did all these things, and thereby transformed a collection of ingenious devices into a problem-solving tool of great power and efficiency.

The mind of Fermat had as many facets as a well-cut diamond and threw off flashes of light in surprising directions. A minor but significant chapter in his intellectual life began when Blaise Pascal, the precocious dilettante of mathematics and physics, wrote to him in 1654 with some questions concerning gambling games that are played with dice. In the ensuing correspondence over the next several months, they jointly developed the basic concepts of the theory of probability. This was the effective beginning of a subject whose influence is now felt in almost every corner of modern life, ranging from such practical fields as insurance and industrial quality control to the esoteric disciplines of genetics, quantum mechanics, and the kinetic theory of gases. However, neither man carried his ideas very far. Pascal was soon caught up in the paroxysms of piety that blighted the remainder of his short life, and Fermat dropped the subject because he had other, more compelling mathematical interests.

The many remarkable achievements sketched here—in analytic geometry, calculus, optics, and the theory of probability—would have sufficed to place Fermat among the outstanding mathematicians of the seventeenth century if he had done nothing else. But to him these activities were all of minor importance compared with the consuming passion of his life, the theory of numbers. It was here that his genius shone most brilliantly, for his insight into the properties of the familiar but mysterious positive integers has perhaps never been equaled. He was the sole and undisputed founder of the modern era in this subject, without any rivals and with few followers until the time of Euler and Lagrange in the next century. Pascal, who called him *le premier homme du monde*—"the foremost man in the world"—wrote to him and said: "Look elsewhere for someone who can follow you in your researches about numbers. For my part, I confess that they are far beyond me, and I am competent only to admire them."

QUESTIONS AND EXERCISES

1. "[Descartes challenged] Fermat to find the tangent to the curve $x^3 + y^3 = 3axy$, and foolishly predicted that he would fail. Descartes was unable to cope with this problem himself." Succeed where Descartes failed and find the equation of the tangent line to the curve at $(3a/2, 3a/2)$.

2. The excerpt, cut off short because the theory of numbers has nothing to do with calculus, goes on to give some of Fermat's discoveries in number theory. Here are two of them, included to illustrate a kind of mathematics quite different from finding tangents and areas:

(a) Fermat asserted that every positive integer is the sum of at most three triangular numbers (1, 3, 6, 10, 15, 21, ...), at most four square numbers (1, 4, 9, 16, 25, 36, ...), at most five pentagonal numbers (1, 5, 12, 22, 35, 50, ...), ... and so on. He was right, but it was not until 1815 that Cauchy could prove that. Check Fermat's result by writing 17, 18, 19, 20, and 21 each as a sum of as few triangular, square, and pentagonal numbers as possible.

(b) Fermat also discovered that if p is a prime number (2, 3, 5, 7, 11, 13, 17, ...) and n is a positive integer not divisible by p, then $n^{p-1} - 1$ is divisible by p. Check Fermat's result for $(p, n) = (5, 4)$, $(3, 8)$, and (using a calculator) $(13, 7)$.

3. "His lack of influence was limited by his lack of interest in publishing his discoveries." Today it seems as if all the people in the world are interested in getting as much recognition and publicity as they deserve, and then some. Why do you think that Fermat was not that way?

4. "Pascal was soon caught up in the paroxysms of piety that blighted the remainder of his short life." Could that be stated in another way? What does it tell you about the author's attitudes?

ON THE SEASHORE

by Eric Temple Bell

No one should study calculus and remain ignorant of who Isaac Newton was and what he did. No one should study *any* science and remain unaware of Newton. No one should study *anything* and not know about Newton. Plato said, "He is unworthy of the name of man who is ignorant of the fact that the diagonal of a square is incommensurable with its side." I say, "He or she is unworthy of the name of educated person who does not know a few things about Isaac Newton." For a variety of reasons, my saying does not have the stately ring of Plato's, but it is nevertheless true. So, if you wish to be worthy of the name of educated person, read the following selection.

The selection is long for two reasons. The first is that Newton is important and thus worth the space. The second is that the author, in the fashion of mathematicians, cannot confine himself to telling only the details of Newton's life; he must also try to explain the fundamental ideas of calculus. This material may be skipped if you already understand completely the idea of derivative. If you do not, here is one more chance for the breakthrough to understanding to take place, one more chance for that great moment when you say to yourself, "Ah, *now* I see. It's so simple! How could anyone not understand that?"

The selection was written more than fifty years ago, a different time when people did not think or write exactly as we do. Also, Professor Bell was not a professional historian, nor did he hesitate to express his opinions, sometimes as facts, when he felt the urge. Thus, the selection is not the last word on Newton, but it is excellent first one.

"I do not know what I may appear to the world; but to myself I seem to have been only like a boy playing on the seashore, and diverting myself in now and then finding a smoother pebble or a prettier shell than ordinary, whilst the great ocean of truth lay all undiscovered before me."

Such was Isaac Newton's estimate of himself toward the close of his long life. Yet his successors capable of appreciating his work almost without exception have pointed to Newton as the supreme intellect that the human race has produced—"he who in genius surpassed the human kind."

Isaac Newton, born on Christmas Day ("oldstyle" of dating), 1642, the year of Galileo's death, came of a family of small but independent farmers, living in the manor house of the hamlet of Woolsthorpe, about eight miles south of Grantham in the county of Lincoln, England. His father, also named Isaac, died at the age of thirty seven before the birth of his son. Newton was a premature child. At birth he was so frail and puny that two women who had gone to a neighbor to get "a tonic" for the infant expected to find him dead on their return. His mother said he was so undersized at birth that a quart mug could easily have contained all there was of him.

Not enough of Newton's ancestry is known to interest students of heredity. His father was described by neighbors as "a wild, extravagant, weak man"; his mother, Hannah Ayscough, was thrifty, industrious, and a capable manageress. After her husband's death Mrs. Newton was recommended as a prospective wife to an old bachelor as "an extraordinary good woman." The cautious bachelor, the Reverend Barnabas Smith, of the neighboring parish of North Witham, married the widow on this testimonial. Mrs. Smith left her three-year-old to the care of his grandmother. By her second marriage she had three children, none of whom exhibited any remarkable ability. From the property of his mother's second marriage and his father's estate Newton ultimately acquired an income of about £80 a year, which of course meant much more in the seventeenth century than it would now. Newton was not one of the great mathematicians who had to contend with poverty.

As a child Newton was not robust and was forced to shun the rough games of boys his own age. Instead of amusing himself in the usual way, Newton invented his own diversions, in which his genius first showed up. It is sometimes said that Newton was not precocious. This may be true so far as mathematics is concerned, but if it is so in other respects a new definition of precocity is required. The unsurpassed experimental genius which Newton was to exhibit as an explorer in the mysteries of light is certainly evident in the ingenuity of his boyish amusements. Kites with lanterns to scare the credulous villagers at night, perfectly constructed mechanical toys which he made entirely by himself and which worked—waterwheels, a mill that ground wheat into snowy flour, with a greedy mouse (who devoured most of the profits) as both miner and motive power, workboxes and toys for his many little girl friends, drawings, sundials, and a wooden clock (that went) for himself—such were some of the things with which this "un-precocious" boy sought to divert the interests of his playmates into "more philosophical" channels. In addition to these more noticeable evidences of talent far above the ordinary, Newton read extensively and jotted down all manner of mysterious recipes and out-of-the-way observations in his notebook. To rate such a boy merely the normal, wholesome lad he appeared to his village friends is to miss the obvious.

The earliest part of Newton's education was received in the common village schools of his vicinity. A maternal uncle, the Reverend William Ayscough, seems to have been the first to recognize that Newton was something unusual. A Cambridge graduate himself, Ayscough finally persuaded Newton's mother to send her son to Cambridge instead of keeping him at home, as she had planned, to help her manage the farm on her return to Woolsthorpe after her husband's death when Newton was fifteen.

Before this, however, Newton had crossed his Rubicon on his own initiative. On his uncle's advice he had been sent to the Grantham Grammar School. While there, in the lowest form but one, he was tormented by the school bully who one day kicked Newton in the stomach, causing him much physical pain and mental anguish. Encouraged by one of the schoolmasters, Newton challenged the bully to a fair fight, thrashed him, and, as a final mark of humiliation, rubbed his enemy's cowardly nose on the wall of the church. Up till this young Newton had shown no great interest in his lessons. He now set out to prove his head as good as his fists and quickly rose to the distinction of top boy in the school. The Headmaster and Uncle Ayscough agreed that Newton was good enough for Cambridge, but the decisive die was thrown when Ayscough caught his nephew reading under a hedge when he was supposed to be helping a farmhand to do the marketing.

While at the Grantham Grammar School, and subsequently while preparing for Cambridge, Newton lodged with a Mr. Clarke, the village apothecary. In the apothecary's attic Newton found a parcel of old books, which he devoured, and in the house generally, Clarke's stepdaughter, Miss Storey, with whom he fell in love and to whom he became engaged before leaving Woolsthorpe for Cambridge in June 1661, at the age of nineteen. But although Newton cherished a warm affection for his first and only sweetheart all her life, absence and growing absorption in his work thrust romance into the background and Newton never married. Miss Storey became Mrs. Vincent.

Before going on to Newton's student career at Trinity College we may take a short look at the England of his times and some of the scientific knowledge to which the young man fell heir. The bull-headed and bigoted Scottish Stuarts had undertaken to rule England according to the divine rights they claimed were vested in them, with the not uncommon result that mere human beings resented the assumption of celestial authority and rebelled against the sublime conceit, the stupidity, and the incompetence of their rulers. Newton grew up in an atmosphere of civil war—political and religious—in which Puritans and Royalists alike impartially looted whatever was needed to keep their ragged armies fighting. Charles I (born in 1608, beheaded in 1649) had done everything in his power to suppress parliament; but in spite of his ruthless extortions and the villainously able backing of his own Star Chamber through its brilliant perversion of the law and common justice, he was no match for the dour Puritans under Oliver Cromwell, who in his time was to back his butcheries and his roughshod march over parliament by an appeal to the divine justice of his holy cause.

All this brutality and holy hypocrisy had a most salutary effect on young Newton's character: he grew up with a fierce hatred of tyranny, subterfuge, and oppression, and when King James later sought to meddle repressively in University affairs, the mathematician and natural philosopher did not need to learn that a resolute show of backbone and a united front on the part of those whose liberties are endangered is the most effective defense against a coalition of unscrupulous politicians; he knew it by observation and by instinct.

To Newton is attributed the saying, "If I have seen a little farther than others it is because I have stood on the shoulders of giants." He had. Among the tallest of these giants were Descartes, Kepler, and Galileo. From Descartes, Newton inherited analytic geometry, which he found difficult at first; from Kepler three fundamental laws of planetary motion, discovered empirically after twenty two years of inhuman calculation; while from Galileo he acquired the first two of the three laws of motion which were to be the cornerstone of his own dynamics. But bricks do not make a building; Newton was the architect of dynamics and celestial mechanics.

As Kepler's laws were to play the role of hero in Newton's development of his law of universal gravitation they may be stated here.

I. The planets move round the Sun in ellipses; the Sun is at one focus of these ellipses. (If S, S' are the foci and P any position of a planet in its orbit, $SP + S'P$ is always equal to AA', the major axis of the ellipse: see Figure 1.)

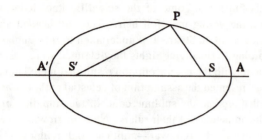

Figure 1

II. The line joining the Sun and a planet sweeps out equal areas in equal times.

III. The square of the time for one complete revolution of each planet is proportional to the cube of its mean [or average] distance from the Sun.

These laws can he proved in a page or two by means of the calculus applied to Newton's law of universal gravitation:

Any two particles of matter in the universe attract one another with a force which is directly proportional to the product of their masses and inversely proportional to the square of the distance between them.

Thus if m, M are the masses of the two particles and d the distance between them (all measured in appropriate units), the force of attraction between them is $\dfrac{kmM}{d^2}$, where k is some constant number (by suitably choosing the units of mass and distance k may be taken equal to 1, so that the attraction is simply $\dfrac{mM}{d^2}$).

For completeness we state Newton's three laws of motion.

I. *Every body will continue in its state of rest or of uniform* [unaccelerated] *motion in a straight line except in so far as it is compelled to change that state by impressed force.*

II. *Rate of change of momentum* ["mass times velocity," mass and velocity being in appropriate units] *is proportional to the impressed force and takes place in the line in which the force acts.*

III. *Action and reaction* [as in the collision on a frictionless table of perfectly elastic billiard balls] *are equal and opposite* [the momentum one ball loses is gained by the other].

The most important thing for mathematics in all this is the phrase opening the statement of the second law of motion, *rate of change*. What is a rate, and how shall it be measured? Momentum, as noted is "mass times velocity." The masses which Newton discussed were assumed to remain constant during their motion—not like the electrons and other particles of current physics whose masses increase appreciably as their velocity approaches a measurable fraction of that of light. Thus, to investigate "rate of change of momentum," it sufficed Newton to clarify *velocity*, which is rate of change of position. His solution of this problem—giving a workable mathematical method for investigating the velocity of any particle moving in any continuous manner, no matter how erratic—gave him the master key to the whole mystery of rates and their measurement, namely the *differential* calculus.

A similar problem growing out of rates put the *integral* calculus into his hands. How shall the total distance passed over in a given time by a moving particle whose velocity is varying continuously from instant to instant be calculated? Answering this or similar problems, some phrased geometrically, Newton came upon the integral calculus. Finally, pondering the two types of problem together, Newton made a capital discovery: he saw that the differential calculus and the integral calculus are intimately and reciprocally related by what is today called "the fundamental theorem of the calculus"—which will be described in the proper place.

In addition to what Newton inherited from his predecessors in science and mathematics he received from the spirit of his age two further gifts, a passion for theology and an unquenchable thirst for the mysteries of alchemy. To censure him for devoting his unsurpassed intellect to these things, which would now be considered unworthy of his serious effort, is to censure oneself. For in Newton's day alchemy *was* chemistry and it had *not* been shown that there was nothing much in it—except what was to come out of it, namely modern chemistry; and Newton, as a man of inborn scientific spirit, undertook to find out *by experiment* exactly what the claims of the alchemists amounted to.

As for theology, Newton was an unquestioning believer in an all-wise Creator of the universe and in his own inability—like that of the boy on the seashore—to fathom the entire ocean of truth in all its depths. He therefore believed that there were not only many things in heaven beyond philosophy but plenty on earth as well, and he made it his business to understand for himself what the majority of intelligent men of his time accepted without dispute (to them it was as natural as common sense)—the traditional account of creation.

He therefore put what he considered his really serious efforts on attempts to prove that the prophecies of Daniel and the poetry of the Apocalypse make sense, and on chronological researches whose object was to harmonize the dates of the Old Testament with those of history. In Newton's day theology was still queen of the sciences and she sometimes ruled her obstreperous subjects with a rod of brass and a head of cast iron. Newton however did permit his rational science to influence his beliefs to the extent of making him what would now be called a Unitarian.

In June, 1661 Newton entered Trinity College, Cambridge, as a subsizar—a student who (in those days) earned his expenses by menial service. Civil war, the restoration of the monarchy in 1661, and uninspired toadying to the Crown on the part of the University had all brought Cambridge to one of the low-water marks in its history as an educational institution when Newton took up his residence. Nevertheless young Newton, lonely at first, quickly found himself and became absorbed in his work.

In mathematics Newton's teacher was Dr. Isaac Barrow (1630-1677), a theologian and mathematician of whom it has been said that brilliant and original as he undoubtedly was in mathematics, he had the misfortune to be the morning star heralding Newton's sun. Barrow gladly recognized that a greater than himself had arrived, and when (1669) the strategic moment came he resigned the Lucasian Professorship of Mathematics (of which he was the first holder) in favor of his incomparable pupil. Barrow's geometrical lectures dealt among other things with his own methods for finding areas and drawing tangents to curves—essentially the key problems of the integral and the differential calculus respectively, and there can be no doubt that these lectures inspired Newton to his own attack.

The record of Newton's undergraduate life is disappointingly meager. He seems to have made no very great impression on his fellow students, nor do his brief, perfunctory letters home tell anything of interest. The first two years were spent mastering elementary mathematics. If there is any reliable account of Newton's sudden maturity as a discoverer, none of his modern biographers seem to have located it. Beyond the fact that in the three years 1664-66 (age twenty one to twenty three) he laid the foundation of all his subsequent work in science and mathematics, and that incessant work and late hours brought on an illness, we know nothing definite. Newton's tendency to secretiveness about his discoveries has also played its part in deepening the mystery.

On the purely human side Newton was normal enough as an undergraduate to relax occasionally, and there is a record in his account book of several sessions at the tavern and two losses at cards. He took his B. A. degree in January, 1664.

The Great Plague (bubonic plague) of 1664-65, with its milder recurrence the following year, gave Newton his great if forced opportunity. The University was closed, and for the better part of two years Newton retired to meditate at Woolsthorpe. Up to then he had done nothing remarkable—except make himself ill by too assiduous observation of a comet and lunar halos—or, if he had, it was a secret. In these two years he invented the method of fluxions (the calculus), discovered the law of universal gravitation, and proved experimentally that white light is composed of light of all the colors. All this before he was twenty five.

A manuscript dated May 20, 1665, shows that Newton at the age of twenty three had sufficiently developed the principles of the calculus to be able to find the tangent and curvature at any point of any continuous curve. He called his method "fluxions"—from the idea of "flowing" or variable quantities and their rates of "flow" or "growth." His discovery of the binomial theorem, an essential step towards a fully developed calculus, preceded this.

The binomial theorem generalizes the simple results like

$$(a + b)^2 = a^2 + 2ab + b^2,$$

$$(a + b)^3 = a^3 + 3a^2b + 3ab^2 + b^3,$$

and so on, which are found by direct calculation; namely,

$$(a + b)^n = a^n + \frac{n}{1}a^{n-1}b + \frac{n(n-1)}{1 \cdot 2}a^{n-1}b^2$$

$$+ \frac{n(n-1)(n-2)}{1 \cdot 2 \cdot 3}a^{n-3}b^3 + \ldots$$

where the dots indicate that the series is to be continued according to the same law as that indicated for the terms written; the next term is

$$\frac{n(n-1)(n-2)(n-3)}{1 \cdot 2 \cdot 3 \cdot 4}a^{n-4}b^4.$$

If n is one of the positive integers 1, 2, 3, ..., the series automatically terminates after precisely $n + 1$ terms. This much is easily proved (as in the school algebras) by mathematical induction.

But if n is not a positive integer, the series does not terminate, and the method of proof is inapplicable. As a proof of the binomial theorem for fractional and negative values of n (also for more general values), with a statement on the necessary restrictions on a, b came only in the nineteenth century, we need merely state here that in extending the theorem to these values of n Newton satisfied himself that the theorem was correct for such values of a, b as he had occasion to consider in his work.

If all modern requirements are similarly ignored in the manner of the seventeenth century it is easy to see how the calculus finally got itself invented. The underlying notions are those of *variable*, *function*, and *limit*. The last took long to clarify.

A letter, say s, which can take on several different values during the course of a mathematical investigation is called a *variable*; for example s is a variable if it denotes the height of a falling body above the earth.

The word *function* (or its Latin equivalent) seems to have been introduced into mathematics by Leibniz in 1694; the concept now dominates much of mathematics and is indispensable in science. Since Leibniz' time the concept has been made more precise. If y and x are two variables so related that whenever a numerical value is assigned to x there is determined a numerical value of y, then y is called a (one-valued, or *uniform*) function of x, and this is symbolized by writing $y = f(x)$.

Instead of attempting to give a modern definition of a *limit* we shall content ourselves with one of the simplest examples of the sort which led the followers of Newton and Leibniz (the former especially) to the use of limits in discussing rates of change. To the early developers of the calculus the notions of variables and limits were intuitive; to us they are extremely subtle concepts hedged about with thickets of semi-metaphysical mysteries concerning the nature of numbers, both rational and irrational.

Let y be a function of x, say $y = f(x)$. *The rate of change of y with respect to x, or, as it is called, the derivative of y with respect to x, is defined as follows.* To x is given any increment, say Δx (read, "increment of x"), so that x becomes $x + \Delta x$, and $f(x)$ or y, becomes $f(x + \Delta x)$. The corresponding increment, Δy, of y is its *new* value *minus* its initial value; namely, $\Delta y = f(x + \Delta x) - f(x)$. As a crude approximation to the rate of change of y with respect to x we may take, by our intuitive notion of a rate as an "average," the result of dividing the increment of y by the increment of x, that is, $\frac{\Delta y}{\Delta x}$.

But this obviously is too crude, as both of x and y are varying and we cannot say that this average represents the rate for *any particular* value of x. Accordingly, we decrease the increment Δx *indefinitely*, till, "in

the limit" Δx approaches zero, and follow the "average" $\frac{\Delta y}{\Delta x}$ all through the process: Δy similarly decreases indefinitely and ultimately approaches zero; but $\frac{\Delta y}{\Delta x}$ does not, thereby, present us with the meaningless symbol 0/0, but with a definite *limiting value*, which is the required rate of change of y with respect to x.

To see how it works out, let $f(x)$ be the particular function x^2, so that $y = x^2$. Following the above outline we get first

$$\frac{\Delta y}{\Delta x} = \frac{((x + \Delta x)^2 - x^2)}{\Delta x}.$$

Nothing is yet said about limits. Simplifying the algebra we find

$$\frac{\Delta y}{\Delta x} = 2x + \Delta x.$$

Having simplified the algebra as far as possible, we *now* let Δx approach zero and see that the limiting value of $\frac{\Delta y}{\Delta x}$ is $2x$. Quite generally, in the same way, if $y = x^n$, the limiting value of $\frac{\Delta y}{\Delta x}$ is nx^{n-1}, as may be proved with the aid of the binomial theorem.

Such an argument would not satisfy a student today, but something not much better was good enough for the inventors of calculus and it will have to do for us here. If $y = f(x)$, the *limiting value* of $\frac{\Delta y}{\Delta x}$ (provided such a value exists) is called the *derivative of y with respect to x*, and is denoted by $\frac{dy}{dx}$. This symbolism is due (essentially) to Leibniz and is the one in common use today; Newton used another ($\dot{y}$) which is less convenient.

The simplest instances of rates in physics are velocity and acceleration, two of the fundamental notions of dynamics. Velocity is rate of change of *distance* (or "position," or "space") with respect to *time*; *acceleration* is the rate of change of *velocity* with respect to *time*.

If s denotes the distance traversed in the time t by a moving particle (it being assumed that the distance is a function of the time), the velocity at the time t is ds/dt. Denoting this velocity by v, we have the corresponding acceleration, dv/dt.

This introduces the idea of a *rate of a rate*, or of a *second derivative*. For in accelerated motion the velocity is not constant but variable, and hence it has a rate of change: the acceleration is a rate of change of the rate of change of distance (both rates with respect to time); and to indicate this *second* rate, or "rate of a rate," we write d^2s/dt^2 for the acceleration. This itself may have a rate of change with respect to the time; this *third* rate is written d^3s/dt^3. And so on for fourth, fifth, ... rates, namely for fourth, fifth, ... derivatives. The most important derivatives in the applications of calculus to science are the first and second.

If we now look back at what was said concerning Newton's second law of motion and compare it with the like for acceleration, we see that "forces" are proportional to the accelerations they produce. With this much we can "set up" the *differential equation* for a problem

O   F(s)   t
—————————————————
        s

which is by no means trivial—that of "central forces": a particle is attracted toward a fixed point by a force whose direction always passes through the fixed point. Given that the force varies as some function of the distance, say as $F(s)$, where s is the distance of the particle at the time t from the fixed point O, it is required to describe the motion of the particle. A little consideration will show that

$$\frac{d^2s}{dt^2} = -F(s),$$

the minus sign being taken because the attraction diminishes the velocity. This is the *differential equation* of the problem, so called because it involves a rate (the acceleration), and rates (or derivatives) are the object of investigation in the *differential* calculus.

Having transformed the problem into a differential equation we are now required to solve this equation, that is, to find the relation between s and t, or, in mathematical language, to solve the differential equation by expressing s as a function of t. This is where the difficulties begin. It may be quite easy to translate a given physical situation into a set of differential equations which no mathematician can solve. In general every essentially new problem in physics leads to types of differential equations which demand the creation of new branches of mathematics for their solution. The particular equation above can however be solved quite simply in terms of elementary functions if $F(s) = 1/s^2$ as in Newton's law of gravitational attraction. Instead of bothering with this particular equation, we shall

consider a much simpler one which will suffice to bring out the point of importance:

$$\frac{dy}{dx} = x.$$

We are given that y is a function of x whose derivative is equal to x; it is required to express y as a function of x. More generally, consider in the same way

$$\frac{dy}{dx} = f(x).$$

This asks, what is the function y (of x) whose derivative (rate of change) with respect to x is equal to $f(x)$? Provided we can find the function required (or provided such a function exists), we call it the *anti-derivative* of $f(x)$ and denote it by $\int f(x)\,dx$—for a reason that will appear presently. For the moment we need only note that $\int f(x)\,dx$ symbolizes a function (if it exists) *whose derivative* is equal to $f(x)$.

By inspection we see that the first of the above equations has the solution $\frac{1}{2}x^2 + c$, where c is a constant (number not depending on the variable x); thus

$$\int x\,dx = \frac{1}{2}x^2 + c.$$

Even this simple example may indicate that the problem of evaluating $\int f(x)\,dx$ for comparatively innocent-looking functions $f(x)$ may be beyond our powers. It does not follow that an "answer" exists at all *in terms of known functions* when an $f(x)$ is chosen at random—the odds against such a chance are an infinity of the worst sort ("non-denumerable") to one. When a physical problem leads to one of these nightmares approximate methods are applied which give the result within the desired accuracy.

With the two basic notions, $\frac{dy}{dx}$ and $\int f(x)\,dx$, of the calculus we can now describe *the fundamental theorem of the calculus* connecting them. For simplicity we shall use a diagram although this is not necessary and is undesirable in an exact account.

Consider a continuous, unlooped curve whose equation is $y = f(x)$ in Cartesian coordinates. It is required to find the area included between the curve, the x-axis and the two perpendiculars AA', BB' drawn to the x-axis from any two points A, B on the curve. The distances OA', OB' are a, b respectively—namely, the coordinates of A', B' are $(a, 0)$, $(b, 0)$. We proceed as Archimedes did, cutting the required area into parallel strips (Figure 2) of equal breadth, treating these strips as

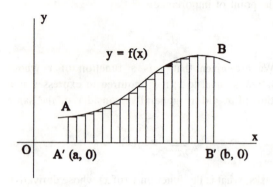

Figure 2

rectangles by disregarding the top triangular bits (one of which is shaded in the figure), adding the areas of all these rectangles, and finally evaluating *the limit of this sum* as the number of rectangles is increased indefinitely. This is all very well, but how are we to calculate the limit? The answer is surely one of the most astonishing things a mathematician ever discovered.

First, find $\int f(x)\,dx$. Say the result is $F(x)$. In this substitute a and b, getting $F(a)$ and $F(b)$. Then subtract the first from the second, $F(b) - F(a)$. *This is the required area.*

Notice the connection between $y = f(x)$, the equation of the given curve; dy/dx, which gives the *slope* of the tangent line to the curve at the point (x, y); and $\int f(x)\,dx$, or $F(x)$, which is the function whose *rate of change* with respect to x is equal to $f(x)$. We have just stated that the *area* required, which is a *limiting sum* of the kind described in connection with Archimedes, is given by $F(b) - F(a)$. Thus we have connected *slopes*, or *derivatives*, with *limiting sums*, or, as they are called, *definite integrals*. The symbol $\int$ is an old-fashioned S, the first letter of the word *Summa*.

Summing all this up in symbols, we write for the area in question $\int_a^b f(x)\,dx$; a is the *lower limit* of the sum, b the *upper limit*; and

$$\int_a^b f(x)\,dx \;=\; F(b) - F(a),$$

in which $F(b)$, $F(a)$ are calculated by evaluating the "*indefinite integral*" $\int f(x)\,dx$, namely, by finding that function $F(x)$ such that its derivative with respect to x, $dF(x)/dx$, is equal to $f(x)$. This is the fundamental theorem of the calculus as it presented itself (in a geometrical form) to Newton and independently also to Leibniz. As a caution we repeat that numerous refine-

ments demanded in a modern statement have been ignored.

Two simple but important matters may conclude this sketch of the leading notions of the calculus as they appeared to the pioneers. So far only functions of a single variable have been considered. But nature presents us with functions of several variables and even of an infinity of variables.

To take a very simple example, the volume, V, of a gas is a function of its temperature, T, and the pressure, P, on it; say $V = F(P, T)$—the actual form of the function F need not be specified here. As T, P vary, V varies. But suppose *only one* of T, P varies while the other is held constant. We are then back essentially with a function of *one* variable, and the derivative of $F(T, P)$ can be calculated with respect to this variable. If T varies while P is held constant, the derivative of $F(T, P)$ with respect to T is called the *partial derivative* (with respect to T), and to show that the variable P is being held constant, a different symbol ∂, is used for this partial derivative, $\dfrac{\partial F(T, P)}{\partial T}$. Similarly, if P varies while T is being held constant, we get $\dfrac{\partial F(T, P)}{\partial P}$. Precisely as in the case of ordinary second, third, ... derivatives, we have the like for partial derivatives; thus $\dfrac{\partial^2 F(T, P)}{\partial T^2}$ signifies the partial derivative of $\dfrac{\partial F(T, P)}{\partial T}$ with respect to T.

The great majority of the important equations of mathematical physics are *partial differential equations*. A famous example is Laplace's equation, or the "equation of continuity," which appears in the theory of Newtonian gravitation, electricity and magnetism, fluid motion, and elsewhere:

$$\frac{\partial^2 u}{\partial x^2} + \frac{\partial^2 u}{\partial y^2} + \frac{\partial^2 u}{\partial z^2} \;=\; 0.$$

In fluid motion this is the mathematical expression of the fact that a "perfect" fluid, in which there are no vortices, is indestructible. A derivation of this equation would be out of place here, but a statement of what it signifies may make it seem less mysterious. If there are no vortices in the fluid, the three component velocities parallel to the axes of x, y, z of any particle in the fluid are calculable as the partial derivatives

$$-\frac{\partial u}{\partial x}, \qquad -\frac{\partial u}{\partial y}, \qquad -\frac{\partial u}{\partial z}$$

of the *same* function u—which will be determined by the particular type of motion. Combining this fact with the obvious remark that if the fluid is incompressible and indestructible, as much fluid must flow out of any small volume in one second as flows into it; and noting that the amount of flow in one second across any small area is equal to the rate of flow multiplied by the area; we see (on combining these remarks and calculating the total inflow and total outflow) that Laplace's equation is more or less of a platitude.

The really astonishing thing about this and some other equations of mathematical physics is that a physical platitude, when subjected to mathematical reasoning, should furnish unforeseen information which is anything but platitudinous. The "anticipations" of physical phenomena mentioned in later chapters arose from such commonplaces treated mathematically.

Two very real difficulties, however, arise in this type of problem. The first concerns the physicist, who must have a feeling for what complications can be lopped off his problem, without mutilating it beyond all recognition, so that he can state it mathematically at all. The second concerns the mathematician, and this brings us to a matter of great importance—that of what are called *boundary-value problems*.

Science does not fling an equation like Laplace's at a mathematician's head and ask him to find the *general* solution. What it wants is something (usually) much more difficult to obtain, a *particular* solution which will not only satisfy the equation but which in addition will satisfy *certain auxiliary conditions* depending on the particular problem to be solved.

The point may be simply illustrated by a problem in the conduction of heat. There is a *general* solution (Fourier's) for the "motion" of heat in a conductor similar to Laplace's for fluid motion. Suppose it is required to find the final distribution of temperature in a cylindrical rod whose ends are kept at one constant temperature and whose curved surface is kept at another; "final" here means that there is a "steady state"—no further change in temperature—at all points of the rod. The solution must not only satisfy the *general* equation, it must also fit the *surface-temperatures*, or the *initial boundary conditions*.

The second is the harder part. For a cylindrical rod the problem is quite different from the corresponding problem for a bar of rectangular cross-section. The theory of *boundary-value problems* deals with the fitting of solutions of differential equations to prescribed initial conditions. It is largely a creation of the past eighty years. In a sense mathematical physics is co-extensive with the theory of boundary-value problems.

The second of Newton's great inspirations which came to him as a youth of twenty two or three in 1666 at Woolsthorpe was his law of universal gravitation (already stated). In this connection we shall not repeat the story of the falling apple.

Most authorities agree that Newton did make some rough calculations in 1666 (he was then twenty three) to see whether his law of universal gravitation would account for Kepler's laws. Many years later (in 1684) when Halley asked him what law of attraction would account for the elliptical orbits of the planets Newton replied at once the inverse square.

"How do you know?" Halley asked—he had been prompted by Sir Christopher Wren and others to put the question, as a great argument over the problem had been going on for some time in London.

"Why, I have calculated it," Newton replied. On attempting to restore his calculation (which he had mislaid) Newton made a slip, and believed he was in error. But presently he found his mistake and verified his original conclusion.

Much has been made of Newton's twenty years' delay in the publication of the law of universal gravitation as an undeserved setback due to inaccurate data. Of three explanations a less romantic but more mathematical one than either of the others is to be preferred here.

Newton's delay was rooted in his inability to solve a certain problem in the integral calculus which was crucial for the whole theory of universal gravitation as expressed in the Newtonian law. Before he could account for the motion of both the apple and the Moon Newton had to find the total attraction of a solid homogeneous sphere on any mass particle outside the sphere. For *every* particle of the sphere attracts the mass particle outside the sphere with a force varying directly as the product of the masses of the two particles and inversely as the square of the distance between them: how are all these separate attractions, infinite in number, to be compounded or added into one resultant attraction?

This evidently is a problem in integral calculus. Today it is given in the textbooks as an example which young students dispose of in twenty minutes or less. Yet it held Newton up for twenty years. He finally solved it, of course: the attraction is the same as if the entire mass of the sphere were concentrated in *a single point* at its centre. The problem is thus reduced to finding the attraction between two mass particles at a given distance apart, and the immediate solution of this is stated as Newton's law. If this is the correct explanation for the twenty years' delay, it may give us some ides of the enormous amount of labor which generations of mathematicians since Newton's day have expended on developing and simplifying the calculus to the point where very ordinary boys of sixteen can use it effectively.

Although our principal interest in Newton centers about his greatness as a mathematician we cannot leave him with his undeveloped masterpiece of 1666. To do so would be to give no idea of his magnitude, so we shall go on to a brief outline of his other activities without entering into detail (for lack of space) on any of them.

On his return to Cambridge Newton was elected a Fellow of Trinity in 1667 and in 1669, at the age of twenty six, succeeded Barrow as Lucasian Professor of Mathematics. His first lectures were on optics. In these he expounded his own discoveries and sketched his corpuscular theory of light, according to which light consists of an emission of corpuscles and is not a wave phenomenon as Huygens and Hooke asserted. Although the two theories appear to be contradictory both are useful today in correlating the phenomena of light and are, in a purely mathematical sense, reconciled in the modern quantum theory. Thus it is not correct to say, as it may have been a few years ago, the Newton was entirely wrong in his corpuscular theory.

The following year, 1668, Newton constructed a reflecting telescope with his own hands and used it to observe the satellites of Jupiter. His object doubtless was to see whether universal gravitation really was universal by observations on Jupiter's satellites. This year is also memorable in the history of the calculus. Mercator's calculations by means of infinite series of an area connected with a hyperbola was brought to Newton's attention. The method was practically identical with Newton's own, which he had not published, but which he now wrote out, gave to Dr. Barrow, and permitted to circulate among a few of the better mathematicians.

On his election to the Royal Society in 1672 Newton communicated his work on telescopes and his corpuscular theory of light. A commission of three, including the cantankerous Hooke, was appointed to report on the work on optics. Exceeding his authority as a referee Hooke seized the opportunity to propagandize for the undulatory theory and himself at Newton's expense. At first Newton was cool and scientific under criticism, but when the mathematician Lucas and the physician Linus, both of Liège, joined Hooke in adding suggestions and objections which quickly changed from the legitimate to the carping and merely stupid, Newton gradually began to lose patience.

A reading of his correspondence in this first of his irritating controversies should convince anyone that Newton was not by nature secretive and jealous of his discoveries. The tone of his letters gradually changes from one of eager willingness to clear up the difficulties which others found, to one of bewilderment that scien-

tific men should regard science as a battleground for personal quarrels. From bewilderment he quickly passes to cold anger and a hurt, somewhat childish resolution to play by himself in future. He simply could not suffer malicious fools gladly.

At last, in a letter of November 18, 1676, he says, "I see I have made myself a slave to philosophy, but if I get free of Mr. Lucas's business, I will resolutely bid adieu to it eternally, excepting what I do for my private satisfaction, or leave to come out after me; for I see a man must either resolve to put out nothing new, or become a slave to defend it." Almost identical sentiments were expressed by Gauss in connection with non-Euclidean geometry.

Newton's petulance under criticism and his exasperation at futile controversies broke out again after the publication of the *Principia*. Writing to Halley on June 20, 1688, he says, "Philosophy [science] is such an impertinently litigious Lady, that a man has as good be engaged to lawsuits, as to have done with her. I found it so formerly, and now I am no sooner come near her again, but she gives me warning." Mathematics, dynamics, and celestial mechanics were in fact—we may as well admit it—secondary interests with Newton. His heart was in his alchemy, his researches in chronology, and his theological studies.

It was only because an inner compulsion drove him that he turned as a recreation to mathematics. As early as 1679, when he was thirty seven (but when also he had his major discoveries and inventions securely locked up in his head or in his desk), he writes to the pestiferous Hooke: "I have for some years last been endeavoring to bend myself from philosophy to other studies in so much that I have long grutched the time spent in that study unless it perhaps be at idle hours sometimes for diversion." These "diversions" occasionally cost him more incessant thought than his professed labors, as when he made himself seriously ill by thinking day and night about the motion of the Moon, the only problem, he says, that ever made his head ache.

Another side of Newton's touchiness showed up in the spring of 1673 when he wrote to Oldenburg resigning his membership in the Royal Society. This petulant action has been variously interpreted. Newton gave financial difficulties and his distance from London as his reasons. Oldenburg took the huffy mathematician at his word and told him that under the rules he could retain his membership without paying. This brought Newton to his senses and he withdrew his resignation, having recovered his temper in the meantime. Nevertheless Newton thought he was about to be hard pressed. However, his finances presently straightened out and he felt better. It may be noted here that Newton was no absent-minded dreamer when it came to a question of money. He was extremely shrewd and he died a rich man for his times. But if shrewd and thrifty he was also very liberal with his money and was already ready to help a friend in need as unobtrusively as possible. To young men he was particularly generous.

The years 1684-86 mark one of the great epochs in the history of human thought. Skillfully coaxed by Halley, Newton at last consented to write up his astronomical and dynamical discoveries for publication. Probably no mortal has ever thought as hard and as continuously as Newton did in composing *Philosophiae Naturalis Principia Mathematica* (Mathematical Principles of Natural Philosophy). Never careful of his bodily health, Newton seems to have forgotten that he had a body which required food and sleep when he gave himself up to the completion of his masterpiece. Meals were ignored or forgotten, and on arising from a snatch of sleep he would sit on the edge of the bed half-clothed for hours, threading the mazes of his mathematics. In 1686 the *Principia* was presented to the Royal Society, and in 1687 was printed at Halley's expense.

A description of the contents of the *Principia* is out of the question here, but a small handful of the inexhaustible treasures it contains may be briefly exhibited. The spirit animating the whole work is Newton's dynamics, his law of universal gravitation, and the application of both to the solar system—"the system of the world." Although the calculus has vanished from the synthetical geometrical demonstrations, Newton states (in a letter) that he used it to *discover* his results and, having done so, proceeded to rework the proofs furnished by the calculus into geometrical shape so that his contemporaries might the more readily grasp the main theme—the dynamical harmony of the heavens.

First, Newton deduced Kepler's empirical laws from his own law of gravitation, and he showed how the mass of the Sun can be calculated, also how the mass of any planet having a satellite can be determined. Second, he initiated the extremely important theory of *perturbations*: the Moon, for example, is attracted not only by the Earth but by the Sun also; hence the orbit of the Moon will be perturbed by the pull of the Sun. In this manner Newton accounted for two ancient observations due to Hipparchus and Ptolemy. Our own generation has seen the now highly developed theory of perturbations applied to electronic orbits, particularly for the helium atom. In addition to these ancient observations, seven other irregularities of the Moon's motion observed by Tycho Brahe (1546-1601), Flamsteed (1646-1719), and others, were deduced from the law of gravitation.

So much for lunar perturbations. The like applies to the planets. Newton began the theory of planetary perturbations, which in the nineteenth century was to lead to the discovery of the planet Neptune and, in the twentieth, to that of Pluto.

The "lawless" comets—still warnings from an angered heaven to superstitious eyes—were brought under the universal law as harmless members of the Sun's family, with such precision that we now calculate and welcome their showy return (unless Jupiter or some other outsider perturbs them unduly), as we did in 1910 when Halley's beautiful comet returned promptly on schedule after an absence of seventy-four years.

He began the vast and still incomplete study of planetary evolution by calculating (from his dynamics and the universal law) the flattening of the earth at its poles due to diurnal rotation, and he proved that the shape of a planet determines the length of its day, so that if we knew accurately how flat Venus was at the poles, we could say how long it takes her to turn completely once round the axis joining her poles. He calculated the variation of weight with latitude. He proved that a hollow shell, bounded by concentric spherical surfaces, and homogeneous, exerts no force on a small body anywhere inside it. The last has important consequences in electrostatics—also in the realm of fiction, where it has been used as the motif for amusing fantasies.

The precession of the equinoxes was beautifully accounted for by the pull of the Moon and the Sun on the equatorial bulge of the Earth causing our planet to wobble like a top. The mysterious tides also fell naturally into the grand scheme—both the lunar and the solar tides were calculated, and from the observed heights of the spring and neap tides the mass of the Moon was deduced. The First Book laid down the principles of dynamics; the Second, the motion of bodies in resisting media, and fluid motion; the Third was the famous "System of the World."

Probably no other law of nature has so simply unified any such mass of natural phenomena as Newton's law of universal gravitation in his *Principia*. It is to the credit of Newton's contemporaries that they recognized at least dimly the magnitude of what had been done, although but few of them could follow the reasoning by which the stupendous miracle of unification had been achieved, and made of the author of the *Principia* a demigod. Before many years had passed the Newtonian system was being taught at Cambridge (1699) and Oxford (1704). France slumbered on for half a century, still dizzy from the whirl of Descartes' angelic vortices. But presently mysticism gave way to reason and Newton found his greatest successor not in England but in France, where Laplace set himself the task of continuing and rounding out the *Principia*.

After the *Principia* the rest is anticlimax. Although the lunar theory continued to plague and "divert" him, Newton was temporarily sick of "philosophy" and welcomed the opportunity to turn to less celestial affairs. James II, obstinate Scot and bigoted Catholic that he was, had determined to force the University to grant a master's degree to a Benedictine over the protests of the academic authorities. Newton was one of the delegates who in 1687 went to London to present the University's case before the Court of High Commission presided over by that great and blackguardly lawyer and Lord High Chancellor George Jeffreys—"infamous Jeffreys" as he is known in history. Having insulted the leader of the delegates in masterly fashion, Jeffreys dismissed the rest with the injunction to go and sin no more. Newton apparently held his peace. Nothing was to be gained by answering a man like Jeffreys in his own kennel. But when the others would have signed a disgraceful compromise it was Newton who put backbone into them and kept from signing. He won the day; nothing of any value was lost—not even honor. "An honest courage in these matters," he write later, "will secure all, having law on our sides."

Cambridge evidently appreciated Newton's courage, for in January, 1689, he was elected to represent the University at the Convention Parliament after James II had fled the country to make room for William of Orange and his Mary, and the faithful Jeffreys was burrowing into dunghills to escape the ready justice of the mob. Newton sat in Parliament till its dissolution in February, 1690. To his credit he never made a speech in the place. But he was faithful to his office and not averse to politics; his diplomacy had much to do with keeping the turbulent University loyal to the decent King and Queen.

Newton's taste of "real life" in London proved his scientific undoing. Influential and officious friends, including the philosopher John Locke (1632-1704) of *Human Understanding* fame, convinced Newton that he was not getting his share of the honors. The crowning imbecility of the Anglo-Saxon breed is its dumb belief in public office or an administrative position as the supreme honor for a man of intellect. The English finally (1699) made Newton Master of the Mint to reform and supervise the coinage of the Realm. For utter bathos this "elevation" of the author of the *Principia* is surpassed only by the jubilation of Sir David Brewster in his life of Newton (1860) over the "well-merited recognition" thus accorded Newton's genius by the English people. Of course if Newton really wanted anything of the sort there is nothing to be said; he had earned the right millions of times over to do anything he

desired. But his busybody friends need not have egged him on.

It did not happen all at once. Charles Montagu, later Earl of Halifax, Fellow of Trinity College and a close friend of Newton, aided and abetted by the everlastingly busy and gossipy Samuel Pepys (1633-1703) of diary notoriety, stirred up by Locke and by Newton himself, began pulling wires to get Newton recognition "worthy" of him.

The negotiations evidently did not always run smoothly and Newton's somewhat suspicious temperament caused him to believe that some of his friends were playing fast and loose with him—as they probably were. The loss of sleep and the indifference to food which had enabled him to compose the *Principia* in eighteen months took their revenge. In the autumn of 1692 (when he was nearly fifty and should have been at his best) Newton fell seriously ill. Aversion to all food and an almost total inability to sleep, aggravated by a temporary persecution mania, brought on something dangerously close to a total mental collapse. A pathetic letter of September 16, 1693 to Locke, written after his recovery, shows how ill he had been.

> Sir,
> Being of opinion that you endeavored
> to embroil me with women and by
> other means, I was so much affected
> with it that when one told me you
> were sickly and would not live, I
> answered, 'twere better if you were
> dead. I desire you to forgive me for
> this uncharitableness. For I am now
> satisfied that what you have done is
> just, and I beg your pardon for having
> hard thoughts of you for it, and for
> representing that you struck at the
> root of mortality, in a principle you
> laid down in your book of ideas, and
> designed to pursue in another book,
> and that I took you to be a Hobbist.
> I beg your pardon also for saying or
> thinking that there was a design to
> sell me an office, or to embroil me.
> > I am your most humble
> > And unfortunate servant,
> > > Is. Newton

The news of Newton's illness spread to the Continent where, naturally, it was greatly exaggerated. His friends, including one who was to become his bitterest enemy, rejoiced at his recovery. Leibniz wrote to an acquaintance expressing his satisfaction that Newton was himself again. But in the very year of his recovery (1693) Newton heard for the first time that the calculus was becoming well known on the Continent and that it was commonly attributed to Leibniz.

The decade after the publication of the *Principia* was about equally divided between alchemy, theology, and worry, with more or less involuntary and headachy excursions into the lunar theory. Newton and Leibniz were still on cordial terms. Their respective "friends," ignorant as Kaffirs of all mathematics and of the calculus in particular, had not yet decided to pit one against the other with charges of plagiarism in the invention of the calculus, and even grosser dishonesty, in the most shameful squabble over priority in the history of mathematics. Newton recognized Leibniz' merits, Leibniz recognized Newton's, and at this peaceful stage of their acquaintance neither for a moment suspected that the other had stolen so much as a single idea of the calculus from the other.

Later, in 1712, when even the man in the street—the zealous patriot who knew nothing of the facts—realized vaguely that Newton had done something tremendous in mathematics (more, probably, as Leibniz had said, than had been done in all history before him), the question as to who had invented the calculus became a matter of acute national jealousy, and all educated England rallied behind its somewhat bewildered champion, howling that his rival was a thief and a liar.

Newton at first was not to blame. Nor was Leibniz. But as the British sporting instinct presently began to assert itself, Newton acquiesced in the disgraceful attack and himself suggested or consented to shady schemes of downright dishonesty designed to win the international championship at any cost—even that of national honor. Leibniz and his backers did likewise, The upshot of it all was that the obstinate British practically rotted mathematically for all of a century after Newton's death, while the more progressive Swiss and French, following the lead of Leibniz, and developing his incomparably better way of merely *writing* the calculus, perfected the subject and made it the simple, easily applied instrument of research that Newton's immediate successors should have had the honor of making it.

In 1696, at the age of fifty four, Newton became Warden of the Mint. His job was to reform the coinage. Having done so, he was promoted in 1699 to the dignity of Master. The only satisfaction mathematicians can take in this degradation of the supreme intellect of ages is the refutation it afforded of the silly satisfaction that mathematicians have no practical sense. Newton was one of the best Masters the Mint ever had. He took his job seriously.

In 1701-2 Newton again represented Cambridge University in Parliament, and in 1703 was elected President of the Royal Society, an honorable office to which he was reelected time after time till his death in 1727. In 1705 he was knighted by good Queen Anne. Probably this honor was in recognition of his services as a money-changer rather than in acknowledgement of his preeminence in the temple of wisdom. This is all as it should be: if "a riband to stick in his coat" is the reward of a turncoat politician, why should a man of intellect and integrity feel flattered if his name appears in the birthday list of honors awarded by the King? Caesar may be rendered the things that are his, ungrudgingly: but when a man of science, *as* a man of science, snaps up the droppings from the table of royalty he joins the mangy and starving dogs licking the sores of the beggars at the feast of Dives. It is to be hoped that Newton was knighted for his services to the money-changers and not for his science.

Was Newton's mathematical genius dead? Most emphatically no. He was still the equal of Archimedes. But the wiser old Greek, born aristocrat that he was—fortunately, cared nothing for the honors of a position which had always been his; to the very last minute of his long life he mathematicized as powerfully as he had in his youth. But for the accidents of preventable disease and poverty, mathematicians are a long-lived race intellectually; their creativeness outlives that of poets, artists, and even of scientists, by decades. Newton was still as virile of intellect as he had ever been. Had his officious friends but let him alone Newton might easily have created the calculus of variations, an instrument of physical and mathematical discovery second only to the calculus, instead of leaving it for the Bernoullis, Euler, and Lagrange to initiate. He had already given a hint of it in the *Principia* when he determined the shape of the surface of revolution which would cleave through a fluid with the least resistance. He had it in him to lay down the broad lines of the whole method. Like Pascal when he forsook this world for the mistier if more satisfying kingdom of heaven, Newton was still a mathematician when he turned his back on his Cambridge study and walked into a more impressive sanctum at the Mint.

In 1696 Johann Bernoulli and Leibniz between them concocted two devilish challenges to the mathematicians of Europe. The first is still of importance; the second is not in the same class. Suppose two points to be fixed at random in a vertical plane. What is the shape of the curve down which a particle must slide (without friction) under the influence of gravity so as to pass from the upper point to the lower in the *least time*? This is the problem of the *brachistochrone* (= "shortest time").

After the problem had baffled the mathematicians of Europe for six months, it was proposed again, and Newton heard of it for the first time on January 29, 1696, when a friend communicated it to him. He had just come home, tired out, from a long day at the Mint. After dinner he solved the problem (and the second as well), and the following day communicated his solutions to the Royal Society anonymously. But for all his caution he could not conceal his identity—while at the Mint Newton resented the efforts of mathematicians and scientists to entice him into discussions of scientific interest. On seeing the solution Bernoulli at once exclaimed, "Ah! I recognize the lion by his paw." (This is not an exact translation of B's Latin.) They all knew Newton when they saw him, even if he did have a moneybag over his head and did not announce his name.

A second proof of Newton's vitality was to come in 1716 when he was seventy four. Leibniz had rashly proposed what appeared to him a difficult problem as a challenge to the mathematicians of Europe and aimed at Newton in particular. Newton received this at five o'clock one afternoon on returning exhausted from the blessed Mint. He solved it that evening. This time Leibniz optimistically thought he had trapped the Lion. In all the history of mathematics Newton has had no superior (and perhaps no equal) in the ability to concentrate all the forces of his intellect on a difficulty at an instant's notice.

The story of the honors that fall to a man's lot in his lifetime makes but trivial reading to his successors. Newton got all that were worth having to a living man. On the whole Newton had as fortunate a life as any great man has ever had. His bodily health was excellent up to his last years; he never wore glasses and he lost only one tooth in all his life. His hair whitened at thirty but remained thick and soft till his death.

The record of his last days is more human and more touching. Even Newton could not escape suffering. His courage and endurance under almost constant pain during the last two or three years of his life add but another laurel to his crown as a human being. He bore the tortures of "the stone" without flinching, though the sweat rolled from him, and always with a word of sympathy for those who waited on him. At last, and mercifully, he was seriously weakened by "a persistent cough," and finally, after having been eased of pain for some days, died peacefully in his sleep between one and two o'clock in the morning of March 20, 1727, in his eighty fifth year. He is buried in Westminster Abbey.

QUESTIONS AND EXERCISES

1. Use the binomial theorem to get the first few terms in the series expansions of

$$(1 - x)^{1/2} \quad and \quad (1 + x^2)^{-1/2}.$$

2. The earth is 93 million miles from the sun and goes around it in a year. It takes Pluto 248 years to go around the sun once. Use Kepler's third law to find out how far from the sun Pluto is.

3. "The crowning imbecility of the Anglo-Saxon breed is its dumb belief in public office or an administrative position as the supreme honor for a man of intellect." Can you think of any examples, besides Newton, of the operation of this phenomenon? The author sounds aggrieved; why do you think that he was vexed?

4. "But for the accidents of preventable disease and poverty, mathematicians are a long-lived race intellectually; their creativeness outlives that of poets, artists, and even of scientists, by decades." Do you think that Bell is right? If he is, what could be an explanation for this?

THE CREATION OF THE CALCULUS

by Morris Kline

Like bread and butter, or epsilons and deltas, Leibniz and Newton always go together, at least when the history of calculus comes up. They both discovered (or, as Professor Kline would have it, "invented") the fundamental ideas of calculus at about the same time. Newton had them first but Leibniz published them first, and the result was a big fight over who should get the credit. Both should. Their approaches were different, and both were necessary for the development of the subject. The excerpt shows, among other things, how hard the ideas of calculus are and how difficult it is to attain an understanding of what *dx, dy,* and *dy/dx really* are. Newton had trouble, Leibniz had trouble, so it is no surprise that millions of calculus students since have also had trouble. Big ideas are hard. The excerpt also shows some of who Leibniz was and what he did; that is not as hard, and everyone who knows a little about Newton should know a little about Leibniz too.

---

4. The Work of Leibniz

Though his contributions were quite different, the man who ranks with Newton in building the calculus is Gottfried Wilhelm Leibniz (1646-1716). He studied law and, after defending a thesis on logic, received a Bachelor of Philosophy degree. In 1666, he wrote the thesis *De Arte Combinatoria* (On the Art of Combinations), a work on a universal method of reasoning; this completed his work for a doctorate of philosophy at the University of Altdorf and qualified him for a professorship. During the years 1670 and 1671 Leibniz wrote his first papers on mechanics, and, by 1671, had produced his calculating machine. He secured a job as an ambassador for the Elector of Mainz and in March of 1672 went to Paris on a political mission. This visit brought him into contact with mathematicians and scientists, notably Huygens, and stirred up his interest in mathematics. Though he had done a little reading in the subject and had written the paper of 1666, he says he knew almost no mathematics up to 1672. In 1673 he went to London and met other scientists and mathematicians, including Henry Oldenburg, at that time secretary of the Royal Society of London. While making his living as a diplomat, he delved further into mathematics and read Descartes and Pascal. In 1676 Leibniz was appointed librarian and councillor to the Elector of Hanover. Twenty-four years later the Elector of Brandenburg invited Leibniz to work for him in Berlin. While involved in all sorts of political maneuvers, including the succession of George Ludwig of Hanover to the English throne, Leibniz worked in many fields and his side activities covered an enormous range. He died neglected in 1716.

In addition to being a diplomat, Leibniz was a philosopher, lawyer, historian, philologist, and pioneer geologist. He did important work in logic, mechanics, optics, mathematics, hydrostatics, pneumatics, nautical science, and calculating machines. Though his profession was jurisprudence, his work in mathematics and philosophy is among the best the world has produced. He kept contact by letter with people as far away as China and Ceylon. He tried endlessly to reconcile the Catholic and Protestant faiths. It was he who proposed, in 1669, that a German Academy of Sciences be founded; finally the Berlin Academy was organized in 1700. His original recommendation had been for a society to make inventions in mechanics and discoveries in chemistry and physiology that would be useful to mankind; Leibniz wanted knowledge to be applied. He called the universities "monkish" and charged that they possessed learning but no judgement and were absorbed in trifles. Instead he urged the pursuit of real knowledge—mathematics, physics, geography, chemistry, anatomy, botany, zoology, and history. To Leibniz the skills of the artisan and the practical man were more valuable than the learned subtleties of the professional scholars. He favored the German language over the Latin because Latin was allied to older, useless thought. Men mask their ignorance, he said, by using the Latin language to impress people. German, on the other hand, was understood by the common people and could be developed to help clarity of thought and acuteness of reasoning.

Leibniz published papers on the calculus from 1684 on, and we shall say more about them later. However, many of his results, as well as the development of his ideas, are contained in hundreds of pages of notes made from 1673 on but never published by him. These notes, as one might expect, jump from one topic to another and contain changing notation as Leibniz's thinking developed. Some are simply ideas that occurred to him while reading books or articles by Gregory of St. Vincent, Fermat, Pascal, Descartes, and Barrow or trying to cast their thought into his own way of approaching the calculus. In 1714 Leibniz write *Historia et Origo Calculi Differentialis*, in which he gives an account of the development of his own thinking. However this was written many years after he had done his work and, in view of the weaknesses of human memory and the greater insight he had acquired by that time, his history may not be accurate. Since his purpose was to defend himself against an accusation of plagiarism, he might have distorted unconsciously his account of the origin of his ideas.

Despite the confused state of Leibniz's notes we shall examine a few, because they reveal how one of the greatest intellects struggled to understand and create. By 1673 he was aware of the important direct and inverse problem of finding tangents to curves; he was also quite sure that the inverse method was equivalent to finding areas and volumes by summations. The somewhat systematic development of his ideas begins with notes of 1675. However, it seems helpful, in order to understand his thinking, to note that in his *De Arte Combinatoria* he had considered sequences of numbers, first difference, second differences, and higher-order differences. Thus for the sequence of squares

$$0, 1, 4, 9, 16, 25, 36,$$

the first differences are

$$1, 3, 5, 7, 9, 11$$

and the second differences are

$$2, 2, 2, 2, 2, 2.$$

Leibniz noted the vanishing of the second differences for the sequence of natural numbers, third differences for the sequence of squares, and so on. He also observed, of course, that if the original sequence starts from 0, the sum of the first differences is the last term of the sequence.

To relate these facts to the calculus he had to think of the sequence of numbers as the y-values of a function and the difference of any two as the difference of two nearby y-values. Initially he thought of x as representing the order of the term in the sequence and y as

representing the value of that term.

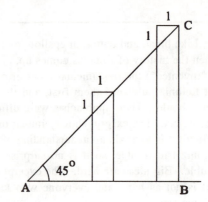

Figure 1

The quantity dx, which he often writes as a, is then 1 because it is the difference of two successive terms, and dy is the actual difference in the values of two successive terms. Then using omn, as an abbreviation for the Latin *omnia*, to mean sum, and using l for dy, Leibniz concludes that omn. $l = y$, because omn. l is the sum of the first differences of a sequence whose terms begin with 0 and so gives the last term. However, omn. yl presents a new problem. Leibniz obtains the result that omn. yl is $y^2/2$ by thinking in terms of the function $y = x$. Thus, as Figure 1 shows, the area of triangle ABC is the sum of the yl (for "small" l) and it is also $y^2/2$. Leibniz says, "Straight lines which increase from nothing each multiplied by its corresponding element of increase form a triangle." These few facts already appear, among more complicated ones, in papers of 1673.

In the next stage he struggled with several difficulties. He had to make the transition from a discrete series of values to the case where dy and dx are increments of an arbitrary function y of x. Since he was still tied to sequences, wherein x is the order of the term, his a or dx was 1; so he inserted and omitted a freely. When he made the transition to the dy and dx of any function, this a was no longer 1. However, while still struggling with the notion of summation he ignored this fact.

Thus in a manuscript of October 29, 1675, Leibniz starts with

$$\text{omn.}\, yl \;=\; \overline{\text{omn.}\,\overline{\text{omn.}\,l}\,\frac{l}{a}},$$

which holds because y itself is omn. l. Here he divides l by a to preserve dimension. Leibniz says that [the last equation] holds, whatever l may be. But, as we saw in connection with Figure 1,

$$\text{omn. } yl = \frac{y^2}{2}.$$

Hence [from the last two equations]

$$\frac{y^2}{2} = \overline{\text{omn.}\overline{\text{omn.}l}\frac{l}{a}}.$$

In our notation, he has shown that

$$\frac{y^2}{2} = \int\left\{\int dy\right\}\frac{dy}{dx} = \int y\frac{dy}{dx}.$$

Leibniz says that this result is admirable.

Another theorem of the same kind, which Leibniz derived from a geometrical argument, is

$$\text{omn.} xl = x\,\text{omn.}\,l - \text{omn.omn.}l,$$

where l is the difference in values of two successive terms of a sequence and x is the number of the term. For us this equation is

$$\int x\,dy = xy - \int y\,dx.$$

Now Leibniz lets l itself be x, and obtains

$$\text{omn.} x^2 = x\,\text{omn.}x - \text{omn.omn.}x.$$

But omn. x, he says, is $x^2/2$ (he has shown that omn. yl is $y^2/2$). Hence

$$\text{omn.} x^2 = x\frac{x^2}{2} - \text{omn.}\frac{x^2}{2}.$$

By transposing the last term he gets

$$\text{omn.} x^2 = \frac{x^3}{3}.$$

In this manuscript of October 29, 1675, Leibniz decided to write $\int$ for omn., so that

$$\int l = \text{omn.}l \quad \text{and} \quad \int x = \frac{x^2}{2}.$$

The symbol $\int$ is an elongated S for "sum."

Leibniz realized rather early, probably from studying the work of Barrow, that differentiation and integration as a summation must be inverse processes; so area, when differentiated, must give a length. Thus, in the same manuscript of October 29, Leibniz says, "Given l and its relation to x, to find $\int l$." Then, he says, "Suppose that $\int l = ya$. Let $l = ya/d$. [Here he puts d in the denominator. It would mean more to us if he wrote $l = d(ya)$.] Then just as $\int$ will increase, so d will diminish the dimension. But $\int$ means a sum, and d, a difference. From the given y we can always find y/d or l, that is, the difference of the y's. Hence one equation may be transformed into the other; just as from the equation

$$\overline{\int c\,\overline{\int T^2}} = \frac{c\,\overline{\int T^3}}{3a^3},$$

we can obtain the equation

$$c\int T^2 = \frac{c\int T^3}{3a^3 d}.$$

In this early paper Leibniz seems to be exploring the *operations* of $\int$ and d and sees that they are inverses. He finally realizes that $\int$ does not raise dimension nor d lower it, because $\int$ is really a summation of rectangles, and so a sum of areas. Thus he recognizes that, to get back to dy from y, he must form the difference of y's or take the differential of y. Then he says, "But $\int$ means a sum and d a difference." This may have been a later insertion. Hence a couple of weeks afterwards, in order to get from y to dy, he changes from dividing by d to taking the differential of y, and writes dy.

Up to this point Leibniz had been thinking of the y-values as values of terms of a sequence and of x usually as the order of these terms, but now, in this paper, says, "All these theorems are true for series in which the differences of the terms bear to the terms themselves a ratio that is less than any assignable quantity." That is, dy/y may be less than any assignable quantity.

In a manuscript dated November 11, 1675, entitled "Examples of the inverse method of tangents," Leibniz uses $\int$ for the sum and x/d for difference. He then says x/d is dx, the difference of two consecutive x-values, but apparently here dx is a constant and equal to unity.

From barely intelligible arguments such as the above, Leibniz asserted the fact that *integration as a summation process is the inverse of differentiation.* This idea is in the work of Barrow and Newton, who obtained area by antidifferentiation, but it is first expressed as a relation between summation and differentiation by Leibniz. Despite this outright assertion, he was by no means clear as to how to obtain an area from what one might loosely write as $\sum y\,dx$—that is, how to obtain an area under a curve from a set of rectangles. Of course this difficulty beset all the seventeenth-century workers. Not possessing a clear concept of a limit, or even clear notions about area, Leibniz thought of the latter sometimes as a sum of rectangles so small and so numerous that the difference between this sum and the true area under the curve could be neglected, and at other times as a sum of the ordinates or y-values. This latter concept of area was common, especially among the indivisibilists, who thought that the ultimate unit of area and the y-value were the same.

With respect to differentiation, even after recognizing that dy and dx can be arbitrarily small quantities, Leibniz had yet to overcome the fundamental difficulty

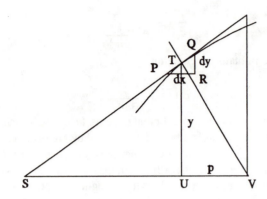

Figure 2

that the ratio dy/dx is not quite the derivative in our sense. He based his argument on the characteristic triangle, which Pascal and Barrow had used. This triangle (Figure 2) consists of dx, dy, and the chord PQ, which Leibniz also thought of *as the curve between P and Q and part of the tangent at T.* Though he speaks of this triangle as indefinitely small, he maintains nevertheless that it is similar to a definite triangle, namely, the triangle STU formed by the subtangent SU, the ordinate at T, and the length of tangent TU. Hence dy and dx are ultimate elements, and their ratio has a definite meaning. In fact, he uses the argument that, from the similar triangles PRQ and SUT, $dy/dx = TU/SU$.

In the manuscript of November 11, 1675, Leibniz shows how he can solve a definite problem. He seeks the curve whose subnormal is inversely proportional to the ordinate. In Figure 2, the normal is TV and the subnormal p is UV. From the similarity of triangles PRQ and TUV, he has

$$\frac{dy}{dx} = \frac{p}{y}$$

or

$$p\,dx = y\,dy.$$

But the curve has the given property

$$p = \frac{b}{y},$$

where b is the proportionality constant. Hence

$$dx = \frac{y^2}{b}\,dy.$$

Then

$$\int dx = \int \frac{y^2}{b}\,dy$$

or

$$x = \frac{y^3}{3b}.$$

Leibniz also solved other inverse tangent problems.

In a paper of June 26, 1676, he realizes that the best method of finding tangents is to find dy/dx, where dy and dx are differences and dy/dx is the quotient. He ignores $dx\,dx$ and higher powers of dx.

By November of 1676, he is able to give the general rules $dx^n = nx^{n-1}$ for integral and fractional n and

$$\int x^n\,dx = \frac{x^{n+1}}{n+1},$$

and says, "The reasoning is general, and it does not depend on what the progressions of the x's may be." Here x still means the order of the terms of a sequence. In this manuscript he also says that to differentiate $\sqrt{a + bz + cz^2}$, let $a + bz + cz^2 = x$, and multiply by dx/dz. This is the chain rule.

By July 11, 1677, Leibniz could give the correct rules for the differential of sum, difference, product, and quotient of two functions and for powers and roots, but no proofs. In the manuscript of November 11, 1675, he had struggled with $d(uv)$ and $d(u/v)$, and thought that $d(uv) = du\,dv$.

In 1680, dx had become the difference of abscissas and dy the difference in the ordinates. He says, "... now these dx and dy are taken to be infinitely small, or the two points on the curve are understood to be a distance apart that is less than any given length. ..." He calls dy the "momentaneous increment" in y as the ordinate moves along the x-axis. But PQ in Figure 2 is still considered part of a straight line. It is "an element of the curve or a side of the infinite-angled polygon that stands for the curve. ..." He continues to use the usual differential form. Thus, if $y = a^2/x$, then

$$dy = -\frac{a^2}{x^2}\,dx.$$

He also says that differences are opposite to sums. Then, to get the area under a curve (Figure 3), he takes the sum of the rectangles and says one can neglect the remaining "triangles, since they are infinitely small compared to the rectangles ... thus I represent in my calculus the area of the figure by $\int y\,dx$ ..." He also gives, for the element of arc,

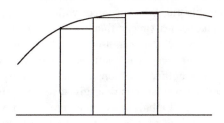

Figure 3

$$ds = \sqrt{dx^2 + dy^2} \ ;$$

and, for the volume of a solid of revolution obtained by revolving a curve around the x-axis,

$$V = \pi \int y^2 dx.$$

Despite prior statements that dx and dy are small differences, he still talks about sequences. He says, "Differences and sums are the inverses of one another, that is to say, the sum of the differences of a series [sequence] is a term of the series, and the difference of the sums of the series is a term of the series, and I enumerate the former thus, $\int dx = x$, and the latter thus, $d\int x = dx$." In fact, in a manuscript written after 1684, Leibniz says his method of infinitesimals has become widely known as the calculus of differences.

Leibniz's first publication on the calculus is in the *Acta Eruditorum* of 1684. In this paper the meaning of dy and dx is still not clear. He says in one place, let dx be any arbitrary quantity, and dy is defined by (see Figure 2)

$$dy : dx = y : \text{subtangent}.$$

This definition of dy presumes some expression for the subtangent; hence the definition is not complete. Moreover, Leibniz's definition of a tangent as a line joining two infinitely near points is not satisfactory.

He also gives in this paper the rules he had obtained in 1677 for the differential of the sum, product, and quotient of two functions and the rule for finding $d(x^n)$. In this last case he sketches the proof for positive integral n but says the rule is true for all n; for the other rules he gives no proofs. He makes applications to finding tangents, maxima and minima, and points of inflection. The paper, six pages long, is so unclear that the Bernoulli brothers called it "an enigma rather than an explanation."

In a paper of 1686 Leibniz gives

$$y = \sqrt{2x - x^2} + \int \frac{dx}{\sqrt{2x - x^2}}$$

as the equation of the cycloid. His point here is to show that by his methods and notation some curves can be expressed as equations not obtainable in other ways. He reaffirms this in his *Historia* when he says that his dx, ddx (second difference), and the sums that are the inverses of these differences can be applied to all functions of x, not excepting the mechanical curves of Vieta and Descartes, which Descartes had said have no equations. Leibniz also says that he can include curves that Newton could not handle even with his method of series.

In the 1686 paper as well as in subsequent papers, Leibniz gave the differential of the logarithmic and exponential functions and recognized exponential functions as a class. He also treated curvature, the osculating circle, and the theory of envelopes. In a letter to Johann Bernoulli of 1697, he differentiated under the integral sign with respect to a parameter. He also had the idea that many indefinite integrals could be evaluated by reducing them to known forms and speaks of preparing tables for such reductions—in other words, a table of integrals. He tried to define the higher-order differentials such as ddy (d^2y) and $dddy$ (d^3y), but the definitions were not satisfactory. Though he did not succeed, he also tried to find a meaning for d^cy where c is any real number.

With respect to notation, Leibniz worked painstakingly to achieve the best. His dx, dy, and dy/dx are, or course, still standard. He introduced the notation $\log x, d^n$ for the nth differential, and even d^{-1} and d^{-n} for $\int$ and the nth iteration of summation, respectively.

In general, Leibniz's work, though richly suggestive and profound, was so incomplete and fragmentary that it was barely intelligible. Fortunately, the Bernoulli brothers, James and John, who were immediately impressed and stirred by Leibniz's ideas, elaborated his sketchy papers and contributed an immense number of new developments we shall discuss later. Leibniz agreed that the calculus was as much theirs as his.

5. A Comparison of the Work of Newton and Leibniz

Both Newton and Leibniz must be credited with seeing the calculus as a new and general method, applicable to many types of functions. After their work, the calculus was no longer an appendage and extension of Greek geometry, but an independent science capable of handling a vastly expanded range of problems.

Both also arithmetized the calculus; that is, they built on algebraic concepts. The algebraic notation and techniques used by Newton and Leibniz not only gave them a more effective tool than geometry, but also permitted many different geometric and physical problems to be treated by the same technique. A major change from the beginning to the end of the seventeenth century was the algebraicization of the calculus. This is comparable to what Vieta had done in the theory of equations and Descartes and Fermat in geometry.

The third vital contribution that Newton and Leibniz share is the reduction to antidifferentiation of area, volume, and other problems that were previously treated as summations. Thus the four main problems—rates, tangents, maxima and minima, and summation—were all reduced to differentiation and antidifferentiation.

The chief distinction between the work of the two men is that Newton used the infinitely small increments in x and y as a means of determining the fluxion or derivative. It was essentially the limit of the ratio of the increments as they became smaller and smaller. On the other hand, Leibniz dealt directly with the infinitely small increments in x and y, that is, with differentials, and determined the relation between them. This difference reflects Newton's physical orientation, in which a concept such as velocity is central, and Leibniz's philosophical concern with ultimate particles of matter, which he called monads. As a consequence, Newton solved area and volume problems by thinking entirely in terms of rate of change. For him differentiation was basic; this process and its inverse solved all calculus problems, and in fact the use of summation to obtain an area, volume, or center of gravity rarely appears in his work. Leibniz, on the other hand, thought first in terms of summation, though of course these sums were evaluated by antidifferentiation.

A third distinction between the work of the two men lies in Newton's free use of series to represent functions; Leibniz preferred the closed form. In a letter to Leibniz of 1676, Newton stressed the use of series even to solve simple differential equations. Though Leibniz did use infinite series, he replied that the real goal should be to obtain results in finite terms, using the trigonometric and logarithmic functions where algebraic functions would not serve. He recalled to Newton James Gregory's assertion that the rectification of the ellipse and hyperbola could not be reduced to the circular and logarithmic functions and challenged Newton to determine by the use of series whether Gregory was correct. Newton replied that by the use of series he could decide whether some integrations could be achieved in finite terms, but gave no criteria. Again, in a letter of 1712 to John Bernoulli, Leibniz objected

to the expansion of functions into series and stated that the calculus should be concerned with reducing its results to quadratures (integrations) and, where necessary, quadratures involving transcendental functions.

There are differences in their manner of working. Newton was empirical, concrete, and circumspect, whereas Leibniz was speculative, given to generalizations, and bold. Leibniz was more concerned with operational formulas to produce a calculus in a broad sense; for example, rules for the differential of a product or quotient of functions, his rule for $d^n(uv)$ (u and v being functions of x), and a table of integrals. It was Leibniz who set the canons of the calculus, the system of rules and formulas. Newton did not bother to formulate rules, even when he could easily have generalized his concrete results. He knew that if $z = uv$, then $\dot{z} = u\dot{v} + v\dot{u}$, but did not point out this general result. Though Newton initiated many methods, he did not stress them. His magnificent applications of the calculus not only demonstrated its value but, far more than Leibniz's work, stimulated and determined almost the entire direction of eighteenth-century analysis. Newton and Leibniz differed also in their concern for notation. Newton attached no importance to this matter, while Leibniz spent days choosing a suggestive notation.

6. The Controversy over Priority

Nothing of Newton's work on the calculus was published before 1687, though he had communicated results to friends during the years 1665 to 1687. In particular, he had sent his tract *De Analysi* in 1669 to Barrow, who had sent it to John Collins. Leibniz visited Paris in 1672 and London in 1673 and communicated with some of the people who knew Newton's work. However, he did not publish on the calculus until 1684. Hence the question of whether Leibniz had known the details of what Newton did was raised, and Leibniz was accused of plagiarism. However, investigations made long after the deaths of the two men show that Leibniz was an independent inventor of major ideas of the calculus, though Newton did much of his work before Leibniz did. Both owe much to Barrow, though Barrow used geometrical methods almost exclusively. The significance of the controversy lies not in the question of who was the victor but rather in the fact that the mathematicians took sides. The Continental mathematicians, the Bernoulli brothers in particular, sided with Leibniz, while the English mathematicians defended Newton. The two groups became unfriendly and even bitter toward each other; John Bernoulli went so far as to ridicule and inveigh against the English.

As a result, English and Continental mathematicians

ceased exchanging ideas. Because Newton's major work and first publication on the calculus, the *Principia*, uses geometrical methods, the English continued to use mainly geometry for about a hundred years after his death. The Continentals took up Leibniz's analytical methods and extended and improved them. These proved to be far more effective; so not only did the English mathematicians fall behind, but mathematics was deprived of contributions that some of the ablest minds might have made.

EXERCISES AND QUESTIONS

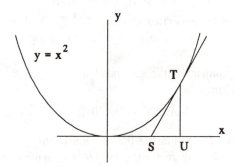

Figure 4

1. In the early days of calculus the subtangent was considerably more important than it is now. In Figure 4, SU is the subtangent of the parabola $y = x^2$.

(a) Show that the length of the subtangent is $x/2$.

(b) Find the length of the subtangent of $y = x^3$.

(c) Generalize: let S denote the subtangent operator and find $S(x^n)$ for $n = 1, 2, 3, \ldots$.

(d) What is $S(\sqrt{x})$?

(e) What is S of the trigonometric functions?

(f) Are there nice formulas for the subtangent of a product and quotient of two functions?

(g) Is it worth going into antisubtangents?

2. "Leibniz noted the vanishing of the second differences for the sequence of natural numbers, third differences for the sequence of squares, and so on." Show that the fourth differences of the sequence of cubes vanishes.

3. "He also observed, of course, that if the original sequence starts from 0, the sum of the first differences is the last term of the sequence." Show that his obser-

vation is correct for

(a) 0, 1, 4, 9, 16, 25

(b) 0, 1, 8, 27, 64, 125

(c) What if the sequence does not start at 0? Modify Leibniz's observation to apply to 3, 7, 13, 21, 31, 43.

4. "Though his profession was jurisprudence, his work in mathematics and philosophy is among the best the world has produced." Ask your favorite philosophy teacher for a brief explanation of Leibniz's contributions to philosophy, and inquire if the author's assessment of them is correct.

5. "He called the universities 'monkish' and charged that they possessed learning but no judgement and were absorbed in trifles." This may have been true in the late seventeenth century; do you think it is true today? If so, should anything be done about it? Does society really need experts on the Mongolian language, Mayan pottery, or, for that matter, the history of calculus?

6. "In a paper of 1686 Leibniz gives

$$y = \sqrt{2x - x^2} + \int \frac{dx}{\sqrt{2x - x^2}}$$

as the equation of the cycloid." Its equation is more usually given parametrically as

$$x = a(t - \sin t)$$
$$y = a(1 - \cos t).$$

Can you show that the two expressions actually represent the same curve?

PRINCIPIA MATHEMATICA

By Isaac Newton

For a change, first will come the excerpt and then the commentary.

LEMMA II

the moment of any genitum *is equal to the moments of each of the generating sides multiplied by the indices of the powers of these sides, and by their coefficients continually.*

I call any quantity a *genitum* which is not made by addition or subtraction of divers parts, but is generated or produced in arithmetic by the multiplication, division, or extraction of the root of any terms whatsoever; in geometry by the finding of contents and sides, or of the extremes and means of proportionals. Quantities of this kind are products, quotients, roots, rectangles, squares, cubes, square and cubic sides, and the like. These quantities I here consider as variable and indetermined, and increasing and decreasing, as it were, by a continual motion or flux; and I understand their momentary increments or decrements by the name of moments; so that the increments may be esteemed as added or affirmative moments; and the decrements as subtracted or negative ones. But take care not to look on finite particles as such. Finite particles are not moments, but the very quantities generated by the moments. We are to conceive them as the just nascent principles of finite magnitudes. Nor do we in this Lemma regard the magnitude of the moments, but their first proportion, as nascent. It will be the same thing, if instead of moments, we use either the velocities of the increments and decrements (which may also be called the motions, mutations, and fluxions of quantities), or any finite quantities proportional to those velocities. The coefficient of any generating side is the quantity which arises by applying the genitum to that side.

Wherefore the sense of the Lemma is, that if the moments of any quantities A, B, C, &c., increasing or decreasing by a continual flux, or the velocities of the mutations which are proportional to them, be called a, b, c, &c., the moment or mutation of the generated rectangle AB will be $aB + bA$; the moment of the generated content ABC will be $aBC + bAC + cAB$; and the moments of the generated powers

$$A^2, A^3, A^4, A^{1/2}, A^{3/2}, A^{1/3}, A^{2/3}, A^{-1}, A^{-2}, A^{-1/2}$$

will be

$$2aA, 3aA^2, 4aA^3, (1/2)aA^{-1/2}, (3/2)aA^{1/2},$$

$$(1/3)aA^{-2/3}, (2/3)aA^{-1/3}, -aA^{-2}, -2aA^{-3}, -(1/2)aA^{-2/3}$$

respectively; and, in general, that the moment of any power $A^{n/m}$ will be $(n/m)aA^{(n-m)/m}$. Also, that the moment of the genitum A^2B will be $2aAB + bA^2$; the moment of the generated quantity $A^3B^4C^2$ will be

$$3aA^2B^4C^2 + 4bA^3B^3C^2 + 2cA^3B^4C;$$

and the moment of the generated quantity A^3/B^2 or A^3B^{-2} will be

$$3aA^2B^{-2} - 2bA^3B^{-3};$$

and so on. The Lemma is thus demonstrated.

Case 1. Any rectangle, as AB, augmented by a continued flux, when, as yet there wanted of the sides A and B half their moments $(1/2)a$ and $(1/2)b$, was $A - (1/2)a$ into $B - (1/2)b$ or

$$AB - \frac{1}{2}Ab - \frac{1}{2}Ba + \frac{1}{4}ab;$$

but as soon as the sides A and B are augmented by the other half-moments, the rectangle becomes $A + (1/2)a$ into $B + (1/2)b$, or

$$AB + \frac{1}{2}Ab + \frac{1}{2}Ba + \frac{1}{4}ab.$$

From the rectangle subtract the former rectangle and there will remain the excess $Ab + Ba$. Therefore with the whole increments a and b of the sides, the increment $Ab + Ba$ of the rectangle is generated. Q. E. D.

Case 2. Suppose AB always equal to G, and then the moment of the content ABC or GC (by Case 1) will be $Gc + cG$, that is (putting AB and $Ab + Ba$ for G and g), $aBC + bAC + cAB$. And the reasoning is the same for contents under ever so many sides. Q. E. D.

Case 3. Suppose the sides A, B, and C, to be always equal among themselves; and the moment $Ab +$

Ba, of A^2, that is, of the rectangle *AB*, will be 2*aA*; and the moment *ABC* + *BAC* + *CAB* of A^3, that is, of the content *ABC*, will be 3*aA*^{2}. And by the same reasoning the moment of any power A^n is naA^{n-1}. Q. E. D.

Case 4. Therefore since 1/*A* into *A* is 1, the moment of 1/*A* multiplied by *A*, together with 1/*A* multiplied by *a*, will be the moment of 1, that is, nothing. Therefore the moment of 1/*A*, or of A^{-1}, is $-a/A^2$. And generally since 1/A^n into A^n is 1, the moment of 1/A^n multiplied by A^n together with 1/A^n into naA^{n-1} will be nothing. And, therefore, the moment of 1/A^n or A^{-n} will be $-na/A^{n+1}$. Q. E. D.

Case 5. And since $A^{1/2}$ into $A^{1/2}$ is *A*, the moment of $A^{1/2}$ multiplied by $2A^{1/2}$ will be *a* (by case 3); and, therefore, the moment of $A^{1/2}$ will be $a/2A^{1/2}$ or $(1/2)aA^{-1/2}$. And generally, putting $A^{m/n}$ equal to *B*, then A^m will be equal to B^n, and therefore maA^{m-1} equal to nbB^{n-1}, and maA^{-1} equal to nbB^{-1}, or $nbA^{-m/n}$; and therefore $(m/n)aA^{(m-n)/n}$ is equal to *b*, that is, equal to the moment of $A^{m/n}$. Q. E. D.

Case 6. Therefore the moment of any genitum $A^m B^n$ is the moment of A^m multiplied by B^n, together with the moment of B^n multiplied by A^m, that is, $maA^{m-1}B^n + nbB^{n-1}A^m$; and that whatever the indices *m* and *n* of the powers be whole numbers or fractions, affirmative or negative. And the reasoning is the same for higher powers. Q. E. D.

Corollary I. Hence in quantities continually proportional, if one term is given, the moments of the rest of the terms will be the same terms multiplied by the number of intervals between them and the given term. Let *A*, *B*, *C*, *D*, *E*, *F* be continually proportional; then if the term *C* is given, the moments of the rest of the terms will be among themselves as $-2A$, $-B$, D, $2E$, $3F$.

Corollary II. And if in four proportionals the two means are given, the moments of the extremes will be as those extremes. The same is to be understood of the sides of any given rectangle.

Corollary III. And if the sum or difference of two squares is given, the moments of the sides will be inversely as the sides.

Did you read that carefully? Was it clear? Most probably it was not clear, since the ideas were presented a form and language that is not familiar to us.

In his first paragraph, Newton is trying to make clear what he means by "moments" of quantities that are changing. He says that moments are "their momentary increments or decrements." That is, a moment is an amount of change in a quantity, the amount that takes place in a moment. However, moments are not finite quantities: "Finite particles are not moments, but the very quantities generated by the moments. We are to conceive them as the just nascent principles of finite magnitudes." It is no wonder that calculus did not sweep through the intellectual world of the seventeenth century like a forest fire, nor that in 1696 there were approximately five people in all the world who truly understood its ideas. "The just nascent principles": Newton knew what he was trying to say, but he couldn't quite put it into words. It is hard to communicate an idea when words fail.

What Newton called a moment we would call a differential. *A*, *B*, and *C* are changing, and *a*, *b*, and *c* are *dA*, *dB*, and *dC*. In the second paragraph, Newton is telling us that

$$d(AB) = B\,dA + A\,dB,$$

$$d(ABC) = BC\,dA + AC\,dB + AB\,dC,$$

and

$$d(A^{n/m}) = \frac{n}{m}A^{\frac{n-m}{m}}\,dA.$$

Then in the following paragraphs, he is telling us why his moments obey those rules. In his Case 1, he says that half an infinitesimal clock tick before *Q* = *AB*,

$$Q = (A - \frac{1}{2}dA)(B - \frac{1}{2}dB)$$

$$= AB - \frac{1}{2}A\,dB - \frac{1}{2}B\,dA + \frac{1}{4}dA\,dB;$$

half an infinitesimal clock tick after,

$$Q = (A + \frac{1}{2}dA)(B + \frac{1}{2}dB)$$

$$= AB + \frac{1}{2}A\,dB + \frac{1}{2}B\,dA + \frac{1}{4}dA\,dB;$$

the difference,

$$A\,dB + B\,dA,$$

is how much Q has changed in the entire infinitesimal clock tick. Talking about infinitesimal clock ticks may be no more clear than talking about nascent principles of finite magnitudes, but there are some ideas that cannot easily be put into words. However, when you truly understand differentials, you will know it and the words will have meaning. Note the cleverness of Newton in setting his infinitesimal clock one-half a tick early: if he had instead found how much Q changed from the start

$$Q = AB$$

to one infinitesimal tick later

$$Q = (A + dA)(B + dB) = AB + A\,dB + B\,dA + dA\,dB,$$

the difference would have been

$$A\,dB + B\,dA + dA\,dB$$

and Newton would have had to explain why the last term could be omitted. He wisely chose not to try that. He was wise because he would not have been able to explain it; the idea of basing derivatives on limits was more than a century in the future.

His succeeding cases are fairly straightforward. In case 2, to find $d(ABC)$, write the changing quantity as a product of two things, $d((AB)C)$, and apply case 1. In case 3, let $A = B = C$ in

$$d(ABC) = BC\,dA + AC\,dB + AB\,dC$$

to get

$$d(A^3) = 3A^2\,dA.$$

In case 4, apply case 1 to get

$$0 = d(1) = d\left(\frac{1}{A}\cdot A\right) = \frac{1}{A}\,dA + A\,d\left(\frac{1}{A}\right)$$

and solve for $d(1/A)$:

$$d\left(\frac{1}{A}\right) = -\frac{1}{A^2}\,dA.$$

Newton realized that there is no need for a quotient rule for differentials as long as you have a product rule:

$$d\left(\frac{A}{B}\right) = d\left(A \cdot \frac{1}{B}\right) = A\,d\left(\frac{1}{B}\right) + \frac{1}{B}\,dA =$$

$$A\left(\frac{-1}{B^2}\right)dB + \frac{1}{B}\,dA = \frac{B\,dA - A\,dB}{B^2}.$$

And so on. The rest of the lemma could similarly be translated into modern notation. The lessons to be learned from this passage, I think, are that important ideas are hard ideas and that expressing them is not very much easier. The number of people who knew calculus in 1700 was very small because the ideas of calculus are difficult to grasp and they had not been presented as clearly as they could have been. It took almost a hundred years for the ideas of calculus to be put into a form that we can look at and think that yes, that is the way that calculus is. It took almost two hundred years for calculus to be part of every scientist's education. Ideas take time to sink in, both for people individually and for a society.

EXERCISES

1. Translate Newton's case 4 into modern notation.
2. Do the same for case 5.
3. Do the same for corollary 1, and prove it. When Newton says that A, B, C, D, E, F are continually proportional, he means that

$$\frac{A}{B} = \frac{B}{C} = \frac{C}{D} = \frac{D}{E} = \frac{E}{F}.$$

When he says that C is given, he means that it is constant, so $dC = 0$. When he says "the moments of the rest of the terms will be among themselves as" he means that the values of the ratios

$$\frac{dA}{dB}, \frac{dB}{dD}, \frac{dD}{dE}, \frac{dE}{dF}$$

are

$$\frac{-2A}{-B}, \frac{-B}{D}, \frac{D}{2E}, \frac{2E}{3F}$$

respectively.

4. Do the same for corollary 3, and prove it. When Newton says that the sum or difference of two squares is given, he means that $A^2 + B^2$ or $A^2 - B^2$ is constant.

THE WORLD'S FIRST CALCULUS TEXTBOOK

The name of it was *Analyse des infiniment petits, pour l'intelligence des lignes courbes* (*Infinitesimal Analysis, with Applications to Curves*), its author was M. le marquis de l'Hospital, and its date of publication was 1696. It really wasn't a textbook in the sense of something used in schools, rather it was meant to explain to people who wanted to know what the new subject was about, what it could do, and how it could do it. The book did its job well, since it lasted a long time. There was a second edition in 1715, and the edition that I have was published in 1781. My edition has some extra notes and commentary by someone named Le Fevre (at the time, a Frenchman's given name was either not important enough to bother mentioning or too important to let anyone know about), but they were added only to improve the reader's understanding. Most of the book, its style and content, is the same as the original edition's. Not many books last almost one hundred years. The average lifespan of a calculus textbook today is a few years at most, and many are born only to die almost immediately.

L'Hôpital (to give his name its modern spelling) as a person has been forgotten, and aside from knowing that his name is attached to l'Hôpital's Rule, hardly anyone knows anything about him. That is too bad, because he sounds like a person it would be good to know better. He was born in 1661, given the triple-barreled name of Guillaume François Antoine as befitted a member of the nobility, and displayed his mathematical talent at an early age, solving a problem about cycloids at the age of fifteen. He was a lifelong lover and supporter of mathematics, publishing several papers in the journals of the day and partially supporting Johann Bernoulli. It was from Bernoulli that he got the rule that bears his name, as well as other material that appears in his book, but there is no question of stealing. Bernoulli was given proper credit as well as money. It is just as well that l'Hôpital did not wait for Bernoulli to publish his writing on calculus, since Bernoulli's work on the integral calculus was not published until 1742 and Bernoulli on the differential calculus appeared only in 1924, rather too late to have much influence on the development of mathematics. L'Hôpital died young, in 1704. Abraham Robinson has written about him

> According to the testimony of his contemporaries, l'Hospital possessed a very attractive personality, being, among other things, modest and generous, two qualities which were not widespread among the mathematicians

of the time.

There should be a historical novel with l'Hôpital as its central character. I would like to know more about him.

His textbook is quite different from the calculus texts that are published today. For one thing, the subject matter is what the title suggests—the analysis of the infinitely small, in order to understand curves. That is what l'Hôpital thought that calculus is for, finding out about the behavior of curves. There are no problems about fields to be fenced in with 500 yards of fence, water leaking out of conical reservoirs, or stones being thrown straight up with initial velocities of 80 feet per second, there are only curves. Curves are what calculus is for. For another, there is not a single derivative in the whole book. The important idea for l'Hôpital, the only necessary idea, was that of the differential. The differential is the infinitely small quantity of the title, whose analysis leads to the understanding of curves. For a third, the only functions l'Hôpital ever uses in examples are algebraic ones—polynomials, their powers, roots, and quotients. There are no logarithms, no exponentials, no sines, and no cosines. Even in the notes added in the 1781 edition everything is algebraic, except for a few logarithms and an infinite series or two. Evidently, the reasons for studying calculus in the eighteenth century were not what they are today.

Here is how the book starts. Proposition 1 is that
$$d(a + x + y - z) = dx + dy - dz.$$

Then come

Proposition 2: $d(xy) = y\,dx + x\,dy$.

Proposition 3: $d\left(\dfrac{x}{y}\right) = \dfrac{y\,dx - x\,dy}{yy}$.

(Descartes wrote xx for x^2 and so does l'Hôpital. Why not? Both notations use exactly the same number of symbols, and xx used to be easier to print.)

Proposition 4: $d(x^r) = rx^{r-1}\,dx$ where r can be any positive or negative integer or rational number. One feature of old calculus books is hard-hitting examples. Modern books would give some easy illustrations of proposition 4, but l'Hôpital gives just four: finding the differentials of

$$(ay - xx)^3, \quad \sqrt{xy + yy}, \quad \sqrt{ax + xx + \sqrt{a^4 + axyy}}\ ,$$

and

$$\frac{(ax + xx)^{1/3}}{\sqrt{xy + yy}}$$

L'Hôpital's answer to the last one is

$$\frac{a\,dx + 2x\,dx}{3\sqrt[3]{ax + xx}}$$

plus

$$\frac{-y\,dx - x\,dy - 2y\,dy}{2\sqrt{xy + yy}}\sqrt[3]{ax + xx}$$

all divided by $xy + yy$.

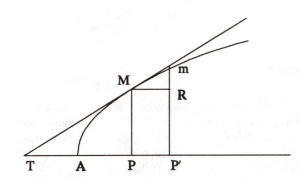

Figure 1

Those four propositions are the content of the first section, "Ou l'on donne les regles du calcul des différences" (Wherein are given the rules for calculating differentials), of the ten sections of the text. The next section, "Usage du calcul des différences pour trouver les tangentes de toutes sortes de lignes courbe" (Using differential calculus to find tangents to various curves), sets the tone of the remainder of the text. Its first problem is

> Soit une ligne courbe *AM* telle que la rélation de la coupée *AP* à l'appliquée *PM*, soit exprimée par une équation quelconque, à qu'il faille du point donné *M* sur cette courbe mener la tangente *MT*.
> (Given any curve, draw a tangent to it at a point *M*.)

(From now on, I will replace l'Hôpital's French with a free English translation.) The solution (see Figure 1) is to take a little right triangle with base dx and height dy: the hypotenuse gives the direction of the tangent line.

In the tradition of all calculus texts, l'Hôpital then gives examples: drawing tangents to

$$ax = y, \quad ay^2 = abx - bx^2,$$
$$ay = bx^m(a + x)^n, \quad \text{and} \quad y^3 - x^3 = axy.$$

There follow also more problems on tangents of the sort that no longer appear in calculus texts because they are too complicated. In a way, this is too bad, since they have gorgeous diagrams, as Figure 2. Modern calculus books, with nothing fancier than a cubic or two, are pale in comparison.

The section on drawing tangents takes forty-one pages of my edition of l'Hôpital; the next section, on

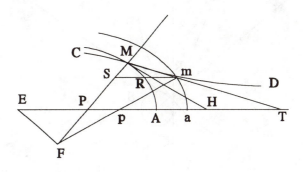

Figure 2

finding maxima and minima, is only eighteen pages long. The reason for the relative shortness is, of course, that once you have said that maxima and minima occur where tangent lines are horizontal (or perhaps vertical), there is nothing more that needs to be added, and you can go directly to examples. That is what l'Hôpital did, and here are his thirteen examples:

1. Find the highest point on the curve $x^3 + y^3 = axy$.

2. Find the lowest point on the curve
$$y - a = a^{1/3}(a - x)^{2/3}.$$

3. Find the maximum distance of a roulette generated by a semicircle from the semicircle's diameter.

4. A geometric problem, leading to finding the maximum of $ax^2 - x^3$.

5. A geometric problem, leading to finding the minimum of
$$\frac{ax + cx - ax - xx}{bx - xx}.$$

6. Find the cone inscribed in a sphere whose surface area is a maximum.

7. Find the parallelipiped, given its volume and the length of one side, with minimum surface area.

8. Find the parallelipiped, given its volume, with minimum surface area.

9. A problem stated geometrically, equivalent to the problem that appears in present-day calculus books about minimizing the time needed to get from a point on one side of a river to a point on the other side, given the speeds of swimming (or rowing) in the river and

walking (or running) on land.

10. Given two points outside a circle, find the point on the circle for which the sum of the two distances from the points to the circle is a minimum.

11. A variation of Example 9.

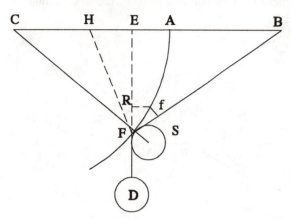

Figure 3

12. Two pulleys are arranged as in Figure 3. What is their equilibrium position?

13. On what day is twilight shortest?

In spite of twilight and the pulleys, the examples are all geometrical.

After maxima and minima, in every modern text come points of inflection, and so it was in l'Hôpital's text. Of course, since he never used derivatives, there was no mention of second derivatives; the section starts with the definition

> the infinitesimal by which the differential of a quantity continually increases or decreases is called the differential of the differential of the quantity, or the second differential.

That is not a definition that we find satisfactory, but l'Hôpital went on to say what it meant and he is soon finding the points of inflection of

$$y = \frac{axx}{xx + aa} \quad \text{and} \quad y - a = (x - a)^{3/5}.$$

That takes us through slightly more than half of the book. The rest of it covers topics that have disappeared from the first course in calculus. In fact, they have disappeared from the second and subsequent courses as well: evolutes, involutes, caustics, roulettes—all sorts of things about pretty curves doing all sorts of pretty things. What is not in the second half of the text is anything about integration or antidifferentiation. What deprived existences our eighteenth-century ancestors

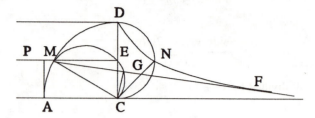

Figure 4

had! No opportunity to evaluate integrals using partial fractions, or to do even the simplest trigonometric substitution! On the other hand, they could contemplate the properties of figures such as Figure 4. It may be that they were no worse off than we are.

The conclusion that I draw from l'Hôpital's text is that what inspired the founders of calculus and drove them on was not a desire to solve physical problems, but rather the fascination of mathematics and mathematical objects. Curves, curves defined by equations! What do they do, how do they do it, why do they act the way they do, where are they going? In l'Hôpital's day those were the questions that gripped the mind and cried out for answers. It happens that calculus and mathematics turn out to be useful for building things and that is nice, but that is not why l'Hôpital, the Bernoullis, Leibniz, and Newton did what they did. They had a different motive. They were after truth. It is a worthy motive. By the way, my conclusion is not one that is universally held so there is no need to agree with it, but it is part of what I have gotten from the world's first calculus textbook. History teaches lessons.

EXERCISES AND QUESTIONS

1. Do l'Hôpital's maximum-minimum problem number 1.

2. Do l'Hôpital's maximum-minimum problem number 7.

3. Find the inflection points on the first of the curves that l'Hôpital used as an example.

4. There is a good chance that your calculus textbook has a title along the lines of *Calculus with Applications to the Physical, Social, and Life Sciences, and to Business Too.* Why are applications so prominent today when they were not in l'Hôpital's time?

5. On page 87 of his text, l'Hôpital says

The differential of $\dfrac{y\,dy}{dx}$ if dx is constant will be

$$\frac{dy^2 + y\,ddy}{dx},$$

and

$$\frac{dx\,dy^2 - y\,dy\,ddx}{dx^2}$$

taking dy as constant.

How would that be said today?

Of *course* the derivative of y with respect to x is written $\dfrac{dy}{dx}$. How else could you possibly write it? You have to have a y to tell what's being differentiated and you have to have an x to tell what it's being differentiated with respect to, and you have to have a d so you know it's differentiation, so that's the natural way to write derivatives. Well, maybe $D_x y$ would do, but only just. Likewise, what could be more natural than $\int_a^b y\,dx$ for the integral of y with respect to x from a to b? You must have the a, b, y, and x and a sign for integration too. The only possible competitor could be something like $I_a^b y\,dx$, or maybe $I\,y_x(a\to b)$. The way that we write calculus is so natural and right that no other way is imaginable.

Not so. Not only are other calculus notations imaginable, they have been used. You probably do not recognize

$$\frac{\overset{n}{\overset{\bullet}{V}}}{\overset{\bullet}{x}{}^{n}} \quad \text{and} \quad \boxed{\begin{array}{c} aa \\ \hline 64x \end{array}}$$

as a derivative and an integral, but that it what they are—

$$\frac{d^{\,n}V}{dx^{\,n}} \quad \text{and} \quad \int \frac{a^2\,dx}{64x}$$

respectively—or were when their authors wrote them in the best and most natural way that they could. Our calculus notation was not handed down to us from above on tablets of stone, nor did it come quickly or easily. The last change in our notation for calculus took place as recently as 1905, more than 200 years after people began writing calculus symbols. In mathematics as in all other areas of human activity, change occurs with what seems like agonizing slowness, obvious and natural improvements can be made only with great effort, and absurdities linger on and on and cannot be gotten rid of. How mathematical notation has evolved is described in Florian Cajori's *A History of Mathematical Notations*, a book (two books, actually, since it comes in two volumes) that was published in 1925 and did its job so well that no one has felt the need to write another on the same subject since. There will be no long excerpt from the book here because it is a work of high scholarship, intended more for reference than for leisure reading. However, what follows is taken directly

from pages 180-262 of volume 2, *Notations Mainly in Higher Mathematics*, the sections on differential and integral calculus.

Newton used dots to write derivatives. He thought of x as a quantity that was changing with time, a *fluent* he called it, and $\dot{x}$ is what he called its rate of change, the *fluxion* of x. The reason that Newton did not do as we do and indicate the variable the derivative was being taken with respect to is that it was always time, and so it didn't need to be explicitly mentioned. A second derivative, a fluxion of a fluxion, Newton wrote as $\ddot{x}$, clearly easier to write than $(\dot{x})^{\cdot}$. A third derivative could be written with three dots, and there was no need for fourth or higher derivatives in the early days of calculus. Dots are fine when you have only one symbol to put them over, but how are you going to indicate the derivatives of $\sqrt{a^2-x^2}$ or $\dfrac{y^2}{b-x}$ using dots? Newton's solution was to write

$$\sqrt{\overset{\bullet}{aa - xx}} \qquad \frac{\overset{\bullet}{yy}}{\overset{\bullet}{b-x}}$$

which looks awkward, as well as being difficult for printers to put into type. To express dy/dx, a rate of change where time was not involved, Newton had to write $\dot{y}:\dot{x}$ (a colon was often used for a divides sign in Newton's day). Dots cause difficulties.

There is no evidence that Newton gave much thought about notation, but there is considerable evidence that Leibniz did. Leibniz wrote, quite correctly,

> In signs one observes an advantage in discovery which is greatest when they express the exact nature of a thing briefly and, as it were, picture it; then indeed the labor of thought is wonderfully diminished.

He tried out various notations, and he wrote letters to people discussing notations. His first notation for differentials was single letters: a, b, or l standing for the differential of x. These are clearly unsatisfactory and liable to cause confusion. He also used z, a bit better since it is closer to x in the alphabet. He tried out $\dfrac{x}{d}$, still better since it has both the x that is being operated on and the d that is doing the operating. He thought further and concluded that dx would be better for a differential, and the way to write the derivative was $\frac{dy}{dx}$. The passage of time has shown that his decision, made in 1675, was the right one. The following, from a letter to Johann Bernoulli, shows Leibniz thinking about notation:

> As regards signs, I see it clearly that it is in the interest of the Republic of Letters and especially of students, that learned men should reach agreement on signs. Accordingly I wish to get your opinion, whether you approve of marking by the sign $\int$ the sum, just as the sign d is displayed for differences; also whether you approve of my designation of ratio as if it were a division, by two dots, for example, that $a:b$ be the same as $\frac{a}{b}$; it is very easily typed, the spacing of the lines is not disturbed. And, as regards proportion, there are some who exhibit such a relation by $a:b :: c:d$; since this really amounts to an equality of quotients, it is sufficient to write, as is my custom $a:b = c:d$ or $\frac{a}{b} = \frac{c}{d}$. Perhaps it will be well to examine other symbols, concerning which more on another occasion.

Newton's notation for antiderivatives was similar to his notation for derivatives. The fluxion of x was $\dot{x}$, and its antifluxion was $\underset{x}{|}$. When Newton needed to indicate an integral of something more than one symbol, the best he could do was to put what was being integrated into a box, as $\frac{aa}{64x}$ was. But, as Cajori says,

> That Newton's notation for integration was defective is readily seen. The $\underset{x}{|}$ was in danger of being mistaken for an abscissa in a series of abscissas x, x', x''; the rectangle was inconvenient in preparing a manuscript and well-nigh impossible for printing, when of frequent occurrence. As a consequence, Newton's signs of integration were never popular, even in

England.

So, English writers who used Newton's notation mostly did without a symbol for integration, getting around the lack by more or less awkward devices.

But a symbol for integration cannot be done without. Leibniz first used "omn.", an abbreviation for *omnia*, meaning "all". So, omn. l was how he first wrote what we would write (and what he would later write) as $\int dx$. (Leibniz had not yet progressed from l to dx as a notation for the differential of x.) In a manuscript dated October 29, 1675, Leibniz wrote

> It will be useful to write $\int$ for omn., as $\int l$ for omn. l, that is, the sum of these l's. Thus one obtains ...

and there is the first-ever appearance of the integral sign. October 29th should clearly be a national calculus holiday, with classes called off to celebrate the inspiration and genius of Leibniz. At the least, no calculus tests should be given on October 29th. Or if a test must be given, it should have no integrals on it. To show that new notations are not born all at once, later in the same manuscript Leibniz wrote a version of the Fundamental Theorem of Calculus:

> If $\int l = ya$, then $l = ya/d$ ...

The d is in the denominator instead of where it belongs, in the numerator.

As Cajori says,

> Perhaps no mathematical symbol has encountered so little competition with other symbols as has $\int$.

Leibniz could have chosen O instead, as a natural abbreviation of omn. But he made the proper choice.

> Why did Leibniz choose the long letter S (*summa*), rather than the letter o of the word *omnia* which he had been using? Was it because the long S stood out in sharper contrast to the other letters and could be more easily distinguished from them?

Yes, most probably. Some writers used $\mathcal{S}$, but that is really the same thing. $\int$ prevailed: it is easier to write, and it is prettier. There is a sensuous pleasure to be gotten in inscribing a well-made integral sign that is hard to match elsewhere in mathematics. A neat $\frac{dy}{dx}$ is satisfactory, but it is the pale satisfaction of a routine job routinely done. Give me a $\int\int xy\,dx\,dy$ any time; the lovely sweep of two integral signs creates a glow that no mere derivative could ever match. Leibniz chose well.

You might think that our notation for definite integrals would follow hard on the heels of the notation for indefinite integrals. If you want to add things up from 0 to 1, what could be more natural than writing $\int_0^1 f(x)\,dx$? It may seem natural to us, but it was not until more than one hundred and fifty years after Newton first wrote integrals that Joseph Fourier first had the idea of putting the limits of integration where we put them today. It seems natural only by hindsight.

Limits of integration were at first indicated only in words. Euler was the first to use a symbol in his integral calculus, of which the following is an illustration:

$$Q = \int \frac{x^{p-1}\,\partial x}{\sqrt[n]{(1-x^n)^{n-q}}} \quad \left[\begin{array}{l} ab\ x^n = \dfrac{1}{2} \\ ad\ x = 1 \end{array}\right].$$

That was in 1768. It is a small change from writing the words "from where $x^n = 1/2$ to $x = 1$" in the line after the integral to putting them in the line with the integral, in abbreviated form. From there it would seem to be only a small step to attach them to the integral sign, but this was not done until 1819, when Fourier off-handedly wrote

We denote in general by the symbol $\int_a^b$ the integral that begins where the variable equals a and is completed where the variable equals b.

The notation was so obviously good that it was eventually universally adopted. But, as usual, even good ideas have to fight to win out. Ohm, after whom Ohm's Law is named, said that his notation for the definite integral, $\int_{b+a} \phi.\,dx$, was better, especially when the integral was complicated, and Peano preferred $S\,(f, a \vdash b)$, though with a bigger and more flamboyant S. Peano was writing this as late as 1903, but by then even he must have known that he was being eccentric. Leibniz and Fourier carried the day.

It was soon after Fourier's notation for definite integrals, in 1823, that a writer had the idea of shorten-

ing $F(b) - F(a)$ to $F(x)\big|_a^b$. Thus, after 150 years, the Fundamental Theorem of Calculus could finally be written in the way that we write it now: if $F'(x) = f(x)$, then

$$\int_a^b f(x)\,dx = F(x)\Big|_a^b.$$

Of course, to write the Fundamental Theorem in that way, you must have the $f(x)$ notation for functions and the prime for differentiation. These were not automatic, nor did they come early in the history of calculus. Early writers used single letters to represent functions, in the same way as we write $y = x^2$ to denote the squaring function. When Leibniz wanted to indicate what variable a function was a function of, he proposed, in a letter to Johann Bernoulli, writing

$$\overline{x\,\rfloor\,1}$$

to denote a function of x. If he had another function that needed writing, it was the same, but with a 2 instead of a 1. He seemed to realize the clumsiness of this notation, since he ended his description with

But in the case of only one function, or only a few of them, the Greek letters suffice, or some such, as you are using.

The first appearance of f for function was in 1734, when Euler wrote

Let $f\left(\dfrac{x}{a} + c\right)$ denote a function of $\dfrac{x}{a} + c$.

It was the great influence of Lagrange's *Théorie des fonctions analytiques* (1797) that established the $f(x)$ notation and the use of primes for differentiation, and also shifted the emphasis in calculus texts from the differential, where it had been, to the derivative, where it is now.

Since in the early days of calculus the idea of limit was not well understood, or the need for it appreciated, it is no surprise that no notation for limits appeared until 1786, when Lhulier wrote "lim." Cauchy did the same in 1823, when he

pointed out that "lim.(sin. x)" has a unique value 0, while "lim.$((1/x))$" admits of two values and "lim.$(($ sin.$(1/x)$ $))$" of an infinity of values, the double parentheses being used to indicate all the values that the inclosed function may take, as x approaches zero.

We no longer allow limits to be double-valued, and certainly not infinitely many-valued, and we no longer write "lim." with the period. Both usages have withered away. Unlike "omn." which turned into $\int$, "lim" has never been replaced with a single symbol, which if not odd is at least a bit inconsistent. However, if "lim" had contracted to one symbol, we would not have the convenience of writing where the variable was going underneath the "lim" as was done by Weierstrass in 1854 and Hamilton in 1853. They did not write it quite as we do: Weierstrass wrote "$n = \infty$" instead of "$n \rightarrow \infty$" and Hamilton did likewise. It was not until 1905 that the arrow was first used. It was generally adopted because of the influence of G. H. Hardy's 1908 text *A Course of Pure Mathematics*, where he wrote

I have followed Mr. J. G. Leathem and Mr. T. J. I'A. Bromwich in always writing

$$\lim_{n \to \infty}, \quad \lim_{x \to \infty}, \quad \lim_{x \to a},$$

and not

$$\lim_{n = \infty}, \quad \lim_{x = \infty}, \quad \lim_{x = a}.$$

This change seems to me one of considerable importance, especially when "∞" is the "limiting value." I believe that to write $n = \infty$, $x = \infty$ (as if anything ever were "equal to infinity"), however convenient it may be at a later stage, is in the early stages of mathematical training to go out of one's way to encourage incoherence and confusion of thought concerning the fundamental ideas of analysis.

Quite right, and with that change the evolution of notation in elementary calculus has come to an end. Or what seems to be an end: perhaps it is just a pause of a few centuries before the introduction of new notation that will be so obviously superior that its users will look back on the primitive days of the twentieth century and smile at the naiveté and simplicity of its inhabitants in not being able to see what, if they only looked, was staring them in the face.

Cajori sums up the evolution of calculus notation as follows:

"In considering the history of the calculus, the view advanced by Moritz Cantor presses upon the mind of the reader with compelling force. Cantor says:

We have felt that we must place the main emphasis on the notation. This is in accordance with the opinion which we have expressed repeatedly that, even before Newton and Leibniz, the consideration of infinitesimals had proceeded so far that a suitable notation was essential before any marked progress could be made.

"Our survey of calculus notations shows that this need was met, but met conservatively. There was no attempt to represent all reasoning in the calculus by specialized shorthand signs so as to introduce a distinct sign language which would exclude those of ordinary written words. There was no attempt to restrict the exposition of theory and application of the calculus to ideographs. Quite the contrary. Symbols were not generally introduced, until their need had become imperative. Witness, for instance, the great hesitancy in the general acceptance of a special notation for the partial derivative.

"It is evident that a sign, to be successful, must possess several qualifications: it must suggest clearly and definitely the concept and operation which it is intended to represent; it must possess adaptability to new developments in the science; it must be brief, easy to write, and easy to print. The number of desirable mathematical symbols that are available is small. The letters of the alphabet, the dot, the comma, the stroke, the bar, constitute the main source of supply. In the survey made in this paper, it was noticed that the forms of the fourth letter of the alphabet, d, D, and ∂, were in heavy demand. This arose from the fact that the words "difference," "differential," "derivative," and "derivation" all began with that letter. A whole century passed before any general agreement was reached among mathematicians of different countries on the specific use which should be assigned to each form.

"The query naturally arises, Could international committees have expedited the agreement? Perhaps the International Association for the Promotion of Vector Analysis will afford an indication of what may be expected from such agencies.

"An interesting feature in our survey is the vitality exhibited by the notation $\dfrac{dy}{dx}$ for derivatives. Typographically not specially desirable, the symbol nevertheless commands at the present time a wider adoption than any of its many rivals. Foremost among the reasons for this is the flexibility of the symbol, the ease with which one passes from the derivative to the differential by the application of simple algebraical processes, the intuitional suggestion to the mind of the reader of the characteristic right triangle which has dx and dy as the perpendicular sides. These symbols readily call to mind ideas which reach to the very heart of geometric and mechanical applications of the calculus.

"For integration the symbol $\int$ has had practically no rival. It easily adapted itself to the need of marking the limits of integration in definite integrals. When one considers the contributions that Leibniz has made to the notation of the calculus, and of mathematics in general, one beholds in him the most successful and influential builder of symbolic notation that the science has ever had."

EXERCISES AND QUESTIONS

1. When Newton wrote l: $\overline{x + a}$ ^{m} he meant $\log(x + a)^m$. Evaluate l: $\overline{x \overset{\bullet}{+} a}$ ^{m} .

2. Write the quotient formula for derivatives as Newton would have.

3. Leibniz wrote $\boxed{2}$ $\overline{r^2 + x^2}$ for $(r^2 + x^2)^2$. What do you think he meant by

(a) $\boxed{3}$ $(AB + BC)$

(b) $\overline{w - e : f \boxed{1: h}}$?

(The answers are in Cajori's *History*, volume 2, page 191.)

4. What do you think Cauchy thaught were the two values of $\lim_{x \to 0} (1/x)$ and the infinity of values of $\lim_{x \to 0} \sin(1/x)$?

5. "Perhaps the International Association for the Promotion of Vector Analysis will afford an indication of what may be expected from such agencies." The IAPVA has disappeared, so the indication is that such agencies fail. However, would it be a good idea to have a national (or international) Notation Bureau to maintain and certify standard notations for mathematics?

THE "WITCH" OF AGNESI (1718-1799)

By Lynn M. Osen

If you pick up any history of mathematics—I will pick up *A History of Mathematics*, by Carl Boyer—and look in the index for names of women, you will not find many. Among the Gs in *A History of Mathematics*, we have Galileo, Galois, Galton, ..., Guldin: 43 names in all, of which 42 are the names of men. I did not pick up, by accident or on purpose, a book by a male chauvinist. Among the Gs in *Mathematical Thought from Ancient to Modern Times* by Morris Kline there are 32 men and 1 woman, and among the Gs in *The Nature and Growth of Modern Mathematics* by Edna Kramer, the proportion is 19 to 1. Why, since women make up about 50% of the human race, are there so few among the mathematicians?

Of course, it is not only in histories of mathematics that the ratio of women to men is less than 50%. Not that I have looked at any, but I am sure that in histories of butchering, baking, and candlestick-making the proportion of female names in the indexes would also be less than one-half. And so would it be in a history, whether it exists or not, of almost every other occupation, because ever since people started to do work for money, the general rule has been for men to tend to concentrate on earning it and for women to be more concerned with the care of home and children.

Whether or not women are underrepresented in mathematics compared with other fields, there can be no argument that their contributions to mathematics have been neglected. Recently several books designed to remedy this defect have been published and the following excerpt is from one of them.

Over this time, when the ancient world was giving way to the medieval, a monstrous tide of misogyny had engulfed Christendom in Europe and did not commence to subside until the Renaissance began. Even in the most enlightened centers there was strong opposition to any form of higher education for females. Most would have denied women even the fundamental elements of education, such as reading and writing, claiming that these were a source of temptation and sin. (To such critics, Hroswitha, the famous nun of Gandersheim, once replied that it was not knowledge itself that was dangerous, but the poor use of it: "Nec scientia scibilis Deum offendit, sed injustitia scientis".)

For the most part, learning was confined to monasteries and nunneries, and these precincts guarded well the sacred mysteries of mathematics, enfranchising only those who subscribed to the religious faith of the ecclesiastics. Such schools generally constituted the only opportunity for education open to girls during the Middle Ages, and in a few of these, women were able to distinguish themselves as scholars.

Perhaps one of the most learned of these was Hroswitha, the well-known nun of the Benedictine Abbey in Saxony during the tenth century. Although she is most often cited for her dramatic compositions and as a writer of history and legend (among the latter,

The Lapse and Conversion of Theophilus was a precursor of the famous legend of Faust), Hroswitha's writings are also an important index to the monastic mathematics of this period, and they reveal a sound intelligence of either Greek or Boethian arithmetic. In her *Sapientia*, for instance, when the emperor Hadrian inquires to know the ages of Sapientia's three daughters (Faith, Hope, and Charity), the reply is that Charity's age is represented by a defective evenly even number; Hope's by a defective evenly odd number; and that of Faith by an oddly even redundant one. It is also worthy of remark that in her writings, Hroswitha mentions four perfect numbers: 6, 28, 496, and 8128.

Hroswitha had both the courage and originality of genius, though not all of her talents were focused on the study and development of mathematics. She was interested in various branches of learning, and her writings were an attempt to provide educational material for the women in her medieval nunnery, a purpose that motivated other scholarly nuns, including Saint Hildegard, Abbess of Bingen on the Rhine. Her capabilities in mathematics and her treatises on science earned recognition, and it has been claimed by some writers that she anticipated Newton by centuries when she wrote that the sun was the center of the firmament and its gravitational pull "holds in place the stars around it, much as the earth attracts the creatures which inhabit it".

After the fall of Constantinople, there was a great influx of scholars from the famous old city on the Bosphorus into the Italian peninsula. These scholars brought with them some of the treasures of science and literature that were to help spark the interesting phenomenon we call the Renaissance. Another notable development significant to learning was the invention of the printing press with movable type. This device helped in the dissemination of knowledge and made the printed page more available to those who were not formally educated.

In Italy some traditions of the relatively free Roman matron had remained alive, but elsewhere on the European continent the status of women changed very slowly, even after the Renaissance. There were occasionally women whose talents and genius were remarkable, but the lives of these women emphasized by contrast the prevalent ignorance of the great mass of women who had little access to instruction even in its most fundamental form.

France and Germany saw a revival of the antifeminist crusade that had stifled women's aspirations in ancient Greece and Rome. The Teutonic mentality did not recognize intelligence in women; Luther was a strong influence in his opposition to the education of females.

In England Henry VIII had destroyed the conventual system, leaving women without any systematic education for a long period. Elizabeth I did nothing for the education of females; where their intellectual progress was enhanced at all during these years, it was due to private tutoring or the protracted efforts of individual women for their right to knowledge. Here, as in most of Europe, women were in many respects even further removed from knowledge than they were during the Dark Ages.

But on the Italian peninsula, where the Renaissance had its origin, some Italian women had made their mark on the academic world, even before the close of the Middle Ages. Some had earned doctorates and had become lecturers and professors in the universities of Bologna and Pavia.

The advent of the Renaissance signaled the return of many Italian women to an active role in the educational movement. One historian wrote of this period in Italian history,

> The universities which had been opened to them at the close of the Middle Ages, gladly conferred upon them the doctorate, and eagerly wel-

comed them to the chairs of some of their most important faculties. The Renaissance was, indeed, the heyday of the intellectual woman throughout the Italian peninsula—a time when women enjoyed the same scholastic freedom as men.

Nor were these women scholars exposed to ridicule. This same historian wrote that the men of those days

> ... were liberal and broad-minded ... who never for a moment imagined that a woman was out of her sphere or unsexed because she wore a doctor's cap or occupied a university chair. And far from stigmatizing her as a singular or strong-minded woman, they recognized her as one who had but enhanced the graces and virtues of her sex by the added attractions of a cultivated mind and a developed intellect. Not only did she escape the shafts of satire and ridicule, which are so frequently aimed at the educated woman of today, but she was called into the councils of temporal and spiritual rulers as well.
>
> Woe betide the ill-advised misogynist who should venture to declaim against the inferiority of the female sex, or to protest against the honors which an appreciative and a chivalrous age bestowed upon it with so lavish a hand. The women of Italy, unlike those of other nations, knew how to defend themselves, and were not afraid to take, when occasion demanded, the pen in self-defense. This is evidenced by numerous works which were written in response to certain narrow-minded pamphleteers or pitiful pedants who would have the activities of women limited to the nursery or the kitchen.

The talent and genius that flowered as a result of this enlightened attitude was enormous: women became famous in the arts, in medicine, literature, philosophy, science, and languages, and there were also important names surfacing in mathematics during the seventeenth and eighteenth century. There were Tarquina Molza, who was taught by the ablest scholars and was honored by the senate of Rome for her accomplishments; Maria Angela Ardinghelli of Naples; Clelia Borromeo of Genoa, who was widely praised and of whom it was said that "no problem in mathematics and mechanics seemed to be beyond her comprehension". There were Elena Cornaro Piscopia, honored by the University of Padua for her proficiency in mathematics; Laura Bassi, primarily known for her work in physics (her work centered on Descartes and Newton, and she was a member of the Bologna Academy of Sciences); and Diamente Medaglia, who wrote a special dissertation on the importance of mathematics in the curriculum of studies for women and is quoted: "To mathematics, to mathematics, let women devote attention for mental discipline".

Some of these women were mathematicians in the most rigorous sense; others worked on the periphery of the discipline, but whatever their proper designation, each had absorbed enough mathematics to make her efforts an exemplary part of that enlightened and powerful age we call the Renaissance, and for this these women have earned a separate recognition, a separate warrant of our attention.

Far more remarkable than any of these women, however, was Maria Gaetana Agnesi, called one of the most extraordinary women scholars of all time. She was born in Milan on May 16, 1718, to a wealthy and literate family; like Hypatia's, her father was a professor of mathematics. Dom Pietro Agnesi Mariami occupied a chair at the University of Bologna, and he, along with Maria's mother, Anna Brivia, very carefully planned the young girl's education so that it was rich and profound.

She was recognized as a child prodigy very early: spoke French by the age of five; and had mastered Latin, Greek, Hebrew, and several modern languages by the age of nine. At around this age, she delivered a discourse in Latin defending higher education for women, a subject that continued to interest her throughout her life.

Maria's teen-age years were spent in private study and in tutoring her younger brothers (she was the oldest of 21 children). During this time, she also mastered the study of mathematics as it had been developed by such masters as Newton, Leibniz, Fermat, Descartes, Euler, and the Bernoulli brothers.

The Agnesi home was a watering place for a select circle of the most distinguished intellectuals of the day, and Maria acted as hostess for her father's carefully chosen assemblies. She participated in the seminars among those gathered in her father's study by presenting theses on the interesting philosophical questions under discussion, and her father encouraged her to engage in disputations with these scholars.

Monsieur Charles De Brosses, the president of the parliament of Burgundy, wrote in his *Lettres sur l'Italie* about one of these seminars to which he and his nephew were invited. He was particularly impressed with Maria's erudite versatility in the discussion of such diverse subjects as

> the manner in which the soul received impressions from corporeal objects, and in which those impressions are communicated from the eyes and ears and other parts of the body on which they were first made, to the organs of the brain which is the general sensorium or place in which the soul receives them; we afterwards disputed on the propagation of light and the prismatic colours. Loppin then discoursed with her on transparent bodies, and curvilinear figures in Geometry, of which last subject I did not understand a word. ... She spoke wonderfully well on all of these subjects though she could not have been prepared before-hand to speak on them, anymore than we were. She is much attached to the Philosophy of Sir Isaac Newton; and it is marvelous to see a person of her age so conversant with the abstruse subjects. Yet, howevermuch I may have been surprised at the extent and depth of her knowledge, I have been much more amazed to hear her speak Latin (a language which she certainly could not often have occasion to make use of) with such purity, ease and accuracy.

De Brosses mentions that this particular party was attended by about thirty people from several different nations of Europe, seated in a circle, questioning Maria. Pietro Agnesi was understandably proud of his accomplished daughter, but these displays were contrary to her shy, bashful nature, and she prevailed upon her father to give them up when she was around twenty years old. At about this time she began to express a desire to enter a convent so that she might spend her life in sequestered study and work with the poor. This request was denied by her father.

Maria never married. She gave most of her time to the study of mathematics, to caring for her younger brothers and sisters, and (after her mother's death) to assuming the duties of the household.

In 1738 she published a collection of complex essays on natural science and philosophy called *Propositiones philosophicae*, based on the discussions of the savants who had gathered in her father's home. Again, these essays expressed her conviction that women should be educated in a variety of subjects.

By the age of twenty, she had entered on her most important work, *Analytical Institutions*, a treatise in two large quarto volumes on the differential and integral calculus. She spent ten years on this work, and her natural talent for mathematics may be reflected in her report that on several occasions during this time, after working all day on a difficult problem that she could not solve, she would arise at night and while in a somnambulistic state, write out the correct solution to the problem.

When her work was finally published in 1748, it caused a sensation in the academic world. Although she had originally begun the project for her own amusement, it had grown, first into a textbook for her younger brothers and then into a more serious effort. It has the distinction of being one of the most important mathematical publications produced by a woman up until that time. It was a classic of its kind and the first comprehensive textbook on the calculus since l'Hôpital's early book. It was also one of the first and most complete works on finite and infinitesimal analysis and was not superseded until Euler produced his great texts on the calculus later in the century.

Agnesi's great service was that she pulled together into her two volumes the works of various mathematicians, including Newton's method of "fluxions" and Leibniz's method of differentials. These and other works concerning analysis were scattered through the writings of various authors, some printed in foreign journals. Maria's scholarship and her facility with languages helped her to collect these into a compendium that saved students the complicated task of seeking out developments and methods formerly dispersed in a variety of sources. Her volumes were translated into French and English and were widely used as textbooks.

The first section of *Analytical Institutions* deals with the analysis of finite quantities and discusses the construction of loci, including conic sections. It also deals with elementary problems of maxima and minima, tangents, and inflections.

The second section is devoted to the analysis of "infinitely small quantities," quantities defined as so small that when compared to the independent variable, the proportion is less than that of any assigned quantity. (If such infinitesimals, called "differences" or "fluxions," are added to or subtracted from the variable, the differ-

ence would not be significant. "Differences" or variables tending to zero, and "fluxions," or finite rates of change, are treated here as essentially the same quantities.)

The third section of Agnesi's work deals with the integral calculus and gives a general idea of the state of knowledge concerning it at the time. She gives some specific rules for integration, and there is a discussion on the expression of a function as a power series. The extent of convergence is not treated.

The last section of the volume discusses "inverse method of tangents" and very fundamental differential equations.

But many of the other important aspects of *Analytical Institutions* have been eclipsed by Agnesi's discussion of a versed sine curve, originally studied by Fermat. This plane cubic curve has the Cartesian equation $xy^2 = a^2(a - x)$. (Agnesi begins with the geometrical principle that if the abscissa of corresponding points on a curve is equal to that of a given semicircle, then the square of the abscissa is to the square of the radius of the semicircle in the same ratio as that in which the abscissa would divide the diameter of the semicircle.) This curve had been studied earlier by Guido Grandi, as well as Fermat. It had come to be called a *versiera*, a word derived from the Latin *vertere*, "to turn," but it was also an abbreviation for the Italian word *avversiera*, or "wife of the devil."

In 1801, when Maria's text was translated into English by John Colson, professor of mathematics at Cambridge, Colson rendered the word *versiera* as witch, and through this or some such mistranslation, the curve discussed by Maria came to be known as the "witch of Agnesi." Subsequently, where mention of this woman is made in modern English textbooks, it is most often by this phrase. The exquisite irony of this term is not lost on those who are familiar with Agnesi's life of selfless service and piety.

Maria's books attracted the attention of the French Academy of Sciences, and a committee was appointed to assess them. A deputy wrote her afterward,

> I do not know of any work of this kind that is clearer, more methodical or more comprehensive. ... There is none in mathematical sciences. I admire particularly the art with which you bring under uniform methods the divers conclusions scattered among the works of geometers and reached by methods entirely different.

Despite this tribute, however, the French Academy

did not admit Agnesi. Its constitution barred females, despite the fact that the very notion of the Academy was introduced to its founder, Richelieu, in the salon of a woman, Madame de Rambouillet.

Fortunately, Italian academics were more liberal, and Maria was elected to the Bologna Academy of Sciences. There were also many other honors: her book had been dedicated to the Empress Maria Theresa, who showed her appreciation by sending Maria a splendid diamond ring and a small crystal casket set with diamonds and precious stones.

But the recognition that pleased her most came from Pope Benedict XIV. He was interested in mathematics, and he recognized the exceptional ability of Maria Agnesi. His letters indicate his respect for her accomplishments, and it was through his invitation that she was given an appointment as honorary lecturer in mathematics at the University of Bologna.

Her name was added to the faculty roll by the senate of the university, and a diploma to this effect was sent her by the pontiff. The diploma was dated 5 October 1750. Sister Mary Thomas a Kempis wrote later that Agnesi's name remained on the university's "Rotuli" until 1795-1796.

There is some difference of opinion among historians as to whether Agnesi accepted this appointment or not. She was urged to do so by many of her contemporaries, including the famous physicist Laura Bassi. Most reviews of her life indicate that she occupied the Chair of Mathematics and Natural Philosophy at Bologna from 1750 to 1752; other writers say she only filled in for her father during his last illness; still others insist that she eschewed the pontiff's offer, preferring instead to remain in her beloved Milan. In retrospect, it would appear that she did accept this position and served at the university until her father's death in 1752, when she decided to return to a quieter life of study and comparative solitude.

She relinquished the ambition to do any further work in mathematics; when, in 1762, the University of Turin asked her for her opinion of the young Lagrange's recent articles on the calculus of variations, her response was that she was no longer concerned with such interests.

True to her deeply religious nature, she began to devote most of her time to charitable projects with the sick at the hospital of Maggiore and with the poor of her parish, San Nazaro.

Sister Mary Thomas a Kempis, whose beautiful article "*The Walking Polyglot*" reviews these charitable efforts of Agnesi's, reports that, "To extend her work more and more she saved on her dresses, on her meals, and on her dear books, she did not hesitate to sell her imperial gifts and even the crown set with precious jewels given her by Pope Benedict XIV".

She turned her home into a refuge for the helpless and the sick, the aged, and the poor. Neglected women were cared for in her own rooms when there were no other facilities. And when the Pio Instituto Trivulzio, a home for the ill and infirm, was opened in 1771, the archbishop asked Maria to take charge of visiting and directing women, particularly the ill. She took on this duty in addition to the burden of maintaining her own small hospital. When these duties became too burdensome, she took up full-time residence at the Institute in 1783, insisting upon paying rent so as not to diminish the capital of the poor. The annals of the Institute call her "an angel of consolation to the sick and dying women until her death at the age of eighty-one years on January 9, 1799".

Agnesi was buried in a cemetery outside the Roman gate of the city walls. She shares a common grave with fifteen old people of the Luogo Pio. There is no elaborate monument over her tomb, nor is one needed. She has been widely honored (and continues to be) for her good works.

On the one-hundredth anniversary of her death, Milan took note of her life: streets in Milan, in Monza, and in Masciago were given her name. A cornerstone has been placed in the facade of the Luogo Pio, the inscription on it proclaiming her "erudite in Mathematics glory of Italy and of her century." A normal school in Milan bears her name, and scholarships for poor girls have been donated in her honor.

Today, almost two hundred years after her lifetime of hard work, her memory is still vital and inspiring.

EXERCISES AND QUESTIONS

1. Discover who the woman whose name began with "G" was who appears in the indexes of histories of mathematics and find out why she appears.

2. What could the ages of Sapientia's daughters have been? To answer that, you need to know that an evenly even number is one that is divisible by 4, that the other even numbers are oddly even numbers, that an evenly odd number is one that is one more than a multiple of 4, and that the other odd numbers are oddly odd. Also, a redundant number is one like 24, whose divisors (1, 2, 3, 4, 6, 12) sum to more than 24, and a defective number is one like 22, whose divisors (1, 2, 11) sum to less than 22. So, for example, 32 is an evenly even defective number and 42 is an oddly even redundant number.

3. "Luther was a strong influence in his opposition

to the education of females." Why the opposition?

4. Sketch a graph of the Witch—

$$y^2 = \frac{a^2(a - x)}{x}$$

and then find, if you can, the area between the curve and its asymptote.

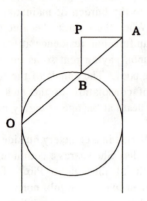

Figure 1

5. "If the abscissa of corresponding points on a curve is equal to that of a given semicircle ..." is not very clear without the aid of a diagram. Here is a geometrical definition of the Witch: in Figure 1, draw any line *OA*, then draw a vertical line from *B* and a horizontal line from *A*; where they intersect gives *P*, a point on the curve. Let the circle have radius 1 and see if the equation that the coordinates of *P* must satisfy is the equation of a Witch.

LEONHARD EULER, 1707-1783

by Jane Muir

Most great mathematicians seem to have led dull lives. Their biographies can be summed up as "X. was born, grew up, studied mathematics, liked it, and worked at mathematics until death." Their lives have no drama. They do not become involved with the politics of the day, or the arts. They do not get rich, they are not noisy drunks, they do not flagrantly abuse their spouses, they do nothing flamboyant or spectacular. Dull lives, you may think, with nothing but work, work, work. But this is only appearance: their lives only seem dull because all we can see is their outsides. Inside, it is different. Inside is mathematics, inside is drama. There is a problem: hard, dark, forbidding, with no way into it. Then comes an idea, something to try, and its working out: will it succeed? Maybe it does not succeed at first—hard and worthwhile problems do not have easy or obvious solutions—so perhaps the idea needs to be changed a little, or maybe a whole new idea is needed. Then more working out. Suspense! Then, sometimes, success—the idea works! The problem is solved! That is drama, classical drama, with conflict, climax, and resolution. Mathematicians have quite enough excitement in their work so that they do not need to look for it elsewhere. That is why their lives seem so dull. Their systems could not stand any more excitement without giving way.

The life of Leonard Euler, a towering figure of eighteenth century mathematics, is an example. He lived, he worked, and he died. But it was a full life. Anywhere you look in mathematics, you see the name of Euler. Whenever you write "sin x", you do it because Euler did it that way. When you differentiate e^x and get e^x it is because Euler used that notation for 2.71828-182845904... raised to the power x. When you are astonished to learn that

$$1 + \frac{1}{4} + \frac{1}{9} + \frac{1}{16} + \frac{1}{25} \ldots$$

is $\pi^2/6$, you are only echoing Euler's astonishment when he first summed the series. There is Euler's ϕ-function in number theory, the Euler-Maclaurin sum formula to change sums into integrals, Euler's polyhedron formula, the Euler force on a beam, and so on. Euler's complete works, when they are finally all published in book form, will take up more than eighty volumes and very large volumes they are. There are twenty-nine on mathematics, thirty-one on mechanics and astronomy, twelve on physics, and the rest on other topics. That is an amazing life's work, but Euler was an amazing person.

The following excerpt gives some of the details of his life and times. It does not give very many details about his mathematics. The reason that the author chose not to give many could have been that what Euler did in mathematics is beyond what ordinary readers could understand. Euler extended calculus beyond what is treated in calculus textbooks and he worked on problems in areas that a typical college student of mathematics never hears of—the calculus of variations, for example, where the Euler-Lagrange equation (one more thing named after Euler) is fundamental. This is why the history of mathematics is not studied by vast masses of people: to understand the history of mathematics, you must understand mathematics, and if the mathematics of the eighteenth century does not mean anything to you then neither can its history. Most people leave school with a sound knowledge of the essence of mathematics as it was in approximately 1550. Students of calculus can add 150 years more, but even so 1700 is a long way from the present. However, even if we do not know the details of Euler's accomplishments, knowing a little about who he was and how he lived can do nothing but good.

The erroneous but popular conception of mathematicians as an eccentric lot is hardly refuted in the persons of Cardano, Pascal, or Newton. As far as sheer conceit, fanaticism, or crotchetiness go, these three are perfect examples. But mathematicians are, basically, human beings, with the same failings and virtues as any other group of humans. If the three above were slightly eccentric, dozens of others were Mr. Normal himself. Leonhard Euler is a case in point. In manner and appearance he gave the impression that had some basis in actuality, for Euler not only came from a small town but was originally slated for a clerical career.

Born in Basel, Switzerland, on April 15, 1707, of Marguerite and Paul Euler, he grew up in a tiny outlying town where his father was the Calvinist minister.

Improbable as it may seem, this small-town boy found his way into some of Europe's most dazzling courts during those glory days of kings—Louis XIV, Catherine and Peter the Great, Frederick the Great of Prussia. All had their own Versailles or replicas thereof, and all vied with each other in cultivating the arts and sciences. In the eighteenth century there were no kings of commerce to support scientific research, education, art, and other worthy causes. This philanthropic role was played by kings of the land, who skimmed the intellectual cream off Europe for their academies and courts.

Euler received his first schooling from his father, who had studied mathematics under one of the Bernoullis, a famous and talented family that produced more than a dozen top-rate mathematicians. While still a young boy Euler entered the University of Basel where he, too, studied under a Bernoulli, Johannes, the brother of his father's teacher.

At Basel his favorite subject was geometry, but after getting his Bachelor's degree at fifteen and Master's at sixteen, he dropped mathematics to study theology and Oriental languages in accordance with his father's wishes. With a little pressure from the Bernoullis, however, Paul Euler soon gave in to his son's pleas to return to the study of mathematics.

At eighteen, Euler published his first mathematical paper, a treatise on the masting of ships, which he submitted in the annual contest held by the French Academy of Science. Although he was competing against Europe's top mathematicians and scientists, many of them two or three times his age, he won second prize. His paper is unimportant as far as the history of mathematics goes, but it does illustrate one aspect of Euler's method. He never checked his results experimentally. Coming from landlocked Switzerland, he knew next to nothing about ships or their sails. But this lack of firsthand experience did not bother him, for

since his conclusions on the height and thickness of masts were "deduced from the surest foundations in mechanics; their truth or correctness could not be questioned."

While Euler was still a schoolboy engrossed in his studies, the free-wheeling, swashbuckling Tsar, Peter the Great, made a tour of Europe. He avoided the fashionable resorts and watering places and concentrated instead on dockyards and universities. Not satisfied by just looking, the six-foot-seven-inch Tsar picked up a hammer and saw and joined the workmen in building ships. In Holland he took a course in anatomy and became the first—and probably only—grand monarch to pull teeth and perform surgery. Everything he saw convinced him more and more of how backward his own country was in comparison with the West, and he returned home determined to lift Russia to the level of France and England. French became the language of the Russian court; nobles were told to shave their outlandish beards; and ladies were introduced to French fashions. Peter built himself a miniature Versailles and a new capital, St. Petersburg, later called Petrograd, and then Leningrad. He founded a medical school and put his own learning of anatomy to use by performing an autopsy on his sister-in-law. And he drew up plans for an Academy in St. Petersburg, which were carried out by his widow, Catherine I.

Two of Euler's friends, Daniel and Nicholas Bernoulli, were invited to come to Russia to teach and study at the new Academy. They went and wrote back about an opening in the physiology department, urging Euler to brush up on the subject and apply for the position. At their own end, they persuaded Catherine I that they knew just the man for the vacancy—Leonhard Euler. Euler quickly crammed in a few courses in medicine and anatomy, applied for the position, was accepted, and at the age of twenty transplanted himself to St. Petersburg, Because of some sort of disorganization or other inscrutable bit of administrative red tape, he was unexpectedly given a position in the mathematics department.

The beginnings of the Academy were inauspicious, if not makeshift. Each professor had to bring with him two students—a policy that resulted in the Academy's opening with twenty teachers and forty pupils.

In addition to the scarcity of pupils, there was a like scarcity of money. Catherine I had died the day Euler reached Russia and her successor's regent, less enlightened about education, considered the Academy an unnecessary drain on the royal exchequer. Within three years Euler found himself on the verge of bankruptcy. He considered giving up his position to join the Russian Navy and was on the verge of doing so when the young

Tsar died and was replaced by the Empress Anna, who was looser with the purse strings although tighter with the whip.

Euler's friend Nicholas Bernoulli died during his first years in Russia and Euler succeeded him as head of the Natural Philosophy department. By 1734 his finances had improved to the point where he could afford to marry. He took as his wife Catharina Gsell, the daughter of a Swiss painter enticed to St. Petersburg by the late Peter the Great. Continuing his work in mathematics, he flooded the scene with writings, many of which attracted a good deal of attention. At the age of twenty-five his first major work—a two-volume book on mechanics—was published and the author hailed as a genius. Several other papers were submitted to the French Academy where he and the Bernoullis seemed to have cornered the market as far as prizes went. Between them, they walked away with twenty-eight prizes, Euler winning twelve.

Quantity as well as quality characterized Euler's work. Undoubtedly, he is the most prolific mathematician who has ever lived. Writing incessantly—short papers, longer treatises, books, all of which required a tremendous amount of study and thought—he produced enough to fill over ninety large volumes. His mind worked like lightning and was capable of intense concentration. But unlike Newton, Euler needed neither quiet nor solitude. Most of his work was done at home in the bedlam created by several small children noisily playing around his desk. Euler remained undisturbed and often rocked a baby with one hand while working out the most difficult problems with the other. He could be interrupted constantly and then easily proceed from where he had left off without losing either his train of thought or his temper.

Research, teaching, writing and his growing family filled his days, but the constant reading and work took their toll. The year after his wedding he received a problem from the French Academy and, devoting three days of intense concentration to it, found a solution. Other mathematicians asked for several months to solve the same problem. Euler's three days' work "threw him into a fever which endangered his life" and, according to the medical diagnosis of the time, was the cause of his losing the sight in his right eye.

At the age of thirty-five, Euler was invited by Frederick the Great of Prussia to come to Berlin to teach and do research. He accepted readily, happy to leave a Russia which was being bathed in blood through the relentless efforts of Empress Anna to clean out spies and traitors.

Frederick, like Peter the Great, was determined to aggrandize his country through art and science, as well as war. A much more polished man than his Russian counterpart, Frederick had managed to acquire a literary veneer despite the hindrances set up by his father. Frederick's father, an ogreish Philistine if ever there was one, considered books and culture as effete and useless and was constantly popping his son into a uniform and thrusting a sword in his hand to make a man of him. Nevertheless, when his father died, Frederick reverted to a cultural life and immediately invited Europe's leading intellectual lights to his court. Voltaire, his idol and "the finest ornament in France," was one of the first to be asked. Frederick tried to get the Bernoullis as well as Euler, but failed.

At Berlin Euler's reputation rose higher than ever through the publication of his *Introduction in Analysis Infinitorum*, and *Instituitiones Calculi Differentialis*. "Through them," said a later scholar, "he became the mathematics teacher of all Europe." The calculus, which had been invented—or at least made public—only a few decades before, was still largely undeveloped. Along with many other mathematicians, Euler believed that it was not a perfect tool but gave correct results because the errors offset each other. Fortified with this belief, he struggled to clarify and simplify the analytical operations. His two books on calculus were not only the best, but the most understandable of the time. Strange as it may seem to us today, they enjoyed a tremendous popularity despite their technical subject matter. The eighteenth century was still an age when no man could consider himself educated without a knowledge of mathematics, for on mathematics all knowledge was based. Its methods set the standard and became the model for every other branch of learning. Indeed, the belief was prevalent that everything—any idea or fact—could be summed up mathematically. Leibniz had even started a grand scheme whereby all ideas were to be reduced to symbols, which could then be handled in the same way as algebraic symbols. This scheme, he believed, would rid the world of wars, for all disputes and differences could be settled peacefully and fairly by simply juggling symbols. The belief that wars are caused only by injustice may be a bit naive, but the general plan of using symbols for ideas is not as farfetched as it might seem. Today an artificial language capable of expressing detailed ideas is being developed for use by electronic computers and a whole branch of modern mathematics, symbolic logic, has been erected on the ruins of Leibniz's scheme.

In Berlin, Euler continued to produce his papers—some being sent to Russian journals, for he still received a salary from the Academy of St. Petersburg; and some being published by the Berlin Academy. The papers fairly gushed forth. He calculated as easily "as men breathe, or as eagles sustain themselves in the wind"; his photographic mind skimmed over the most difficult problems and came up with solutions in a flash—throwing "new light on nearly all parts of pure or abstract mathematics."

Euler worked in every field of mathematics: analysis (he was called "analysis incarnate"), algebra, geometry, and number theory. In all of these branches he consolidated and united the work that had been done before. He supplied the missing links and went on to develop the unfinished theories of others. He extended the applications of analytic geometry, for instance, to three dimensions, where he found general equation forms for planes and the different solids. A line, as shown in the chapter on Descartes, is of the form $ax + by + c = 0$. In the three-dimensional system, the general equation for the corresponding figure, a plane, is $ax + by + cz + d = 0$; the general equation for a sphere, $x^2 + y^2 + z^2 = d^2$; also bears a resemblance to that of the circle. Compare it with $x^2 + y^2 = c^2$, the equation of a circle.

If the general equation in geometry (plane) has two variables (x and y), and if three-dimensional geometry (solid) has three variables (x and y and z), then four-dimensional geometry should have four variables, and n-dimensional geometry n variables. At least that was the thought that struck some mathematicians a hundred and fifty years later. True, there is no fourth dimension, realistically speaking, but there is one *mathematically* speaking, for we can give equations that describe a figure in four dimensions. Today this fourth dimension is usually taken as *time*—and who has not heard of its use in Einstein's theory of relativity? Later it will be shown how this idea of four dimensions had to be combined with something else before it could be used to describe the universe and its many time-space worlds. Einstein could never have formulated his theory without building on what others had done. Like Newton, he, too, stood on the shoulders of giants: Descartes and Euler were two of them.

Euler was one of hundreds who tackled the problem known as "Fermat's last theorem": to prove that there are no positive integers such that $a^n + b^n = c^n$, where n is more than 2. A problem of this type falls into the category of number theory, a branch of mathematics where the problems are so simple that even an amateur can understand them, but where the solutions are so difficult that they tax the best minds. For instance, the example above contains the *particular* problem of proving that no positive integers exist, the sum of whose cubes equals another cube. We know that there are many integers whose squares added together equal a square ($3^2 + 4^2 = 5^2$, $5^2 + 12^2 = 13^2$, etc.), but what about cubes? No one has ever found two cubes equal to a third, but that does no *prove* that they do not exist. Of course the problem is even more general. It involves proving that there are no two integers which raised to *any* power above the second can be added together to produce a third number of the same power.

The problem was—and still is—especially intriguing because of the following notation made by Fermat in the margin of a book: "It is impossible to partition a cube into two cubes, or a biquadrate [fourth power] into two biquadrates, or generally any power higher than a square into two powers of like degree. I have discovered a truly wonderful proof of this, which, however, this margin is too narrow to hold." Because Fermat lacked a few inches of space, mathematicians have wasted reams of paper in trying to find this "truly wonderful proof." Nobody has ever found it. Euler set his mind to it, too, and failed, although he did succeed in proving the impossibility when n equals 3 or 4.

Euler was the great organizer in mathematics, and in organizing the subject, he supplied many missing links. His work in analytic geometry has been mentioned. In trigonometry, he invented the calculation of sines and put the whole subject on an algebraic rather than a geometric basis. Until he came along, trigonometry consisted of a number of unrelated formulas which even as far back as Archimedes' time were used to find the lengths of the sides of triangles. Trigonometric functions had been invented but not recognized as ratios. Euler tied the whole thing together into one consistent whole—just as it is taught in schools today.
...

And what about mathematics? Do its truths exist independently of man, and if so, how much is accessible to human reason? Or has man himself created mathematics whole cloth out of the fabric of his mind, weaving together postulates with logic? Is mathematics, after all, only a technique and not absolute Truth? Neither Euler nor Voltaire nor anyone else though of asking these questions. Mathematics had been synonymous with Truth for so long that its position was unassailable. It was its offspring, science, that had to meet these philosophical hurdles—hurdles have been given in some detail because later mathematics, too, had to meet them, and used the same arguments against the same objections. The contests between Euler and Voltaire, fought on the battlefield of religion and science, were later transferred to the battlefield of mathematics. Euler did not live to see the second conflict, which was just as well, for he had enough arguments in his lifetime to keep him busy—arguments that he always lost. "Our friend Euler," wrote one of his admirers, "is a great mathematician but a bad philosopher ... . It is incredible that he can be so shallow and childish in metaphysics."

Antagonistic religious beliefs were not the only thing that plagued Euler in Berlin. The members of the Academy indulged in constant intrigues and feuds—and Euler was expected to take sides. Wearily, he chose the path of least resistance—to support what the Academy's president, Maupertuis, wanted. Maupertuis, on the other hand, was an aggressive fighter who was seldom between feuds. He quarreled continually with Voltaire, members of his own Academy, and other scientists, always dragging Euler along with him. When Maupertuis finally became ill—"owing to an excess of brandy," remarked Voltaire—and died, Euler was the most likely candidate for the position of president. But Frederick did not like Euler and was reluctant to appoint him. Euler was too bourgeois, too obtrusively pious, too unsophisticated and unpolished. Frederick cruelly ridiculed "that great cyclops of a geometer" and constantly tried to get the men whose intellects decorated the court to admit that mathematics was really not very important.

Frederick's distaste for Euler may have sprung in part from the fact that Frederick was never very good at mathematics and resented someone who was better. Also, he preferred men like Voltaire—literary, witty, sophisticated men whom he could meet on his own level to discuss "the immortality of the soul, freedom, and Plato's hermaphrodites." He could show his poems to Voltaire—Frederick's great ambition was to be a writer—and receive effusions of praise from his idol. By Euler he could only be told that the sides of a triangle are proportional to the sines of the opposite angles.

Frederick decided to bypass Euler and invite the French mathematician d'Alembert, to be president. D'Alembert, however, unlike his benefactor, had a sense of propriety and justice. He refused the invitation, saying that it would be absurd to put anyone over Euler.

Meanwhile, Euler had kept up his relationship with the Russian academy. He continued to send them papers for their journal and entertained visiting Russian mathematicians and students. The esteem in which he was held by Russia—contrasting with his low status in Prussia—was illustrated during the 1760 invasion of Berlin by Russia. The invaders assigned two men to protect Euler's town house from looters. Unfortunately, his country house was inadvertently destroyed by the invading army; but when the Tsarina heard about it, she immediately sent 4,000 crowns indemnity.

The difference in treatment was not lost in Euler. He finally decided to leave Berlin, his home for the past twenty-four years, and to return to Russia.

When he arrived in St. Petersburg, the grateful Catherine the Great presented him with a completely furnished house, her own royal cook, and saw to it that his sons were appointed to good positions. Although eight of Euler's thirteen children had died, his household now comprised eighteen people counting in-laws and grandchildren.

Shortly after his return to Russia, Euler developed a cataract in his remaining eye and went completely blind. Now, the average blind mathematician is about as useful as a blind painter—but Euler was not average. He continued his work—increasing it if anything, and in the next few years submitted over four hundred treatises to the Russian Academy. He composed by dictating to a secretary who knew next to nothing about mathematics (he was a tailor by trade) and by writing formulas in chalk on a large slate. The secretary merely copied them down in the manuscript.

Euler's remarkable memory stood him in good stead during his years of blindness. He calculated long and difficult problems in his head—sometimes to as many as fifty places. Equally as remarkable, he completely memorized the *Aeneid* and could recite the whole thing, word for word, noting where each page ended and the next began.

In 1771, another disaster struck. His house caught fire and the poor blind man was trapped, unable to escape through the smoke and flames which he could feel but not see. A servant, Peter Grimm, dashed into his room and carried him out to safety. Euler's manuscripts were also saved, but everything else burned.

Later the same year his bad luck took a temporary change for the better. He had a successful operation to remove the cataract and after five years of total darkness could see again. But a few weeks later, infection set in, accompanied by days of almost unbearable agony. When it was over, Euler was once again in total darkness.

The faith that the sophisticated Berliners had ridiculed and a strong innate optimism carried Euler through these many trials. When his wife died in 1776, the indomitable mathematician courted her half-sister and married her a few months later. He was sixty-nine at the time.

It is said that Euler calculated as long as he breathed. Old—past seventy—blind, and slightly deaf, he continued to produce his prodigious works. At Catherine the Great's request, he turned his hand to writing a book on elementary algebra—nothing was beneath him and his books for beginners show the same superb organization and elegant style as his more advanced writings.

Euler loved mathematics and rejoiced over each new discovery, whether it was his own or not. He never seemed to mind if his theories were criticized or exploded nor did he indulge in quarrels over priority of discovery—it was the mathematics and not his reputation that was important. When Euler had been in Berlin, a twenty-three-year-old boy sent him a method of solving a certain type of problem in calculus—a method Euler himself had only recently discovered. Euler sent his compliments and encouragement to the young man, informing him that he was delaying his own publication of the method "so as not to deprive you of any part of the glory which is your due." The young man, Joseph-Louis Lagrange, went on to become one of Europe's most eminent mathematicians. Euler's generosity in helping Lagrange become recognized is unmatched in a field where jealous feuding, backbiting, and selfish credit-grabbing have been the standard practice of many of its greatest men. ...

In his seventy-seventh year, on November 18, 1783, Euler was sitting at the table having tea with one of his grandchildren when he suffered a stroke. "I am dying," he cried, and minutes later became unconscious. He died a few hours afterward.

QUESTIONS AND EXERCISES

1. Do as Euler did and multiply out
$$(1 - x)(1 - x^2)(1 - x^3)(1 - x^4) \ldots .$$

What do you get? Do you notice anything about the coefficients? If you do not, you can look up what Euler noticed by looking for a book that includes his pentagonal number theorem.

2. Fermat thought that all of the numbers $2^{2^n} + 1$ were prime because they are prime when $n = 0, 1, 2, 3,$ or 4. The Fermat number with $n = 5$ is composite, though, as Euler first showed. Use a machine of one sort or another to discover at least one of its factors.

3. How well could you do mathematics if you were blind? Find out by trying to solve the following problems with your eyes closed:
(a) $1 + 2 + 3 + 4 + 5 + 6 + 7 = ?$
(b) $314 - 159 = ?$
(c) $35 \times 6 = ?$
(d) $188490/6 = ?$
(e) $2x + 13 = 3x + 4: x = ?$
(f) $x^2 - 10x + 9 = 0: x = ?$
(g) $x^2 - 10x - 10 = 0: x = ?$

4. "By Euler he could only be told that the sides of a triangle are proportional to the sines of the opposite angles." What equations say the same thing as those words? Would Euler really have told that to Frederick? What was the author trying to convey?

5. The November, 1983 issue of *Mathematics Magazine* (volume 56, number 5) is devoted entirely to Euler, and it contains a glossary listing forty-two mathematical objects to which the name of Euler is attached. Find the issue of the magazine and see how many of them make sense to you.

6. "And what about mathematics? Do its truths exist independently of man, and if so, how much is accessible to human reason? Or has man himself created mathematics whole cloth out of the fabric of his mind, weaving together postulates with logic?" That is, is mathematics discovered, or invented? That is a big question, and not all mathematicians answer it the same way. Most would probably come down in favor of discovery, but there are many who are convinced that invention is correct. Instead of trying to answer it, try to answer this one first: is the question answerable? That is, is there any way that it would be possible to decide which view is correct?

7. "Frederick was determined to aggrandize his country through art and science, as well as war." That was in the eighteenth century: how do nations seek to aggrandize themselves today?

TRANSITION TO THE TWENTIETH CENTURY

by Howard Eves

After Newton and Leibniz, after the discovery of calculus, then what? Quite a lot. Calculus was a new idea, a very large new idea that took minds as powerful as Newton's and Leibniz's to find, but new ideas, even very large ones, can be understood, appreciated, and applied by people with minds less powerful. So it was with calculus. Calculus *worked*. It was a method, a new method, for dealing quantitatively with problems of motion and change. No such method existed before. The theorems of geometry are about objects that just sit there. Triangles do not fly, circles do not roll, and pyramids do not stir off their bases. Lines may *seem* to shoot off into the distance, but that is because our vision is limited. Actually, lines are *there*, in their infinitely long entireties at all times, eternally unmoving. It is the same with algebra as with geometry. The x's and y's of algebra do not move either. They stay in their places, while the methods of algebra strip their disguises from them so that the numbers that have been hiding behind them are revealed. So, $3x + 4 = 5$, does it? You cannot hide from Algebra, x. Algebra will find out what you are. Algebra knows that you are 1/3. Note that x was not *changing*. It remained motionless, though perhaps shivering a little, as Algebra exerted its inexorable power on its equation. Calculus is different. Calculus is about motion, change, and variation. There is plenty of motion, change, and variation in the world. There are infinitely many questions about things that are moving, changing, and varying that can be asked. Calculus can answer some of them. Geometry and algebra could not, but calculus can. People saw that, and got to work.

There was an explosion. Not one that you could see or hear, but an explosion nevertheless. Problems were solved, and as is always the way in mathematics, whenever a problem is solved new problems suggest themselves. More problems were solved, and more people solved them, as the new tool of calculus was used, refined, and made more powerful. It was a great time to be a mathematician, since new results were all around, waiting to be found. It was a time of exuberance, when Euler could write

$$-1 = 1 + 2 + 4 + 8 + 16 + \dots .$$

Today, in more sedate times, no one would think of writing that or anything like it. It was a time of flowering, a time when mathematics enjoyed huge prestige—had not Newton explained the universe?—a time whose like we will not see again. The following excerpt presents a few of the accomplishments made in those wonderful years.

You may notice that you do not completely understand all of the accomplishments. That is because in the eighteenth century mathematics was getting away from things that could be understood by anyone with a decent education and into areas where only people with quite a bit of mathematical training could understand what has been done. Calculus is the reason why this happened in the eighteenth century. The sixteenth century was when the solution to the cubic polynomial equation was found, and anyone who has ever solved $x^2 + 2x + 3 = 0$ can understand what has been done, even if the details of how to solve

$$x^3 + 2x^2 + 3x + 4 = 0$$

(the solution turns out to be

$$x = -\left(\frac{5}{9}\sqrt{6} + \frac{35}{27}\right)^{1/3} - \left(\frac{5}{9}\sqrt{6} - \frac{35}{27}\right)^{1/3} - \frac{2}{3}$$

and two other complex roots) are not at their fingertips. The early seventeenth century was when Descartes had the idea of combining algebra and geometry into analytic geometry, and anyone who has ever drawn the graph of $x^2 + 2x + 3 = 0$ can understand what has been done. The late seventeenth century was when calculus burst upon the world, and anyone who has ever used derivatives to find the minimum value of $x^2 + 2x + 3$ can understand, at least a little, what has been done. The eighteenth century is different. When you read that

> Euler employed the idea of an integrating factor in the solution of differential equations, was one of the first to develop a theory of continued fractions, contributed to the fields of differential geometry and the calculus of variations, and considerably enriched number theory

you may not have a grasp on what Euler did. What's an integrating factor? How can you continue a fraction? What is differential geometry about? Is the calculus of variations different from plain calculus? How can you have a theory about numbers? These are questions that even people with decent educations probably do not know how to answer. However, it can be helpful, sometimes, to read things that we cannot understand fully at the time.

---

1. Introduction

The calculus, aided by analytic geometry, was the greatest mathematical tool discovered in the seventeenth century. It proved to be remarkably powerful and capable of attacking problems quite unassailable in earlier days. It was the wide and astonishing applicability of the discipline that attracted the bulk of the mathematical researchers of the day, with the result that papers were turned out in great profusion with little concern regarding the very unsatisfactory foundations of the subject. The processes employed were justified largely on the ground that they worked, and it was not until the eighteenth century had almost elapsed, after a number of absurdities and contradictions had crept into mathematics, that mathematicians felt it was essential that the basis of their work be logically examined and rigorously established. The painstaking effort to place analysis on a logically rigorous foundation was a natural reaction to the pell-mell employment of intuition and formalism of the previous century. The task proved to be a difficult one, its various ramifications occupying the better part of the next hundred years. A result of this careful work in the foundations of analysis was that it led to equally careful work in the foundations of all branches of mathematics and to the refinement of many important concepts. Thus the function idea itself had to be clarified, and such notions as limit, continuity, differentiability, and integrability had to be very carefully and clearly defined. This task of refining the basic concepts of mathematics led, in turn, to intricate generalizations. Such concepts as space, dimension, convergence, and integrability, to name only a few, underwent remarkable generalization and abstraction. A good part of the mathematics of the first half of the twentieth century has been devoted to this sort of thing, until now

generalization and abstraction have become striking features of present-day mathematics. But some of these developments have, in turn, brought about a fresh batch of paradoxical situations. The generalizations to transfinite numbers and the abstract study of sets have widened and deepened many branches of mathematics, but, at the same time, they have revealed some very disturbing paradoxes which appear to lie in the innermost depths of mathematics. Here is where we seem to be today, and it may be that the second half of the twentieth century will witness the resolution of these critical problems.

In summarizing the last paragraph we may say, with a fair amount of truth, that the eighteenth century was largely spent in exploring the new and powerful methods of the calculus, that the nineteenth century was largely devoted to the effort of establishing on a firm logical foundation the enormous but shaky superstructure erected in the preceding century, and that the first half of the twentieth century has, in large part, been spent in generalizing as far as possible the gains already made, and that at present many mathematicians are becoming concerned with even deeper foundational problems. This general picture is complicated by the various sociological factors that affect the development of any science. Such matters as the growth of life insurance, the construction of the large navies of the eighteenth century, the economic and technological problems brought about by the opening of the industrial revolution on continental Europe, the present world-war atmosphere, and today's concentrated effort to conquer outer space, have led to many practical developments in the field of mathematics. A division of mathematics into "pure" and "applied" has come about, research in the former being carried out to a great extent by those specialists who have become interested in the subject for its own sake, and in the latter by those who remain attached to immediately practical uses.

We shall now fill in some of the details of the general picture sketched above.

2. The Bernoulli Family

The bulk of the mathematics of the eighteenth century found its genesis and its goal in the fields of mechanics and astronomy, and it was not until well into the nineteenth century that mathematical research generally emancipated itself from this viewpoint.

The principal contributions to mathematics in the eighteenth century were made by members of the Bernoulli family, Abraham De Moivre, Brook Taylor, Colin Maclaurin, Leonhard Euler, Alexis Claude Clairaut, Jean-le-Rond d'Alembert, Johann Heinrich Lambert, Joseph Louis Lagrange, and Gaspard Monge.

One of the most distinguished families in the history of mathematics and science is the Bernoulli family of Switzerland, which, from the late seventeenth century on, has produced a remarkable number of capable mathematicians and scientists. The family record starts with the two brothers, Jakob Bernoulli (1654-1705) and Johann Bernoulli (1667-1748). These two men gave up earlier vocational interests and became mathematicians when Leibniz' papers began to appear in *Acta eruditorum*. They were among the first mathematicians to realize the surprising power of calculus and to apply the tool to a great diversity of problems. From 1687 until his death, Jakob occupied the mathematical chair at Basel University. Johann, in 1697, became a professor at Groningen University, and then, on Jakob's death in 1705, succeeded his brother in the chair at Basel University, to remain there for the rest of his life. The two brothers, often bitter rivals, maintained an almost constant exchange of ideas with Leibniz and with each other.

Among Jakob Bernoulli's contributions to mathematics are the early use of polar coordinates, the derivation in both polar and rectangular coordinates of the formula for the radius of curvature of a plane curve, the study of the catenary curve with extensions to strings of variable density and strings under the action of a central force, the study of a number of other higher plane curves, the discovery of the so-called *isochrone*—or curve along which a body will fall with uniform vertical velocity (it turned out to be a semicubical parabola with a vertical cusptangent), the determination of the form taken by an elastic rod fixed at one end and carrying a weight at the other, the form assumed by a flexible rectangular sheet having two opposite edges held horizontally fixed at the same height and loaded

with a heavy liquid, and the shape of a rectangular sail filled with wind. He also proposed and discussed the problem of isoperimetric figures (planar closed paths of given species and fixed perimeter which include a maximum area), and was thus one of the first mathematicians to work in the calculus of variations. He was also one of the early students of mathematical probability; his book in this field, the *Ars conjectandi*, was posthumously published in 1713. There are several things in mathematics which now bear Jakob Bernoulli's name; among these are the *Bernoulli distribution* and *Bernoulli theorem* of statistics and probability theory, the *Bernoulli equation* met by every student of a first course in differential equations, the *Bernoulli numbers* and *Bernoulli polynomials* of number-theory interest, and the *lemniscate of Bernoulli* encountered in any first course in calculus. In Jakob Bernoulli's solution to the problem of the isochrone curve, which was published in *Acta eruditorum* in 1690, we meet for the first time the word integral in a calculus sense. Leibniz had called the integral calculus *calculus summatorius*; in 1696 Leibniz and Johann Bernoulli agreed to call it *calculus integralis*. Jakob Bernoulli was struck by the way the equiangular spiral reproduces itself under a variety of transformations and asked, in imitation of Archimedes, that such a spiral be engraved on his tombstone, along with the inscription "Eadem mutata resurgo" ("I shall arise the same, though changed").

Johann Bernoulli was an even more prolific contributor to mathematics than was his brother Jakob. Though he was a jealous and cantankerous man, he was one of the most successful teachers of his time. He greatly enriched the calculus and was very influential in making the power of the new subject appreciated in continental Europe. It was his material that the Marquis de l'Hospital (1661-1704), under a curious financial agreement with Johann, assembled in 1696 into the first calculus textbook. It was in this way that the familiar method of evaluating the indeterminate form 0/0 became incorrectly known, in later calculus texts, as *l'Hospital's Rule*. Johann Bernoulli wrote on a wide variety of topics, including optical phenomena connected with reflection and refraction, the determination of orthogonal trajectories of families of curves, rectification of curves and quadrature of area by series, analytical trigonometry, the exponential calculus, and other subjects. One of his more noted pieces of work is his contribution to the problem of the *brachystochrone*—the determination of the curve of quickest descent of a weighted particle moving between two given points in a gravitational field; the curve turned out to be an arc of an appropriate cycloid curve. The problem was also discussed by Jakob Bernoulli. The cycloid curve is also the solution to the problem of the *tautochrone*—the determination of the curve along which a weighted particle will arrive at a given point of the curve in the same time interval no matter from what initial point of the curve it starts. The latter problem, which was more generally discussed by Johann Bernoulli, Euler, and Lagrange, had earlier been solved by Huygens (1673) and Newton (1687) and applied by Huygens in the construction of pendulum clocks.

Johann Bernoulli had three sons, Nicolaus (1695-1726), Daniel (1700-1782), and Johann II (1710-1790), all of whom won renown as eighteenth-century mathematicians and scientists. Nicolaus, who showed great promise in the field of mathematics, was called by the St. Petersburg Academy, where he unfortunately died, by drowning, only eight months later. He wrote on curves, differential equations, and probability. A problem on probability, which he proposed from St. Petersburg, later became known as the *Petersburg paradox*. The problem is: if *A* receives a penny should head appear on the first toss of a coin, 2 pennies if head does not appear until the second toss, 4 pennies if head does not appear until the third toss, and so on, what is *A*'s expectation? Mathematical theory shows that *A*'s expectation is infinite, which seems a paradoxical result. The problem was investigated by Nicolaus' brother Daniel, who succeeded Nicolaus at St. Petersburg. Daniel returned to Basel seven years later. He was the most famous of Johann's three sons, and devoted most of his energies to probability, astronomy, physics, and hydrodynamics. In probability he devised the concept of *moral expectation*, and in his *Hydrodynamica*, of 1738, appears the principle of hydrodynamics that bears his name in all present-day elementary physics texts. He wrote on tides, established the kinetic theory of gases, studied the vibrating string, and pioneered in partial differential equations. Johann (II), the youngest of the three sons, studied law but spent his later years as a professor of mathematics at the University of Basel. He was particularly interested in the mathematical theory of heat and light.

There was another eighteenth-century Nicolaus Bernoulli (1687-1759), a nephew of Jakob and Johann, who achieved some fame in mathematics. This Nicolaus held, for a time, the chair of mathematics at Padua once filled by Galileo. He wrote extensively on geometry and differential equations. Later in life he taught logic and law.

Johann Bernoulli (II) had a son Johann (III) (1744-1807) who, like his father, studied law but then turned to mathematics. When barely 19 years old, he was called as a professor of mathematics to the Berlin Academy. He wrote on astronomy, the doctrine of

chance, recurring decimals, and indeterminate equations.

3. De Moivre, Taylor, Maclaurin

Abraham De Moivre (1667-1754) was born in France but lived most of his life in England, becoming an intimate friend of Isaac Newton. He is particularly noted for his work *Annuities upon Lives*, which played an important role in the history of actuarial mathematics, his *Doctrine of Chances*, which contained much new material on the theory of probability, and his *Miscellanea analytica*, which contributed to recurrent series, probability, and analytic trigonometry. De Moivre is credited with the first treatment of the probability integral

$$\int_0^\infty e^{-x^2}dx = \frac{\sqrt{\pi}}{2},$$

and of (essentially) the normal frequency curve

$$y = ce^{-hx^2}, \quad c \text{ and } h \text{ constants,}$$

so important in the study of statistics. The misnamed *Stirling's formula*, which says that for very large *n*

$$n! \approx (2\pi n)^{1/2}e^{-n}n^n,$$

is due to De Moivre and is highly useful for approximating factorials of large numbers. The familiar formula

$$(\cos x + i\sin x)^n = \cos nx + i\sin nx, \quad i = \sqrt{-1},$$

known by De Moivre's name and found in every theory of equations textbook, was familiar to De Moivre for the case where *n* is a positive integer. The formula has become the keystone of analytic trigonometry.

Rather interesting is the fable often told of De Moivre's death. According to the story De Moivre noticed that each day he required a quarter of an hour more sleep than on the preceding day. When the arithmetic progression reached 24 hours De Moivre passed away.

Every student of the calculus is familiar with the name of the Englishman Brook Taylor (1685-1731) and the name of the Scotsman Colin Maclaurin (1698-1746), through the very useful Taylor's expansion and Maclaurin's expansion of a function. It was in 1715 that Taylor published (with no consideration of convergence) his well-known expansion theorem,

$$f(a + h) = f(a) + hf'(a) + \frac{h^2}{2!}f''(a) + \dots .$$

In 1717 Taylor applied his series to the solution of numerical equations as follows: let *a* be an approximation to a root of *f(x) = 0*; set

$$f(a) = k, \quad f'(a) = k', \quad f''(a) = k'', \quad \text{and} \quad x = a + h;$$

expand $0 = f(a + h)$ by the series; discard all powers of h above the second; substitute the values of k, k', k'', and then solve for h. By successive applications of the process, closer and closer approximations can be obtained. Some work done by Taylor in the theory of perspective has found recent application in the mathematical treatment of photogrammetry, the science of surveying by means of photographs taken from an airplane.

Maclaurin was one of the ablest mathematicians of the eighteenth century. The so-called Maclaurin expansion is nothing but the case where $a = 0$ in the Taylor expansion above and was actually given by James Stirling 25 years before Maclaurin used it in 1742. Maclaurin did very notable work in geometry, particularly in the study of higher plane curves, and he showed great power in applying classical geometry to physical problems. Among his many papers in applied mathematics is a prize-winning memoir on the mathematical theory of tides.

4. Euler, Clairaut, d'Alembert

Leonhard Euler was born in Basel, Switzerland, in 1707, and he studied mathematics there under Johann Bernoulli. In 1727 he accepted the chair of mathematics at the new St. Petersburg Academy formed by Peter the Great. Fourteen years later he accepted the invitation of Frederick the Great to go to Berlin to head the Prussian Academy. After 25 years in this post Euler returned to St. Petersburg, remaining there until his death in 1783 when he was 76 years old.

Euler was a voluminous writer on mathematics; indeed, the most prolific writer in the history of the subject; his name is attached to every branch of the study. It is of interest to note that his amazing productivity was not in the least impaired when, about 1768, he had the misfortune to become totally blind.

Euler's work represents the outstanding example of eighteenth-century formalism, or the manipulation, without proper attention to matters of convergence and mathematical existence, of formulas involving infinite processes. For example, if the binomial theorem is applied formally to $(1 - 2)^{-1}$ we find

$$-1 = 1 + 2 + 4 + 8 + 16 + \dots,$$

a result which caused Euler no wonderment! Also, by adding the two series

$$x + x^2 + x^3 + \dots = \frac{x}{1 - x}$$

and

$$1 + \frac{1}{x} + \frac{1}{x^2} + \dots = \frac{x}{x - 1},$$

Euler found that

$$\dots + \frac{1}{x^2} + \frac{1}{x} + 1 + x + x^2 + \dots = 0.$$

The eighteenth-century effort to inject rigor into mathematics was brought about by an accumulation of absurdities such as these.

Euler's contributions to mathematics are too numerous and, in general, too advanced to expound here, but we may note some of his contributions to the elementary field. First of all, we owe to Euler the conventionalization of the following notations:

$f(x)$	for functional notation,
e	for the base of natural logarithms,
a, b, c	for the sides of a triangle ABC,
s	for the semiperimeter of triangle ABC,
Σ	for the summation sign,
i	for the imaginary unit, $\sqrt{-1}$.

To Euler is also due the very remarkable formula

$$e^{ix} = \cos x + i \sin x$$

which, for $x = \pi$, becomes

$$e^{i\pi} + 1 = 0,$$

a relation connecting five of the most important numbers in mathematics. By purely formal processes, Euler arrived at an enormous number of curious relations, like

$$i^i = e^{-\pi/2},$$

and he succeeded in showing that any nonzero real number r has an infinite number of logarithms (for a given base), all imaginary if $r < 0$ and all imaginary but one if $r > 0$. In college geometry we find the *Euler line* of a triangle, in college courses in the theory of equations the student sometimes encounters *Euler's method* for solving quartic equations, and in even the most elementary course in number theory one meets *Euler's theorem* and the *Euler ϕ-function*. The beta and gamma functions of advanced calculus are credited to Euler, though they were adumbrated by Wallis. Euler employed the idea of an integrating factor in the solution of differential equations, was one of the first to develop a theory of continued fractions, contributed to the fields of differential geometry and the calculus of variations, and considerably enriched number theory. In one of his smaller papers occurs the relation $V - E + F = 2$, already known to Descartes, connecting the number of vertices V, edges E, and faces F of any simple closed polyhedron. In another paper he investigates *orbiform curves*, or curves which, like the circle, are convex ovals of constant width. Several of his papers are devoted to mathematical recreations, such as unicursal and multicursal graphs (inspired by the seven bridges of Königsberg), the re-entrant knight's path on a chess board, and Graeco-Latin squares. He also published extensively in areas of applied mathematics, in particular to lunar theory, the three-body problem of celestial mechanics, the attraction of ellipsoids, hydraulics, ship-building, artillery, and a theory of music.

Euler was a masterful writer of textbooks, in which he presented his material with great clarity, detail, and completeness. These texts enjoyed a marked and a long popularity, and to this day make very interesting and profitable reading. One cannot but be surprised at Euler's enormous fertility of ideas, and it is no wonder that so many of the great mathematicians coming after him have admitted their indebtedness to him.

Claude Alexis Clairaut was born in Paris in 1713 and died there in 1765. He was a youthful mathematical prodigy, composing in his eleventh year a treatise on curves of the third order. This early paper, and a singularly elegant subsequent one on the differential geometry of twisted curves in space, won him a seat in the French Academy of Sciences at the illegal age of 18. In 1736 he accompanied Pierre Louis de Maupertuis (1698-1759) on an expedition to Lapland to measure the length of a degree of one of the earth's meridians. The expedition was undertaken to settle a dispute as to the shape of the earth. Newton and Huygens had concluded, from mathematical theory, that the earth is flattened at the poles. But about 1712, the Italian astronomer and mathematician Giovanni Domenico Cassini (1625-1712), and his French-born son Jacques Cassini (1677-1756), measured an arc of longitude extending from Dunkirk to Perpignan, and obtained a result that seemed to support the Cartesian contention that the earth is elongated at the poles. The measurement made in Lapland unquestionably confirmed the Newton-Huygens belief, and earned Maupertuis the title of "earth flattener." In 1743, after his return to France, Clairaut published his definitive work, *Théorie de la figure de la Terre*. In 1752 he won a prize form the St. Petersburg academy for his paper *Théorie de la Lune*, a mathematical study of lunar motion which cleared up some, to then, unanswered questions. He applied the process of differentiation to the differential equation

$$y = px + f(p), \quad p = \frac{dy}{dx},$$

now known in elementary textbooks on differential

equations as *Clairaut's equation*, and he found the singular solution, but this process had been used earlier by Brook Taylor. In 1759 he calculated, with an error of about a month, the 1759 return of Halley's comet.

Clairaut had a brother who died when only 16, but who at 14 read a paper on geometry before the French Academy and at 15 published a work on geometry. The father of the Clairaut children was a teacher of mathematics, a correspondent of the Berlin Academy, and a writer on geometry.

Jean-le-Rond d'Alembert (1717-1783), like Alexis Clairaut, was born in Paris and died in Paris. As a newly-born infant he was abandoned near the church of Saint Jean-le-Rond and was discovered there by a gendarme who had him hurriedly christened with the name of the place where he was found. Later, for reasons not known, the name d'Alembert was added.

There existed a scientific rivalry, often not friendly, between d'Alembert and Clairaut. At the age of 24, d'Alembert was admitted to the French Academy. In 1743 he published his *Traité de dynamique*, based upon the great principle of kinetics that now bears his name. In 1744 he applied his principle in a treatise on the equilibrium and motion of fluids, and in 1746 in a treatise on the causes of winds. In each of these works, and also in one of 1747 devoted to vibrating strings, he was led to partial differential equations, and he became a pioneer in the study of such equations. With the aid of his principle he was able to obtain a complete solution of the baffling problem of the precession of the equinoxes. D'Alembert showed interest in the foundations of analysis, and in 1754 he made the important suggestion that a sound theory of limits was needed to put analysis on a firm foundation, but his contemporaries paid little heed to his suggestion. It was in 1754 that d'Alembert became permanent secretary of the French Academy. During his later years he worked on the great French *Encyclopédie*, which had been begun by Denis Diderot and himself.

5. Lambert, Lagrange, Monge

A little younger than Clairaut and d'Alembert was Johann Heinrich Lambert (1728-1777), born in Mulhouse (Alsace), then part of Swiss territory. It was in 1766 that Lambert wrote his investigation of the parallel postulate entitled *Die Theorie der Parallellinien*. Lambert was a mathematician of high quality. As a son of a poor tailor he was largely self-taught. He possessed a fine imagination and he established his results with great attention to rigor. In fact, Lambert was the first to prove rigorously that the number π is irrational. He showed that if x is rational, but not zero, then $\tan x$

cannot be rational; since tan $\pi/4 = 1$, it follows that $\pi/4$, or π cannot be rational. We also owe to Lambert the first systematic development of the theory of hyperbolic functions and, indeed, our present notation for these functions. Lambert was a many-sided scholar and contributed noteworthily to the mathematics of numerous other topics, such as descriptive geometry, the determination of comet orbits, and the theory of projections employed in the making of maps (a much-used one of these projections is now named after him). At one time he considered plans for a mathematical logic of the sort once outlined by Leibniz.

The two greatest mathematicians of the eighteenth century were Euler and Joseph Louis Lagrange (1736-1813), and which of the two is to be accorded first place is a matter of debate that often reflects the varying mathematical sensitivities of the debaters. Lagrange was born in Turin, Italy. In 1766, when Euler left Berlin, Frederick the Great wrote to Lagrange that "the greatest king in Europe" wished to have at his court "the greatest mathematician of Europe." Lagrange accepted the invitation and for twenty years held the post vacated by Euler. A few years after leaving Berlin, and in spite of the chaotic political situation in France, Lagrange accepted a professorship at the newly established École Normale, and then at the École Polytechnique, schools which became famous in the history of mathematics inasmuch as many of the great mathematicians of modern France have been trained there and many held professorships there. Lagrange did much to develop the high degree of scholarship in mathematics that has been associated with these institutions.

Lagrange's work had a very deep influence on later mathematical research, for he was the earliest mathematician of the first rank to recognize the thoroughly unsatisfactory state of the foundations of analysis and accordingly to attempt a rigorization of the calculus. The attempt, which was far from successful, was made in 1797 in his great publication *Théorie des fonctions analytiques contenant les principes du calcul différential*. The cardinal idea here was the representation of a function $f(x)$ by a Taylor series. The derivatives $f'(x)$, $f''(x)$, ... were then defined as the coefficients of h, $h^2/2!$, ... in the Taylor expansion of $f(x + h)$ in terms of h. The notation $f'(x)$, $f''(x)$, ... , very commonly used today, is due to Lagrange. But Lagrange failed to give sufficient attention to matters of convergence and divergence. Nevertheless, we have here the first "theory of functions of a real variable." Two other great works of Lagrange are his *Traité de la résolution des équations numériques de tous degrés* and his monumental *Mécanique analytique*; the former was written late in the

century and gives a method of approximating the real roots of an equation by means of continued fractions, the latter (which has been described as a "scientific poem") dates from Lagrange's Berlin period and contains the general equations of motion of a dynamical system known today as *Lagrange's equations*. His work in differential equations (for example, the method of variation of parameters), and particularly in partial differential equations, is very notable, and his contributions to the calculus of variations did much for the development of that subject. Lagrange had a penchant for number theory and wrote important papers in this field also, such as the first published proof of the theorem that every positive integer can be expressed as the sum of not more than four squares. Some of his early work on the theory of equations later led Galois to the theory of groups. In fact, the important theorem of group theory that states that the order of a subgroup of a group G is a factor of the order of G, is called *Lagrange's theorem*.

Whereas Euler wrote with a profusion of detail and a free employment of intuition, Lagrange wrote concisely and with attempted rigor. Lagrange was "modern" in style and can be characterized as the first real analyst.

The last outstanding mathematician of the eighteenth century whom we shall consider is Gaspard Monge (1746-1818), who in 1794 became a professor of mathematics at the École Polytechnique, which he had vigorously helped to establish. He is particularly noted as the elaborator of descriptive geometry, the science of representing three dimensional objects by appropriate projections on the two dimensional plane. A work of his entitled *Application de l'analyse à la géométrie* ran through five editions and was one of the most important of the early treatments of the differential geometry of surfaces. Monge was a gifted teacher and his lectures inspired a large following of able geometers.

We conclude our very brief survey of eighteenth-century mathematics by noting that while the century witnessed considerable further development in such subjects as trigonometry, analytic geometry, calculus, theory of numbers, theory of equations, probability, differential equations, and analytic mechanics, it witnessed also the creation of a number of new subjects, such as actuarial science, the calculus of variations, higher functions, descriptive geometry, and differential geometry.

EXERCISES AND QUESTIONS

1. Use Stirling's formula to get an approximation

to $\dbinom{2n}{n}$.

2. "In 1717 Taylor applied his series to the solution of numerical equations as follows ..." Use Taylor's method to get a numerical value for $3^{1/3}$. (Take $f(x) = x^3 - 3$ and $a = 3/2$.) Check your result with your calculator.

3. (a) Play the Petersburg paradox game ten times and see how much you win. It was less than infinitely much, was it not? Therein lies the paradox.

(b) Of course, your opponent in the Petersburg game does not have infinite resources and at some point the coin flipping would have to stop. Suppose that the coin can be flipped at most three times, with you getting four pennies for either *TTT* or *TTH*. Then what is your expected gain? What if there can be five flips? How about twenty? Is the paradox now less paradoxical?

4. "When the arithmetic progression reached 24 hours De Moivre passed away." This anecdote is absolutely incredible and must have been made up by someone. Why would someone make up such a story? Why would other people repeat it?

5. Make a list of the problems solved or considered by Jakob Bernoulli. *Why* did he solve or consider them?

6. What in the world is a *Graeco-Latin square*?

7. "Such matters as the growth of life insurance ... have led to many practical developments in the field of mathematics." Life insurance is a subject that most people don't think of when asked about applications of calculus to life. Look up Makeham's Law in a book on actuarial mathematics and use differentiation to find out at what age the human population is decreasing most rapidly.

ARE VARIABLES NECESSARY IN CALCULUS?

By Karl Menger

Newton and Leibniz wrote derivatives and integrals in quite different ways, and the notation for calculus did not become standardized until the time of Euler, almost one hundred years after Newton and Leibniz. It takes time for notations to evolve, just as it does for species. In fact, the parallels between the evolution of living things and the notations of mathematics are very close. Animals that are well adapted to their environment tend to reproduce more than animals that are less fit and pass their characteristics along to the next generation. Notations that make doing mathematics easy tend to be used more than those that make things more difficult, so notations that are not as well adapted to the mathematical environment tend to die out. That is why we no longer write

$$4 \text{ L M } 6 \text{ Q P } 7$$
$$3 \text{ Q P } 4 \text{ L M } 5$$

$$12 \text{ C M } 18 \text{ Q Q P } 21 \text{ Q}$$
$$16 \text{ Q M } 24 \text{ C P } 28 \text{ L}$$
$$\text{M } 20 \text{ L P } 30 \text{ Q M } 35$$

$$67 \text{ Q P } 8 \text{ L M } 12 \text{ C M } 18 \text{ Q Q M } 35$$

when we want to multiply $4x - 6x^2 + 7$ by $3x^2 + 4x - 5$. (The notation is that of G. Gosselin, *De Arte Magna*, Paris, 1577.) Our notation is superior: there are fewer symbols to know the meaning of, and the symbols used for exponents and operations are not both capital letters and are thus distinguished from each other. So, the notation of Gosselin has become extinct.

Species tend to evolve until they fit their environment very well, or well enough to get along, and then they stop. Horseshoe crabs, sharks, and palm trees have been the same for millions of years. Not the same individuals, of course, but a palm tree a million years ago looked just like a palm tree today. Similarly, a 100-year-old calculus book can be recognized instantly as being a calculus book. Calculus notation has been the same for about two hundred years, not quite as long as millions but long enough so that it seems unlikely that it will change very much in the future. When evolution stops, it does not mean that the thing evolving has become perfect. Sharks and horseshoe crabs could probably be improved on, and tree designers might be able to produce blueprints for superpalms. However, new and improved species do not appear because of the inertia and conservatism built into nature. If something serves well enough, evolution sees no reason why it should be replaced. All of us are born with appendices that we could very well do without. So it is with the notation of calculus, and other non-living things. Everyone knows that the design of the QWERTYUIOP typewriter keyboard is inefficient and the Dvorak keyboard is superior, but keyboards have not changed because it would be too much effort to retrain everyone's fingers so that the arrangement

$$\text{P Y F G C R L}$$
$$\text{A O E U I D H T N S}$$
$$\text{Q J K X B M W V Z}$$

became natural. Inertia is one of the most powerful forces there is. Calculus notation is no more perfect than the typewriter keyboard is. For example, if

$$u = f(x, z) \text{ and } z = g(x, y),$$

what is u_x? The answer is, it depends. If the independent variables are x and y it is one thing, while if they are x and z it is another. The meaning of the symbol is ambiguous, and a system of notation with ambiguous symbols is not perfect. But it works well enough, so it does not change.

This does not mean that efforts to improve should not be made. Sharks and palm trees are left to their own devices and show little inclination to alter themselves, but the notation of calculus can be acted on by people. After learning how to integrate things, many students have thought, "Why do we have to write that dx all of the time?" Some have asked their teachers, and gotten more or less unsatisfactory answers. "If you leave it off, you'll get a point taken off your score" is a common one, as is "That's just the way it's written." A little better is "It stands for one of the little bits of x that you're adding up," as is "When we do integration by substitution you'll see why we need it." However, the *real* answer, kept from students so as not to unsettle them, is "You *don't* have to write the dx." In the following paper, Karl Menger says that

$$\int_0^\pi \sin = 2$$

is better than

$$\int_0^\pi \sin x \, dx = 2$$

and that

$$S \cos \sim \sin$$

is preferable to

$$\int \cos x \, dx = \sin x + C.$$

He had a point, several points in fact. The notation of calculus does have illogicalities and inconsistencies that could be corrected. However, Menger's effort of reform went nowhere. After his paper was published, he wrote a calculus text that used his notation, but it did not sweep the nation. Just as with the typewriter keyboard, the effort needed to change to a new system was just too much for people to make even if the new system was superior. Inertia claimed another victim.

The chance for a revolution in calculus notation seems to be very small. But you never know. Any year, there may appear a mutant palm tree so well adapted to its environment that it will choke out oaks and maples and the continent will be covered with a forest of waving palm fronds. Any time, some one may find a new way of writing calculus so superior to the way it is written now that everyone will say, "Yes, that's obviously the right way to do it. Why didn't I think of that myself?" and the old notation will disappear. You never know what will happen. That is one reason why the future is so interesting.

---

1. The definite integral. We begin with a case in which the variables are superfluous beyond any doubt. By virtue of their definitions, the numbers

$$\int_0^\pi \sin x \, dx, \quad \int_0^\pi \sin z \, dz, \quad \int_0^\pi \sin \gamma \, d\gamma$$

are identical. Hence it does not make any difference which letter we use for the variable. But then why write the dummy part at all? We shall simply write

$$\int_0^\pi \sin.$$

A geometric consideration confirms this view. In a cartesian coordinate system, the sine function represents a curve, the sine curve

$$y = \sin x, \quad w = \sin z, \quad \text{or} \quad \delta = \sin \gamma$$

according to whether the points are denoted by (x, y), (z, w), or (γ, δ). The numbers 0 and π determine an arc on the curve. The $\int$ sign indicates that we form the area under this arc. How we denote the points has no bearing on the area. We have

$$\int_0^\pi \sin = 2$$

and similarly

$$\int_{-1}^1 \sinh = 0, \quad \int_1^2 \log = 2\log 2 - 1.$$

If we wish to define

$$\int_0^1 e^x \, dx$$

without dummy variables we experience the first

difficulty. For e^x denotes the value which the exponential function associates with the number x rather than the exponential function itself. But we remember that if in e^x we have to replace x with a long expression such as

$$-\frac{1}{2(1-\rho^2)}\left(\frac{x^2}{\sigma_1^2} - \frac{2\rho xy}{\sigma_1\sigma_2} + \frac{y^2}{\sigma_2^2}\right),$$

then typographical difficulties force us to print

$$\exp\left[-\frac{1}{2(1-\rho^2)}\left(\frac{x^2}{\sigma_1^2} - \frac{2\rho xy}{\sigma_1\sigma_2} + \frac{y^2}{\sigma_2^2}\right)\right]$$

where the symbol, exp, is used for the exponential function in the same way that the symbol, log, is used for the logarithmic function or the symbol, sin, for the sine function. In this notation we can write

$$\int_0^1 \exp = e - 1.$$

In eliminating the dummy part of

$$\int_0^1 x^n \, dx$$

we are confronted with the complete lack of a symbol for the nth power function which associated the number x^n with x. If we feel that this important function deserves a symbol and we denote it by $^n\!\prod$, then we can write

$$\int_0^1 \,^n\!\prod = \frac{1}{n+1}.$$

We do have a symbol of the less important nth root function. The typical polynomial has neither a name nor a symbol. The polynomial whose value for x is

$$a_0 + a_1 x + + \ldots + a_n x^n$$

might be denoted by $p_{a_0, a_1, \ldots, a_n}$. Polynomials of special importance may, of course, be denoted by special symbols. For instance, $p_{-1/2, 0, 3/2}$ is the second Legendre polynomial and is usually denoted by P_2. The frequent occurrence of the function that associates with x the number $+\sqrt{1 - k^2 x^2}$ might warrant the introduction of a symbol. The fact that the graph of the function is the upper half of an ellipse with eccentricity $\sqrt{(1 - k^2)}$ suggests the symbol

$$\text{ell}_{(1 - k^2)^{1/2}}$$

for this function. In particular,

$$y = \sqrt{(1 - x^2)}$$

is the equation of a semi-circle, and we might write cir instead of ell_0. We should then have

$$\int_0^1 \text{cir} = \frac{\pi}{4}.$$

Instead of the constant polynomial p_c, which associates c with every x, we shall simply write c. For instance,

$$\int_0^\pi 2 = 2\cdot\pi, \qquad \int_0^\pi \sin 2 = \sin 2\cdot\pi.$$

All in all, what we need in the calculus of definite integrals are symbols for the most important functions rather than variables.

2. Substitution and the identity function. Instead of

$$\int_0^\pi \sin 2x \, dx = 0$$

we might write

$$\int_0^\pi \sin p_{0,2} = 0$$

where $\sin p_{0,2}$ denotes the function obtained by substituting the polynomial $p_{0,2}$ into the sine function. This notation for substitution is modelled after the symbol, log sin, in the classical theory where log sin x denotes the value for x of the function obtained by substituting the sine into the logarithmic function. Denoting substitution by juxtaposition is unambiguous if we consistently use a dot in denoting a product. For instance, while $\sin p_{0,2}$ denotes the function whose value for x is $\sin 2x$ we write $\sin \cdot p_{0,2}$ for the function whose value for x is $\sin x \cdot 2x$. Similarly we distinguish between the functions log sin and log $\cdot$ sin assuming the values log sin x and log $x \cdot \sin x$, respectively.

But polynomials of the form $p_{0,c}$ are so frequently studied and the function $1/c \cdot p_{0,c}$, that is, the polynomial

$$p_{0,1} = \,^1\!\prod,$$

is of such paramount importance that it seems to deserve a short symbol of its own. For $p_{0,1}$ is the function which associates x with x; the function which may be substituted into any function f without changing f; the function into which any function may be substitutes without being changed. It is, in other words, the identity function, and the lack of a current symbol for this function strikingly illustrates how little heed we give to the algebraic aspects of calculus. We shall denote the identity function by j. The above integral reads

$$\int_0^\pi \sin(2 \cdot j) = 0.$$

Denoting substitution by juxtaposition we can express the properties of j as follows

$$fj = jf = f \text{ for every } f, \text{ in particular, } jj = j.$$

Incidentally, these equalities show that it would be incorrect to describe the introduction of j for the identity function as "just writing the letter j instead of the letter x." For could we in classical notation write

$$fx = xf = f \quad \text{or} \quad f(x) = x(f) = f \,?$$

In particular, could we write

$$xx = x \quad \text{or} \quad x(x) = x \,?$$

As a matter of fact, the only way of transcribing the simple formula $fj = jf = f$ into the classical notation is by way of an implication of the following form: if $j(x) = x$ for every x, then $f(j(x)) = j(f(x)) = f(x)$ and, in particular, $j(j(x)) = j(x) = x$.

The function j also enables us to define pairs of inverse functions, such as sin and arcsin, exp and log. We call g and g^* inverse functions if $gg^* = j$. (The functions g and g' are reciprocal if $g \cdot g' = 1$.)

As long as we refrain from introducing a special symbol for the identity function, we are comparable to virtuosos in multiplication without a symbol for the number 1.

3. Differential calculus. It is obvious also that the formulae of differential calculus can be written without variables. If Df denotes the derivative of f, then the basic formulae read as follows:

I. $D(f + g) = Df + Dg$;

II. $D(f \cdot g) = fDg + gDf$;

III. $D(fg) = (Df)g \, Dg$.

In III, the term $D(fg)$ denotes the derivative of the function obtained by substituting g into f, the term $(Df)g$, the function obtained by substituting g into the derivative of f. By conventions about the scope of the symbol D, we could dispense with parentheses in either one of the two expressions.

If we set

$$f = \text{logabs, and } g = \sin$$

where logabs is the function assuming the value $\log |x|$

for x, then we have

$$Df = \text{rec and } Dg = \cos$$

where rec is the function $^{-1}\prod$ associating with x the reciprocal number $1/x$, and we obtain from formula III

$$D(\text{logabs sin}) = \text{rec sin cos} = \cot.$$

4. The calculus of antiderivatives. We shall denote by Sf any antiderivative of f, that is, any function having the derivative f. Hence we have

A. $D(Sf) = f$.

Moreover, we shall write

B. $f \sim g$ if and only if $Df = Dg$.

Obviously, the relation $\sim$ is reflexive, symmetrical, and transitive. We further readily prove

C. $Sf \sim g$ if and only if $f = Dg$,

D. $S(Df) \sim f$.

In this notation, the classical results concerning antiderivatives can be written without variables. For instance,

$$S \cos \sim \sin, \ S \exp \sim \exp, \ S \text{ rec} \sim \text{logabs},$$

$$S \log \sim (j - 1) \log.$$

5. Changing variables. The crucial test for a notation without variables is integration by substitution. For, traditionally, this method is treated as a change of variables. In our notation we have, first of all,

(1) $(Sh)g \sim S[hg \, Dg]$.

For, by B of Section 4, this formula is equivalent to

$$D\{(Sh)g\} = D\{S[hg \, Dg]\}$$

and this last equality is true since both expressions are equal to $hg \, Dg$; the expression on the right side by virtue of A of Section 4; the expression on the left side since by virtue of Rules III of Section 3 and A of Section 4 we have

$$D\{(Sh)g\} = \{D(Sh)\}g \, Dg = hg \, Dg \,.$$

This completes the proof of (1). If on both sides of (1) we substitute an inverse of g, that is, a function g^* such that $gg^* = j$, then, in view of $(Sh)j = Sh$, we obtain

$$Sh \sim [S(hg\ Dg)]g^*$$

which is the formula for integration by substitution. The formula clearly indicates the four steps that we have to take in integrating h by the substitution of g, namely,

1) the substitution of g into h;
2) the multiplication by Dg;
3) the integration of the product;
4) the substitution of g^* into the antiderivative.

For instance, let h be rec cir, that is, the function associating the number

$$\frac{1}{\sqrt{1 - x^2}}$$

with x. If we wish to find an antiderivative of h by the substitution of the function $g = \sin$ for which $Dg = \cos$ and $g^* = \arcsin$, then, noting that cir sin $=\cos$ and $S1 \sim j$, we obtain

$$S \text{ rec cir} \sim [S(\text{rec cos cos})]\ \arcsin \sim (S1)\ \arcsin$$

$$\sim j\ \arcsin = \arcsin.$$

Thus we do not need variables in order to "change variables."

It is important to realize that we can apply the method described above even if we refrain from introducing new symbols for special functions beyond j for the identity function. For instance, if we denote the function associating $1/\sqrt{1 - x^2}$ with x by f and note that

$$\frac{1}{\sqrt{1 - \sin^2 t}} = \sec t$$

or, without variables, that $f \sin = \sec$, then we still have

$$Sf \sim [S f \sin D \sin]\ \arcsin \sim [S \sec \cos]\ \arcsin$$

$$\sim [S1]\ \arcsin \sim j\ \arcsin = \arcsin.$$

...

10. Conclusions. While variables are not necessary for the presentation of fundamental results of calculus, there remain two questions. To what extent are variables necessary in proving these results? And, are variables not desirable even in formulating the theorems?

Since most students learn calculus as a tool, and since books on physics, engineering, statistics, mathematical economics, etc., are written in the classical notation, it is clear that, in initiating students into calculus, we have to use the classical notation. Yet I feel that the possibility of a consistent notation without variables should influence our teaching, namely, in the direction of reducing the use of variables. I further think that, at least in a few cases, we should mention the alternative form and, in particular, make the student aware of the possibility of a consistent notation which dispenses with dummy variables. I even believe that the ability to grasp, say, integration by substitution without variables is a gauge for a student's real understanding of calculus.

In proving formulae, we shall make use of variables although perhaps again at a diminishing rate. In proving, for instance, Formula III of Section 3 we show that if for a number x_0 the three numbers

$$Dg, \quad (Df)g, \quad D(fg)$$

are meaningful, then the third is the product of the first two. (In fact, we prove even more.) This result may be interpreted in the following form: At a place where the three terms of formula III are meaningful, the formula is true. Many formulae can be interpreted in the sense that they are true provided that every term is meaningful. For elementary functions, one could even develop an algebra of their domains of definition accompanying the algebras of their substitution and differentiation.

Another point brought out by these developments is that the application of the limit concept can be confined to the proof of very few basic formulae from which all the other formulae can be obtained by some algebra.

EXERCISES AND QUESTIONS

1. Multiply $x^2 - 2x + 3$ by $2x - 5$ using Gosselin's notation and only Gosselin's notation. That is, do not multiply in the modern fashion and translate to the other notation, but try to think as Gosselin would have. Then check to see if your result is correct.

2. Let $u = 2x + 3z$ and $z = 4x + 5y$. Show that u_x is 14 if the independent variables are x and y and 2 if the independent variables are x and z.

3. How would you write the formula for integration by parts in Menger's notation?

4. Menger later advocated using j^n for the function whose value at x is x^n.

(a) Write $(d/dx)(3x^3 - x^2 + 2)$ in Menger's notation, and carry out the differentiation operation.

(b) What is S rec j^3 ?

(c) What is S j^3 rec ?

(d) How would you express, in ordinary notation, rec $f = f$ rec ?

5. Write cir in terms of j.

6. What would be the advantages that would follow the initial period of confusion (lasting, say, twenty-five years) if

(a) everyone adapted the Dvorak typewriter keyboard

(b) everyone adapted the Menger calculus notation? How do the two proposed improvements differ and how are they similar?

MISTEAKS

by Barry Cipra

These days there are support groups for almost everything. There are support groups for people with any number of diseases or disabilities, support groups for the relatives of people with diseases or disabilities, support groups for people with mental problems, support groups for people with almost any kind of problem. My local newspaper prints a list of support groups every week, and it contains groups for agoraphobia, AIDS, alcoholism, Alzheimer's disease, apnea, arthritis, asthma, ... The common denominator of all these groups is that they are for *victims*. There are so many of them that it might seem that all victims have groups available for them to join. Not so. At least one group of victims gets no support. That group is students of calculus. To be a victim of calculus is by no means as serious as being a victim of one of the things on my newspaper's list—calculus, at least, tends to go away with time—but it nevertheless can cause sleeplessness, digestive upsets, pain, and tears. Calculus students need help.

Calculus students fight their battle with the subject almost alone. Alone, they struggle with the textbook, often so heavy that it is hard to lift, not to mention read. Alone, they struggle with the problems, often failing to attain the answer in the back of the book and not knowing why. Alone, they struggle with examinations, trying to solve problems with no answer in the back of the book to look at and under the pressure of time as well. Alone, they look at their returned examinations, sometimes filled with large red Xs, sometimes with comments like "–5", "–10", or even "–20", sometimes with no other explanations. It is almost too much to bear alone. Calculus students need support groups.

They are not likely to get them, however. The reason, I think, is that society is insufficiently aware of their suffering and does not see them as victims. College, society thinks, is a carefree time, an idyllic existence, free of responsibilities and cares. Society, however, has never had to find out how fast the height of water in a leaky conical reservoir is changing when there are only ten minutes left to complete a test. So much for society! Lacking support groups, calculus students need all the help they can get.

The author of the next excerpt has tried to give some. His book is unusual in being directly addressed to calculus students. He wants to help. Read the excerpt and see if he succeeds. If you think that he does, find a copy of the book and read it all—it is only sixty-five pages long. It can't hurt, and may help.

Most of us are prone to making certain mistakes, or certain kinds of mistakes, over and over again. Some people always mix up the minus sign when they differentiate $\sin x$ and $\cos x$. Others multiply when they should divide, as in

$$\int x^2 dx = 3x^3 + C.$$

Personally, I tend to make mistakes in addition, such as adding up students' test scores:

$$88 + 72 + 81 + 83 = 314,$$

320 being the cut-off for a B.

These are the so-called "stupid mistakes" everyone complains about making. It would be nice if there were some sure-fire way of dealing with them.

Unfortunately, the handiest hint I can think of for remedying such errors is Socrates' maxim: Know thyself. If you know you're going to make a mistake, you may not be able to avoid it, but at least you can catch yourself right away, and (presumably) correct it. So go ahead and write down

$$\frac{d}{dx} \cos x = \sin x,$$

but, if you know you tend to get this wrong, ask yourself, Did I get it right this time? You can always stick in the minus sign, and no one'll be the wiser.

The mistakes you make will be as unique as your fingerprints or your handwriting (unless you're copying from someone else's paper; I once had the dubious pleasure of nailing a cheater who had copied verbatim from another student's test paper, including a mistake so distinctive that the odds against the cheater making it independently were surpassed only by the odds against the cheater passing the course on his own). Nevertheless, there are errors that are common enough that, perhaps by pointing them out here, we can take steps toward their control. What follows then, though by no means an exhaustive survey of the common errors of calculus, is at least an introduction. I have chosen six categories of common mistakes: 1) missing minus signs; 2) disappearing parentheses; 3) lost coefficients; 4) dropped or otherwise damaged exponents; 5) fractional inversion (sounds pretty foreboding, doesn't it?); 6) uncontrollable computations.

1. Missing Minus Signs

Aside from the controversy over what to do with sin x and cos x, there is always the Chain Rule to be dealt with, and problems of denominators. As far as I know, there is still a reward out for the correct differentiation of

$$\frac{1}{1 - (1 - x^{-2})^{-3}}.$$

(There's a bonus for integrating it as well.) The best thing to do in a situation like this is to call for help. If help is unavailable (or recalcitrant), I suggest *counting* the minus signs, being sure to include one for the denominator. Alternatively, you can try determining if the function is increasing or decreasing. This probably won't help, but it's better than nothing.

Also, it goes without saying (but we'll say it again anyhow): Area, volume, and stuff like that are never negative.

2. Disappearing Parentheses

Parentheses are a way of keeping straight what goes with what. When you leave them off you run the risk of doing your test score serious harm. For instance

$$\int \frac{2x \, dx}{(x^2 + 1)^2} = -1/x^2 + 1 + C$$

looks all right, if you remember what you really mean is

$$\frac{-1}{(x^2 + 1)} + C,$$

but most likely you'll eventually convert it to $(-1/x^2) + 1 + C$ (at which point the C should really scoop up that 1, but never mind). This gives strange answers: the integral

$$\int_0^1 \frac{2x \, dx}{(x^2 + 1)^2}$$

looks perfectly well behaved, yet $(-1/x^2) + 1$ *blows up* at the lower limit. There seems to be an infinite amount of area beneath this unassuming curve!

Parentheses also have a tendency to disappear in differentiating:

$$\frac{d}{dx}(x^2 + x - 1)^4 = 4(x^2 + x - 1)^3 \, 2x + 1.$$

Actually, disappearing parentheses is a problem that may itself soon disappear. Most students who leave out parentheses do so because they don't fully understand why the parentheses are there. But as students become more accustomed to working with computers, where oftentimes a program won't even run if you don't stick in enough parentheses, they will (teachers hope) be impressed early with the necessity of stating things precisely. [The computer age in general may tend to make this book obsolete, but I doubt it (at least I hope not!). The human potential for error is boundless; it's something we can always count on. Computers may eventually take all our derivatives, do all our integrals, and graph all our functions (there are already pocket calculators that do this stuff), but we'll still be setting up the problems and pushing the buttons, and we'll keep on doing those things wrong. Computers allow us to handle bigger and more complicated problems; our mistakes will likewise get bigger and more complicated. In fact, as we remove ourselves further and further from the computational drudgery of mathematics, letting machines handle all that, it becomes increasingly important to ask the question, What does this answer mean? Can this nonsense be correct? The computer won't be able to answer that; all it will say is what it has always said: Garbage in, garbage out. *We* are ultimately responsible for the mistakes made by our misguided machines. How's that for pompous sermonizing?]

3. Lost Coefficients

This happens when you differentiate—

$$\frac{d}{dx}(x^4 + 5x^3 - x + 1) = 4x^3 + 3x^2 - 1,$$

when you integrate—

$$\int (x^4 + 5x^3 - x + 1) \, dx =$$

$$\frac{1}{5}x^5 + \frac{1}{4}x^4 - \frac{1}{2}x^2 + x + C,$$

or simply when you recopy a line—

$$x^4 + x^3 - x + 1.$$

(In case you missed it, the coefficient 5 has been ignored, as if it were never there.) Lost coefficients can be hard to detect. If it's a large enough number, or if it's a funny number like π or e, the "What Did You Expect?" method is helpful. (Whatever happened to that $942\pi^{47}$, anyway?) If the coefficient is a variable, as in

$$\frac{d}{dx}(ax + 7)^3 = 3a(x + 7)^2,$$

checking for dimensions can identify the problem. It's the small constants, 2, 3, and 4 (not to mention -1), that cause the most trouble. You either have to check over your work very carefully (which never seems to work), or wait until you get nonsense (negative area, etc.) for a final answer, which then obligates you to go back and dig up your mistake, wherever it is. (This assumes that the problem has some eventual meaning to it, which not every calculus problem does, especially on tests. Also, it occasionally happens—let's admit it—that you actually did the problem correctly, in which case you'll never be able to find the mistake.)

4. Dropped Exponents

This error is frequently seen in company with the preceding mistake. As you copy and recopy a formula, something like this may happen:

$$x^5 - 4x^4 + 3x^2 - x + 1$$

$$= x^5 - 4x^4 + 3x^3 - x^2 + 1$$

$$= x^5 + 4x^4 + x^3 - x - 1$$

$$= x^5 + 4x^3 + x^2 - x + 1$$

$$= 1 + x - x^2 + 4x^3 + x^4$$

$$= \text{etc.}$$

Fractional and negative exponents can suffer an even worse fate:

$$\frac{1 + (1 + x^2)^{-3/2}}{x^{5/2} - 1} = \frac{1 + (1 + x^2)^{3/2}}{x^{1/2} - 1}$$

$$= \frac{1 + (1 + x^3)/2}{x^{1/2} - 1}$$

$$= \left(1 + \frac{(1 + x)^3}{2}\right)(x^{1/2} - 1)^{-1}$$

$$= \left(1 + \left(\frac{1 + x}{2}\right)^3\right)(x^2 - 1)^*$$

$$= \left(1 + \left(\frac{1 + x}{2}\right)\right)^3 (x^2 - 1)^4$$

and so forth. One might wonder whatever happened to the square roots, or the power of 5 (in the $x^{5/2}$)? If you can assign dimensions to things (which is hard to do here, since x has to be dimensionless in order to get added to 1), you're in good shape. Otherwise, reread the section on lost coefficients.

5. Fractional inversion

Fractions are the last straw in a great many people's mathematical training. People who would rush into a burning building to save a child, who make confident decisions to buy this stock or that (junk) bond, who write informed articles on the global village, are all too often reduced to fear and trembling—not to mention loathing—when faced with a math problem that has "things in the denominator."

Even those of us who made it past fractions still have our problems with them. What does it mean to divide one fraction by another? Or into another? (And is there a difference?) And even if we "understand" all this, we still make mistakes:

$$\int \frac{1}{300} t^3 dt = \frac{1}{75} t^4 + C.$$

Reason: When you integrate t^3 to t^4, you *divide* by 4. That's just what we did: 300 divided by 4 is 75.

It's very common in calculus for students (and teachers as well, though presumably not quite as much) to multiply when they should have divided, or divide when they should have multiplied:

$$\frac{d}{dx}\left(1 + \frac{x}{10}\right)^{1/3} = \frac{10}{3}\left(1 + \frac{x}{10}\right)^{-2/3}$$

$$\frac{d}{dx}(1 + 10x)^{1/3} = \frac{1}{30}(1 + 10x)^{-2/3}$$

$$\int\left(1 + \frac{x}{10}\right)^{1/3} dx = \frac{4}{30}\left(1 + \frac{x}{10}\right)^{-2/3} + C.$$

This happens especially when you're first learning to integrate. You're used to differentiating, where you multiply, so you keep doing that:

$$\int \sin ax \, dx \;=\; a \cos ax + C.$$

(Later, having done nothing but integrals for what seems like forever, you realize that you can no longer differentiate correctly.) And then there are fractional powers. People who would never write

$$\int x^3 \, dx \;=\; 4x^4 + C$$

will invariably write

$$\int x^{1/3} \, dx \;=\; \frac{4}{3}x^{4/3} + C.$$

I really don't have any suggestions here. Dimensions help, I suppose, when you can assign them. Beyond that, you're on your own. Good luck.

6. Uncontrollable computations

Teachers may be cruel, but they are usually not perverse. Whatever else this may mean, it certainly means the following: On a one-hour calculus exam, you are not supposed to wind up doing fifty-nine minutes of computational arithmetic. Thus if you're asked to differentiate $(x^2 + 5)^{15}$, you are *not* supposed to start expanding the polynomial. Even a problem such as to differentiate

$$(x^4 - 1)(x^2 + 1)^2$$

is best done as

$$4x^3(x^2 + 1)^2 + 2x(x^4 - 1)(x^2 + 1),$$

and left at that—why should you expand things out if the teacher didn't? (Furthermore, what'll happen to you if you make a mistake in the expansion?)

For some problems a certain amount of computation is unavoidable, but even then you shouldn't let it get out of hand. By way of example, let's start with the innocuous test problem

$$\int \frac{(x+1)(x+5)}{(x-1)(x+2)(x+3)} \, dx.$$

Of course this integral must be done by that absolute misery, partial fractions:

$$\frac{(x+1)(x+5)}{(x-1)(x+2)(x+3)}$$

$$= \frac{A}{(x-1)} + \frac{B}{(x+2)} + \frac{C}{(x+3)},$$

so

$$(x+1)(x+5) = A(x+2)(x+3) + B(x-1)(x+3)$$
$$+ C(x-1)(x+2),$$

or

$$x^2 + 5x + 5 = A(x^2 + 5x + 6) +$$
$$B(x^2 + 2x + 3) + C(x^2 - x + 2),$$

which gives us three equations in three unknowns:

$$A + B + C = 1$$
$$5A + 2B - C = 5$$
$$6A - 3B + 2C = 5.$$

From here on out, I'll just show the steps as they might appear on a test paper, without explanation, rhyme or reason—see if you can figure out what I've been thinking.

$$6B - 2C = 1$$
$$28B - 17C = -5$$
$$102B - 34C = 17$$
$$56B - 34C = -10$$

$$158B = 7 \qquad B = \frac{7}{158}$$

$$6\frac{7}{158} - 2C = 1$$
$$\frac{42}{158} - 2C = 1$$

$$-2C = 1 - \frac{45}{158} = \frac{158-45}{158} = \frac{113}{158}$$

$$C = -\frac{113}{316}$$

$$A + \frac{7}{158} - \frac{113}{316} = 1$$

$$A = 1 - \frac{7}{158} + \frac{113}{316}$$

$$= \frac{158 \times 316 - 7 \times 316 + 158 \times 113}{158 \times 316}$$

$$= \frac{49928 - 212 + 17854}{49928}$$

$$A = \frac{62570}{49982}$$

It's obvious there's an error somewhere. No

teacher would ever put a problem on a test whose answer involves so much computation and such large numbers. (Perhaps I should amend that: No *reasonable, humane* teacher would do such a thing. Your teacher may be different.) This is supposed to be a problem on a *calculus* exam, not an arithmetic test. I should expect the problem to be computationally easy, not messy; I should have been suspicious as soon as I got a denominator 158 for *B*. As it is, notice I never did finish the problem—I never integrated anything. Why? *Because I ran out of time trying to do all those stupid multiplications!*

Of course, in "real life" (those quotations marks again!) the problems you are handed are not artificially constructed so as to be computationally easy. One must distinguish between the classroom and reality. But even so, anytime you find yourself doing an inordinate amount of arithmetic—or any other kind of unpleasant work—you should stop for a moment and ask yourself if what you're doing is really necessary, what's the point of it, is the problem really this hard, isn't there some easier way? You may find some surprising answers.

Only one exercise: Look into your soul—and your old math papers, if you haven't destroyed them—and ask yourself, What kind of fool am I?

QUESTIONS AND EXERCISES

1. Do

$$\int \left(1 + \frac{x}{10}\right)^{1/3} dx$$

right.

2. Show that the derivative of

$$\frac{1}{1 - (1 - x^{-2})^{-3}}$$

is

$$-\frac{6x^5(x^2 - 1)^2}{(3x^4 - 3x^2 + 1)^2}.$$

3. "Computers may eventually take all our derivatives, do all our integrals, and graph all our functions." In fact, a computer program is responsible for the derivative in problem 2 and I did not check it by hand.

(a) How much confidence can you have that it is correct?

(b) Is there any way to know that the machine has not made some gross error? If so, how?

(c) Should students of calculus have to learn how to do derivatives and integrals? Why should humans

have to do what machines can do better?

(d) Show that the integral in the last section of the selection is

$$-\ln(x + 3) + \ln(x + 2) + \ln(x - 1) + C.$$

To find that took me, from start to finish, thirty-seven seconds. How long did it take you? I used a machine: was my time better spent than yours?

4. Suppose that you are supervising a mathematical project (aiming missiles, say, or designing a nuclear reactor) in which no mistakes are permissible (because the missiles would wipe out Houston, or the reactor would explode). How could you make sure that your team would make none?

MATHEMATICAL OBJECTIVES

by R. M. Winger

Why study calculus, or mathematics? The following selection gives one answer.

The practical man, who frequently finds time between business and golf to lament the sins of the schools, is likely to insist that education be directed toward a definite goal. A considerable number of the students, on the other hand, still exemplify the refrain of the old song "I don't know where I'm going, but I'm on the way." Our pedagogical friends who have accepted the weighty but voluntary task of rebuilding the curriculum have adopted a popular catch-word: before any course may be considered for the new curriculum, the expert must first ascertain its *objectives*—although his vague ideas of the objectives of education itself may defy formulation. "What are the objectives of your course in trigonometry?" one of these zealots will demand, in a manner that implies that the quaking victim is expected to "stand and deliver."

What are the objectives of the courses in arithmetic, algebra, geometry and calculus? To teach the children mathematics—what, are you answered Mr. Critic? Alas no, for whatever objective may be proposed, that of teaching subject matter seems to be obsolete. We have gone far—perhaps forward, perhaps backward, possibly along the arc of a circle—since Chief Justice Taft, accepting the Kent professorship of law at his Alma Mater, announced in effect that it should be his purpose to acquaint the youth of Yale with the Constitution of the United States. Even our efficient school authorities, in the occasional inventories of their educational wares—with a view to reducing the overhead—look with misgivings upon any commodity however staple that fails to record the standard turnover. They too would examine the objectives of mathematical instruction.

Now a mere enumeration of objectives is futile unless the objectives be worthy of attainment. I propose accordingly the following line of defense on behalf of the mathematician. Trigonometry, college algebra, analytic geometry, and calculus are required subjects in most engineering schools, since they are the indispensable equipment of a trained engineer, be he civil, electrical, mechanical, or aeronautical. It would be presumptuous for a comparative layman to attempt a vindication of the colleges of engineering—that task belongs rather to the engineering profession.

Again, the same subjects are normally required or strongly recommended for major students in physics and chemistry while some of them are suggested for biologists and geologists. Physics, chemistry and biology in turn are required for entrance to the best medical schools. These several branches of natural science have been gaining ground in the past fifty years until they are now universally recognized as suitable studies in a program of liberal education. Let those who question them apply to the scientists. The medical schools alluded to are of course the schools of scientific or orthodox medicine, which have been the object of attack from the allied ranks of a multitude of dissenting cults. Whether a public university should take sides in favor of one to the exclusion of all the others is a problem outside the realm of mathematics—let those concerned consult the doctors of medicine.

Plato made geometry the entrance requirement to his philosophical academy. For argued Plato, "Geometry will create the mind of philosophy" since "that knowledge at which geometry aims is of the eternal and not of the perishing and transient." Pythagoras likewise and his school believed that geometry and number held the key to the riddle of the universe. But these men lived centuries ago and their ideas may need revision in the modern world. Much of the domain of ancient philosophy has been usurped by the sciences, including mathematics; witness the philosophical implications of the work of Einstein, Eddington, Weyl, Whitehead, Russell and a multitude of others. The skeptic who seeks an apology for such philosophy as remains under the parent name is respectfully referred to the disciples of Plato and Kant on the university faculties.

Mathematics is a kind of language—a "divine shorthand" as one enthusiast expresses it—the most precise and abridged yet evolved, and truly international in scope. A recent book on pedagogy goes so far as to treat mathematics and grammar in the same chapter because of the common elements in the two sciences. Now language has been regarded as a desirable accomplishment of scholars since the days of the Vedas, before Homer sang or Demosthenes thundered. That this honored tradition has lost some of its former momentum however, is apparent from the decadence of the classics in recent years. Nevertheless, I am confident that the departments of language can offer at least some rhetorical reasons for the continuation of linguistic courses.

Occasionally a mathematical memoir is classed as news. At rare intervals, an article from the pen of a less notable mathematician is to be found in the Sunday supplements. Whether it is the function of a university thus to assist in the creation of news or merely to train students in the art of writing it, is a point that schools of journalism might wish to debate.

Mathematicians agree that mathematics possesses qualities of beauty, analogous to those of poetry, music, sculpture and other forms of art. The trinity of the good, the true and the beautiful have enjoyed a sanction as general as it is ancient. But the justification of esthetics—if justification be demanded—is rather the province of the faculties of fine arts.

Mathematics furnishes the most accurate and adequate view of infinity to be found in any subject. In many religions, various infinite attributes are ascribed to the Deity, so that mathematics might throw some light on the nature of divinity. Indeed, Plato asserted that "God continually geometrizes." Whether, in a university supported by taxation, in a country where the separation of church and state is a cardinal principle, there is a place for the teaching of religion is perhaps a question—but let the theologians, not the mathematicians, make answer.

Mathematics is fast becoming essential to the study of economics and the calculations of modern business. Life insurance is doubtless the most scientific as it is one of the largest and most important of business enterprises—and actuarial science is merely the application of mathematical principles to the statistical problem of life and death. He would be a rash man indeed who would question the propriety of erecting on every college campus suitable shrines, dedicated to the worship of the Almighty Dollar. Need a mere mathematician lift his feeble voice to swell the mighty hallelujah chorus?

The foregoing will peradventure appear to the objective trailer as a naive response, blinded as he is by the naiveté of his inquiry. Let us resort to the parable of the ancient sage who was accustomed to "explain" to the simple-minded questioners that the earth rested on the back of a huge elephant. The more sophisticated searchers, who were curious about the support of the elephant, were silenced by the assurance that the elephant stood on an immense rock.

We have now shifted the burden of justifying mathematics upon the ample shoulders of the elephant herd of the natural sciences, engineering, economics,

language, philosophy, esthetics and religion. What do the elephants stand on? Why, the solid rock of human needs. And the rock? Ah, I was expecting that. Then abandoning parables, pachyderms and pedestals, let it be perennially and externally proclaimed that the study of mathematics fosters careful, accurate, sustained thinking, stimulating the while thinking itself. It strengthens the reason, develops the power of generalization, cultivates the imagination, and brings one face to face with chaste but naked truth. Was it Spinoza who said in substance that if mathematics—unlike history and politics—had not been independent of personal interest, the world should never have known truth?

In short I hold with the musing poet

> Flower in the crannied wall,
> I pluck you out of the crannies,
> I hold you here, root and all, in my hand,
> Little flower—but *if* I could understand
> What you are, root and all, and all in all,
> I would know what God and man is.

And this is as valid a defense of mathematics as of botany. Of no other subject can it be affirmed so completely as of mathematics

> Nothing useless is or low,
> Each thing in its place is best
> And what seems but idle show
> Strengthens and supports the rest.

QUESTIONS

1. What the author (who was a professional mathematician, by the way) seems to be saying in answer to the question, "Why study calculus?" is, "Don't ask *me*, ask those people over there." How satisfactory an answer do you think that is? Can you think of other jobs or professions where it would be suitable?

2. At the end the author throws in, either for good measure or to make his case for mathematics stronger, the assertion that mathematics fosters accurate thinking, stimulates and strengthens reasoning, and cultivates the imagination. Do you think that this assertion is true? Opinions aside, how could its truth or falsity be determined? Truth or falsity aside, I think that most people would agree with the assertion, on no grounds other than intuition. Am I thinking correctly? If so, how does that intuition arise, and how far can it, or any intuition, be trusted?

3. About when do you think that the selection was written? The answer is $1900 + x$ where x is the larger root of $x^2 - 35x + 174 = 0$. Perhaps you thought that it was written later than that, and it could be that you thought, influenced by the reference to the calculus movie *Stand and Deliver*, that it was written just recently. Whether you thought that or not, could the selection have been recent? If so, why is it that the same questions are being asked, and the same answers being given, in $1900 + x$ and $199y$? Are we making no progress at all?

"Why study calculus?" This is a question that every student of calculus should have asked, either out loud or only in thought, at least once. If you are a student of calculus and have never asked it, then you are either (a) not interested in calculus *at all* or (b) already beaten down by the system into doing whatever you are told without asking why. There is nothing much that can be done if you are in category (a). No one can find everything interesting and it is perfectly all right if calculus for you is a task to be gotten through on the way to something more important. Nevertheless, read on to see one way that the question could be answered if you had asked it. Nor is there much that can be done if you are in category (b). Systems are designed to beat people down. However, it is too bad if you have been defeated so quickly: many people hold out for years, and some never give in.

The question can be answered at several levels. There is no need to spend time on the superficial answers. "I'm taking calculus because it's a requirement." "I had calculus in high school, so I figured it'd be an easy course." "I've always gotten pretty good grades in mathematics, so I'm taking some more." "I've got to take *something*—I need 120 hours of credit to get out of here—and calculus is as good as anything else." Complete and satisfactory answers, giving perfectly valid reasons for being enrolled in a calculus class (it is not true that you must have noble and commendable reasons for all of your actions, or even for your noble and commendable actions), but there is nothing much that can be said about them. They do not get at the deeper questions about why calculus classes exist to be taken, or why schools exist that teach calculus.

A more fundamental question is, why study *anything*? Until a satisfactory answer can be given, there is no sense in worrying about studying something as specialized as calculus. Yes indeed, why study anything? Studying is not necessary. Animals don't study anything, and they do all right. Take squirrels, for example. Why not be a squirrel? Squirrels seem to have good lives: they do what they want to, when they want to. Their lives have freedom and variety. (Squirrels are *never* bored.) They have nuts to eat, trees to climb, and squirrels of the opposite sex to chase or be chased by. Squirrels don't have worries about the future. Squirrels don't have to work. Squirrels don't have to learn calculus. Squirrels have great lives! Why not be a squirrel?

I'll tell you why not. Observe squirrels. Squirrels are constantly twitching. They are always looking over their shoulders. Squirrels are in a continual state of panic. *Everything* surprises a squirrel. Squirrels spend a lot of their time running away from things. Squirrels are constantly threatened. A squirrel's life is one of random potential disasters, repeated and repeated and repeated without end. I wouldn't want to be a squirrel.

Well, you might say, if you don't want to be a squirrel because you can't take the stress, how about being a sheep? Sheep have calm and contented lives with plenty of grass to eat, all their needs taken care of, and unlimited time to think sheep thoughts. You would have nothing to worry about, no anxieties about the future, no family problems, never a care about money, no need to fret over the progress of your career or be apprehensive about who will take care of you when you get old. A life of tranquillity, a good life! Why not be a sheep?

Life as a sheep is not for me. Have you ever looked deeply into the eyes of a sheep? What you see there is a look of bafflement, of incomprehension, of confusion and unknowing. Sheep do not know what is happening to them or why it is happening, and they are puzzled. Sheep are aware, though very dimly, that they do not have the answers to anything and they yearn, with a sheeplike yearning that can never be satisfied, for someone to tell them in terms that they can understand what the world is about and what it is doing to them. I wouldn't want to be a sheep.

In any event, we are stuck with our humanity which carries with it the gifts of self-consciousness and rationality, not given in the same measure to squirrels and sheep. Or are they instead burdens and afflictions? If we do not surrender to delusions, either imposed from the outside or generated from within, we can use them to look around and see the world whole and clear. What we see does not always lead to joy. Consider the following excerpt.

Vanity of vanities, saith the Preacher, vanity of vanities; all is vanity. What profit hath a man of all his labor which he taketh under the sun? One generation passeth away, and another generation cometh: but the earth abideth forever.

The thing that hath been, it is that which shall be; and that which is done is that which shall be done: and there is no new thing under the sun. Is there anything whereof it may be said, See, this is new? it hath been already of old time, which was before us. There is no remembrance of former things; neither shall there be any remembrance of things that are to come with those that shall come after.

For what hath man of all his labor, and of the vexation of his heart, wherein he hath labored under the sun? For all his days are sorrows, and his travail grief; yea, his heart taketh nor rest in the night.

For that which befalleth the sons of men befalleth beasts; even one thing befalleth them: as the one dieth, so dieth the other; yea they have all one breath; so that a man hath no preeminence above a beast: for all is vanity. All go unto one place; all are of the dust, and all turn to dust again.

I considered all the oppressions that are done under the sun: and beheld the tears of such as were oppressed, and they had no comforter; and on the side of their oppressors there was power; but they had no comforter. Wherefore I praised the dead which are already dead more than the living which are yet alive. Yea, better is he than both they, which hath not yet been, who has not seen the evil work that is done under the sun. I considered all travail, and every right work, that for this a man is envied of his neighbor. This is also vanity and vexation of spirit.

Better is the sight of the eyes than the wandering of desire: this is also vanity and vexation of spirit. That which hath been is named already, and it is known that it is man: neither may he contend with him that is mightier than he. Seeing there be many things that increase vanity, what is man the better? For who knoweth what is good for man in this life, all the days of his vain life which he spendeth as a shadow?

I saw under the sun, that the race is not to the swift, nor the battle to the strong, neither yet bread to the wise, nor yet riches to men of understanding, nor yet favor to men of skill: but time and chance happeneth to them all.

Truly the light is sweet, and a pleasant thing it is for the eyes to behold the sun: but if a man live for many years, and rejoice in them all; yet let him remember the days of darkness; for they shall be many. All that cometh is vanity.

The evil days come, and the years draw nigh, when thou shalt say, I have no pleasure in them. The sun and the light of day shall darken and the clouds return with the rain. The keepers of the house shall tremble, and the strong men shall bow themselves, and the grinders cease because they are few, and those that look out of the windows are darkened, and the doors shall be shut in the streets, the sound of the mill is low; when they shall be afraid of that which is high, and fears shall be in the way, when the almond-blossoms whiten, and the locust shall be a burden, and desire shall fail: because man goeth to his long home, and the mourners go about the streets. The silver cord is loosed, the golden bowl is broken, the pitcher is shattered at the fountain, and the wheel broken at the well. Then shall the dust return to the earth as it was. Vanity of vanities, saith the preacher; all is vanity.

COMMENTS

Any comment must be pale in comparison to that passage, whose power is immense. Surely life cannot be that grim and bleak. Surely not all is empty. Surely the Speaker must have missed something, surely the Speaker's vision must be somehow limited or flawed.

Yes, it is limited and yes, the Speaker did miss something. What the Speaker missed, what would have made all the difference, was a course in calculus.

Let me explain. The Speaker looked about and saw that nothing changes. People come and people go, but they forever do the same things. They want to be rich, they want to be powerful, they want, want, want, but what is the point of all their wanting? They all die, and what of their riches and power then? Those that replace them will die also. Emptiness, all is emptiness. That was an accurate reflection of the time and place in which the Speaker lived, the Near East something over two thousand years ago. What was lacking there and then was science and mathematics. The Greeks had invented mathematics as a deductive system a very few centuries before and science and the scientific method lay many centuries in the future. Mathematics and science are antidotes to the despair that comes from thinking that in spite of mighty efforts nothing changes or will ever change. It is true that in government, in economics, in morals, the wheel of existence goes around and around, endlessly turning, and there is nothing new under the sun. We overthrow kings and replace them with democracies that change into dictatorships; we have free markets until we see that a little regulation would make conditions better and we then regulate more and more until we see that no regulation would be better still; we forbid divorce to minimize the damage to families and society and then we make divorce easy for exactly the same reason; around and around the wheel of existence goes. However, it is not the case that nothing ever changes and there is nothing new under the sun. In science and mathematics, progress is made. In science and mathematics, the arrow points forward only and does not bend into a circle. The advances made in this century in physics and medicine are astonishing, and a person would have to be devoid of curiosity—and hence not completely human—not to wonder what was going to happen next. Whatever happens next will be new. It will not be emptiness.

It is better to be human than to be a squirrel or a sheep, and one of the higher callings of humanity is to find out why the world is as it is. That is a purpose that is not chasing the wind. That is a reason for studying calculus.

THE FABULOUS FOURTEEN OF CALCULUS

by Charles A. Jones

Sines and cosines, exponentials and logarithms, arctangents and hyperbolic secants: functions like those are not as nice as polynomials. If you want to know the value of a polynomial there is no difficulty: substitute a number, do some computation, and there you are. If you have had a properly old-fashioned mathematical education, you can use the magic of synthetic division to make the computations easier: to find the value of $x^4 - 3x^3 + 5x^2 - 7x + 9$ when $x = 3$, write

1	−3	5	−7	9	3
	3	0	15	24	
1	0	5	8	33	

and there you have the answer, 33, at the cost of only a few seconds' labor. Those other functions are different, and stranger. To find the value of ln 365, substitution will not do. A calculator will yield 5.8999 with very little labor on our part, but think of the labor that went into creating the calculator. It is of a different order of magnitude entirely from synthetic division. Before there were calculators there were tables, but if we had neither tables nor calculator, we would be helpless in the face of something so simple-looking as sin 1.

Polynomials are the puppy dogs of mathematics: friendly, open, and eager to help. You may not be able to see their tails wagging nor feel their tongues giving you an adoring lick, but polynomials love people. The transcendental functions—that is the name for the sine, the exponential, and their relatives—are more like cats. They live with us, and do useful things, but on their terms and not ours. They are aloof, they have secrets, they do not jump with joy and try to please when they see a person coming.

Why can't all functions be polynomials? The world would be *such* a nicer place. However, the world has not been constructed entirely for the convenience of humans, there are cats, mosquitoes, and poisonous snakes (each with a mathematical counterpart—e. g., that there is an even prime number is a mathematical mosquito), and we must live with them all. The following selection shows why transcendental functions cannot be puppydogs.

The functions studied in calculus are built from algebraic operations plus the following fourteen functions: the six trigonometric functions, the natural logarithm function, the six inverse trigonometric functions, and the exponential function, e^x. The common feature of these fourteen functions is that they are defined in geometrical, not computational, ways. This paper discusses why no computational definitions are possible.

The real numbers are divided into several categories: there are integers, rational numbers, irrational numbers, and transcendental numbers. I assume the integers are familiar to the reader. The *rational numbers* are those real numbers p/q where both p and q are integers and $q \neq 0$. The *irrational numbers* are the real numbers that are not rational; a common example of an irrational number is $\sqrt{2}$ (see Exercise 1). The *algebraic numbers* are those numbers which are the zeros of polynomials with integer coefficients. For example, any rational number p/q is algebraic since p/q is a zero of $f(x) = qx - p$. (To see this, substitute $x = p/q$ and see that $f(p/q) = 0$.) In addition to all the rationals, *some* of the irrational numbers are algebraic; e. g., $\sqrt{2}$ is a zero of $x^2 - 2$. The real numbers which are not algebraic are called *transcendental*. Two well-known transcendental numbers are π and e. Figure 1 summarizes the above discussion.

You may notice that the figure shows the transcendentals as the largest collection. This may seem unusual since we don't usually use many of the transcendentals, but it turns out that there are *far more* transcendentals than nontranscendentals (algebraic numbers); (see Note 1).

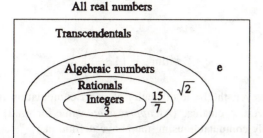

Figure 1

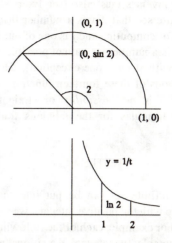

Figure 2

The transcendental numbers π and e have been computed to millions of decimal places, but they can never be "known exactly." You never *compute* transcendentals in any straightforward fashion since, by definition, there are no algebraic formulas for computing them starting from integers.

The functions studied in calculus are built from the algebraic operations of $+$, $*$, $-$, $\div$, and raising to rational powers, for example

$$\frac{x^2 + x\sqrt{x^3 - 1}}{\left(4x - \dfrac{2}{x}\right)^{8/3}},$$

and from the fourteen functions already mentioned, for example

$$17xe^{x^3} + \ln\left(\sin^{-1}\sqrt{x + 3}\right).$$

The purpose of this section is to indicate how these fourteen functions are vastly different from functions built from algebraic operations only. From your own experience, you may think these fourteen functions are strange and hard to deal with; if you do, you are absolutely correct.

There is one major problem with these fourteen functions; they are hard to compute. For example, try to compute $\sin(2)$ exactly. The definition of sine suggests you start at $(1, 0)$ and go 2 units along the unit circle in the counterclockwise direction, then stop and report back your current y-coordinate. (Compare this with evaluating $2x^3 + 3x - 1$ when $x = 2$, which quickly yields the result of 21.) The logarithm function is no better. By definition, $\ln(2)$ is the area under the curve $y = 1/t$ between $t = 1$ and $t = 2$. (See Figure 2.)

In fact, seven of our fourteen functions (the six trigonometric functions and the logarithm function) have geometrical, not computational definitions. The remaining seven functions are defined in an even less satisfactory way as the *inverses* of the first seven. If you feel that defining a function as the inverse of another function which is impossible to compute is an unsatisfactory state of affairs, your point of view is well taken. If you feel like demanding that your instructor or textbook do something to remedy this sorry state of affairs, you are out of luck. Two facts doom your case. First, these fourteen functions cannot be discarded—they are among the most important functions in mathematics, both theoretically and practically. Second, the following theorem due to the German mathematician F. Lindemann proves there is no (algebraic) computational scheme available for these fourteen functions.

A version of *Lindemann's Theorem* (1882): If $x \neq 0$, and x is an algebraic number, then e^x is transcendental. In other words, if you supply a nonzero x that you can compute algebraically (x will be algebraic), the result e^x, will *not* be a number you can compute.

You may have noticed that Lindemann's Theorem mentions only the exponential function, not the other thirteen functions. However, all of our fourteen functions have a similar property: with at most one exception, given an algebraic x, the functions evaluated at x yield transcendental numbers. To see why this is so, we will use the fact that Lindemann's Theorem holds when using complex variables. In complex variables, e^x, $\sin(x)$, and $\cos(x)$ are related algebraically by $e^{bi} = \cos(b) + i\sin(b)$, and from this one can show (but I'm not going to) that sine and cosine also have our property. Once sine and cosine have the property, the other four trigonometric functions follow suit. Finally, since e^x and the six trigonometric functions have the property, their inverses must also (see Exercise 3).

We therefore see that if x is a number that we can have any hope of computing and f is one of our fourteen functions, $f(x)$ is a number we cannot possible compute algebraically (with at most one exception). Hence, the strange definitions of these fourteen functions are here to stay. There is *no possibility* of straightforward computational formulas for the fabulous fourteen of calculus.

NOTES

1. If an infinite set can be put into one-to-one correspondence with the positive integers, we say the set is *countable*. For example, each of the following sets of numbers is countable: the integers, the rational numbers, and the algebraic numbers. The real numbers and the transcendental numbers are not countable (called *uncountable*); there are far more numbers in these sets than in the countable sets.

The following facts illustrate these ideas.

a. If you could somehow pick a real number at random (every real number equally likely to be picked), the probability of choosing an algebraic number is 0 (hence the probability of choosing a transcendental number is 1).

b. If you repeated your experiment of choosing a random real number a trillion times, the probability that at least one of your numbers is algebraic is 0 (i. e., they would all be transcendental with probability 1).

2. The functions described in this paper as "the functions studied in calculus" are known as the *elementary functions*.

3. The well-known trigonometrical values, such as $\sin(\pi/6) = 1/2$, do not violate our property since $\pi/6$ is *not* algebraic. This follows from π being transcendental by the following argument: if $\pi/6$ were algebraic, $\pi/6$ would be a zero of some polynomial, $p(x)$, with integer coefficients. But then π would be a zero of $p(x/6)$. Also, π is a zero of $q(x) = 6^{(\text{degree of } p)}p(x/6)$ since $q(x)$ is just a constant times $p(x/6)$. However, q has integer coefficients since the 6's in the denominators of $p(x/6)$ are all cancelled by $6^{(\text{degree of } p)}$. Similar arguments show that any nonzero rational number times π is not transcendental. Thus, all of the familiar angles from trigonometry (except 0) are transcendental (e. g., $\pi/4$, $-5\pi/6$, $17\pi/6$, 2π, etc.).

4. Some readers may have seen some of our fourteen functions defined using limits instead of geometry. For example,

$$\sin x = x - \frac{x^3}{3!} + \frac{x^5}{5!} - \dots$$

or

$$e^x = \lim_{n \to \infty} \left(1 + \frac{x}{n}\right)^n.$$

However, the basic difficulty of *exact* computation for nonzero x remains. (An *approximate* value for sin(2) is easily computable using the above definition.)

5. A reference for many of the topics in this paper is Alan Baker, *Transcendental Number Theory*, Cambridge University Press, 1975. Pages 1-8 contain an excellent treatment of transcendental numbers, the transcendence of π and e, and Lindemann's Theorem. A mathematics student who has completed abstract algebra should be able to work through these results and understand the proofs.

EXERCISES

1. Show that $\sqrt{2}$ is irrational by giving reasons for the steps in the following outline.

Preliminary facts: An integer n is even means n is a multiple of 2. An integer n is even if and only if n^2 is even. (If n is even, n^2 is a multiple of 4.)

Proof by contradiction: Assume there exists two integers p, q so that $p/q = \sqrt{2}$ and assume p/q is in lowest terms. (In particular, p and q are not both even.)

$\dfrac{p^2}{q^2} = 2$.

$p^2 = 2q^2$.

p^2 is even.

p^2 is a multiple of 4.

q^2 is even

Hence p and q are both even.

Contradiction; so no such p, q exist.

2. Each of the fourteen functions in this paper has our property of sending algebraic numbers to transcendental numbers with at most one exception. For e^x, $x = 0$ is the exception since $e^0 = 1$ and 1 is also algebraic. For each of the thirteen other functions, find the exception. *Hint:* For our fourteen functions, seven have $x = 0$ as the exception, three have $x = 1$ as the exception, and four have no exceptions.

3. Show that if f has the property of sending algebraic numbers to transcendental numbers, then f^{-1} must also. *Hint:* Suppose a is algebraic and $f^{-1}(a)$ is also algebraic. What is $f(f^{-1}(a))$?

IMPOSSIBILITY

by Charles A. Jones and Nathan Root

Language changes, and the meaning of words blurs with time. *Impossible* used to mean ... well, impossible: as my dictionary puts it, "not capable of existing or happening." But meaning blurs and strong words invariably become weaker. *Unique* used to mean ... well, unique: one of a kind, the only example there is of something. Now *unique* has come to mean merely unusual, and things can be "almost unique" or even, heaven help us, "somewhat unique." *Impossible* has also weakened. My dictionary's second meaning for the word is "having little likelihood of happening or being accomplished" and, given the way that language evolves, it may be the first meaning in my dictionary's next edition. Then the language will have to create a new word to take on the old first meaning of *impossible*, and there is no telling what it will be.

The reason that a new word will be created is that there is a need for it, because in mathematics *impossible* still means impossible and it always will. When I say "It's impossible for the sum of two even integers to be odd" I do not mean that there is very little likelihood that the sum of two even integers will turn out to be odd, or that finding two such even integers is such a difficult task that no one has yet succeeded at it, I mean it's impossible. Period. I can prove it. Let a and b be even integers. Then $a = 2m$ and $b = 2n$ for some integers m and n. Thus, $a + b = 2m + 2n = 2(m + n)$ and that is divisible by 2. Thus $a + b$ is even, because it is divisible by 2, no matter what a and b are. $a + b$ isn't odd. It's impossible.

In mathematics, we can have absolute certainty. The Pythagorean Theorem about squares of legs and hypotenuses of right triangles is true, always has been, and always will be. It will never need revision. 17 is a prime number, once and for all. Mathematics provides eternal truth! This does not occur elsewhere. The law of gravity has been verified sufficiently many times that it is called a law and everyone is confident that it will remain in force, but we cannot be certain. Economists cannot agree on what the laws of economics are. Anything involving people is wildly uncertain. Mathematics, beautiful mathematics, gives us unchanging truth.

It is important to realize the difference between the two senses of impossible. "It is impossible to send a person to Sirius and back" is a true statement, but at some time in the future it may be false. "It is impossible to trisect angles with straightedge and compass alone" is a true statement that will never be false. The people who trisect angles with straightedge and compass alone (and there are quite a few of them) do not understand the difference, and some of them have had their lives blighted, or even ruined, as a consequence. Angles cannot be trisected because it has been proved that they cannot be trisected, and a mathematical proof is forever. Unlike the proof that it is impossible for the sum of two even integers to be odd, the proof that the trisection is impossible is too hard for most people to understand but, understood or not, a proof is a proof and that is that. When something in mathematics is impossible, it is impossible in the first, and best, sense of the word. Mathematics provides eternal truth.

The next selection explains why it is impossible to construct, with straightedge and compass alone, the side of a cube that has double the volume of a given cube and also why it is impossible to find a nice function f such that $f'(x) = e^{x^2}$.

What does it mean to say that a task is impossible? It means that one cannot accomplish the goal of the task that one has set. This is not to say that the person trying to do it was not smart or clever enough and that someone else may come along later and discover how to do it: it is to say that there is no way to do it and it cannot be done, by anyone, ever.

In this paper we describe five famous impossible tasks in mathematics: (1) doubling a cube, (2) trisecting an angle, (3) squaring a circle, (4) finding a formula to solve a fifth degree polynomial equation, and (5) finding an antiderivative of any elementary function. The first three tasks are from geometry and concern straightedge and compass constructions. The fourth task is algebraic and the fifth is from calculus. Even though they are from different areas of mathematics, the proofs of their impossibility use a common set of tools from abstract algebra: fields, field extensions, and (for some) automorphisms of fields.

An interesting question about the five tasks is: for what instances are they impossible? Tasks (1) and (3) are always impossible: no cube can be doubled and no circle can be squared. In contrast, task (2) can sometimes be carried out: we can trisect some angles, such as 90°, though such angles are exceptional. For task (4) there is a very powerful impossibility result. Not only is determining a general formula for solving fifth degree polynomial equations impossible, but there are specific examples of polynomials, such as $2x^5 - 10x + 5$, which do not have a formula for any of their zeros. (A formula for the zeros is one that uses rational numbers, addition, subtraction, multiplication, division, and the taking of roots of any degree, i. e. square root, cube root, 9/5 root, and so on.) Note that this polynomial has exactly three real zeros, as can be shown using the intermediate value theorem and Rolle's theorem. By algebraic theory there are two more zeros, which are complex numbers. Values of all of the zeros can be found only by using numerical methods. Task (5) is more complicated to analyze. Some functions, such as e^x, are easy to antidifferentiate, but others, like e^{x^2}, are impossible. What does this impossibility result mean? By the Fundamental Theorem of Calculus we know that e^{x^2} *does have* an antiderivative, just as $2x^5 - 10x + 5$ has five zeros, it is that we cannot express the antiderivative using the usual functions and operations from calculus, just as we cannot express the zeros of $2x^5 - 10x + 5$ in the manner that we would like. The technical terminology is that the antiderivative of e^{x^2} is not an elementary function.

We will now examine in detail the task of doubling a cube. Before we proceed, we need to make sure that the problem is understood. The problem is this: given the side of a cube, is it possible to construct, with a straightedge and compass only, the side of a cube that has double the volume of the original cube? The key lies in the phrase "with a straightedge and compass only"; what exactly does this mean? Simply put, it means that you can draw lines through two points (though you cannot measure distances), circles around points, and mark the intersection of two lines, two circles, or a line and a circle.

To show that doubling a cube is impossible we approach the problem as follows. First we show what numbers are constructible with straightedge and compass alone, then we show that to double a cube would

require the construction of a nonconstructible number. What does it mean to construct a number? We start with a line segment of unit length; that is, we draw a line segment with our straightedge and call its length one. To construct the number 2, we need to construct a line segment of length 2, twice the length of the unit segment. Since this is an elementary geometric construction, the number 2 is a constructible number. We can generalize to a definition of constructible: given a line segment of length one we say that a number x is constructible if, using a straightedge and compass alone, we can construct a line segment of length x.

Once we have this definition, we ask, is it possible to construct every number? Given the title of this paper, you would guess that the answer is *no*, but how can that be proved? First, it is not hard to show (though we will not show it here) that any number of the form p/q, where p and q are positive integers, is constructible. That is, we can construct all the positive rational numbers. Next, we ask if any other numbers, such as $\sqrt{2}$, are constructible. We answer this question in the following discussion describing exactly what numbers are constructible. We start by showing that square roots of rational numbers are constructible. Then, by extending this argument, we show that in addition to the four arithmetic operations we can take square roots, so that the sum of two constructible numbers is also constructible, as is their product, and as is the square root of either. Then we examine the argument to show that we can add *only* the taking of square roots.

To begin the argument, we show that the square root of a rational number is constructible. Given that we can construct rational numbers, we can locate any point in the x-y plane with rational coordinates. Conversely, if we can construct a point (a, b) then we can construct a and b. So, constructing points and constructing segments are equivalent, and we can turn our attention to the construction of points. Points are constructed by the intersections of lines and circles. There are three types of intersections: (1) line with line, (2) line with circle, and (3) circle with circle. Given the constraints that we have placed on our lines and circles—that each line passes through two known points with rational coordinates and the center of each circle is a point with rational coordinates and the square of its radius is rational—it is easy to show (and is recommended as an exercise) that we can represent lines as $ax + by + c$ and circles as $x^2 + y^2 + dx + ey + f$ where a, b, c, d, e, f are *rational* numbers. Applying these equations to the three types of intersections we see that the new points that we add are simply the solutions to the system of equations depending on which case we look at. We examine the cases individually. (See Figure 1.)

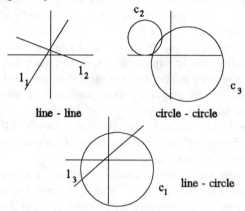

line - line   circle - circle

line - circle

Figure 1

Case 1: Line Intersect Line. Here we have two lines, l_1 and l_2, that intersect in a point. This point must have rational coefficients since if we solve for x in the equation of l_1 and then substitute the equation of l_2 we just get another linear equation with rational coefficients. So we see that the intersection of lines yields us no new points and hence no new constructible numbers.

Case 2: Line Intersect Circle. Here we have a line represented by $ax + by + c$ and a circle represented by $x^2 + y^2 + dx + ey + f$. In this case, if we solve for x in the equation of the line and then substitute it into the equation of the circle we see that we arrive at a quadratic. However, applying the quadratic formula requires the use of the square root operation. Therefore, we can add the square root of rational numbers by looking at the case where a circle and a line intersect.

Case 3: Circle Intersect Circle. Here we have two circles, c_2 and c_3, where c_2 is represented by $x^2 + y^2 + d_2x + e_2y + f_2$ and c_3 is represented by $x^2 + y^2 + d_3x + e_3y + f_3$ However, by looking at the difference of the two equations for c_2 we see that we get $(d_2 - d_3)x + (e_2 - e_3)y + (f_2 - f_3)$. Since this is a line we have that the intersection of two circles is equivalent to the intersection of a line and a circle. That is, the solution to the equations

$$x^2 + y^2 + d_2x + e_2y + f_2,$$

$$x^2 + y^2 + d_3x + e_3y + f_3$$

is the same as the solution to the equations

$$x^2 + y^2 + d_2x + e_2y + f_2,$$

$$(d_2 - d_3)x + (e_2 - e_3)y + (f_2 - f_3).$$

Therefore Case 3 can be reduced to Case 2, so Case 3, like Case 2, also adds the square root of rational numbers.

So we have shown that we may add the square root of a rational number. If we repeat the argument using constructible numbers instead of rational numbers, we see that we may add the square root of any constructible number. It is easy to see that if a and b are constructible then so is $a + b$ and $a - b$ and using the same proof for the construction of all the rational numbers we can show that the same holds true for ab and a/b. Finally, the above argument shows that if a is constructible than so is $\sqrt{a}$. If we now examine the above argument carefully we see that we can only add square roots, since that is the only operation besides the four basic arithmetic operations that is employed in the argument. So, we cannot take cube roots, fifth roots, sixth roots, ..., though we can construct numbers such as

$$\sqrt{2}, \quad \sqrt[4]{2}, \quad \sqrt{4 + \sqrt{7}}, \quad \sqrt{3 + \sqrt[4]{1 + \sqrt{2}}}.$$

So we have completely determined exactly which numbers are constructible. The fact that cube roots and so on are not constructible can be very precisely shown using the tools of abstract algebra.

Now that we have determined exactly which numbers are constructible, we are ready to consider the problem of doubling a cube. Let the given cube have a side of length 1, so to double the volume the new cube must have a side of length $\sqrt[3]{2}$. So to construct it we must construct the number $\sqrt[3]{2}$. However, $\sqrt[3]{2}$ is a cube root and hence we cannot construct it. Therefore, doubling a cube is impossible.

Now let us consider antiderivatives. Given a function, say f, what does it mean to say that it is possible to find an antiderivative for it? The heart of the question is in the word *possible*. We can define it in terms of its opposite, *impossible*. If something is impossible then no one can do it, so an impossible antiderivative is one that no one can find, while a possible antiderivative is one that can be found. However, before we can proceed we must define exactly what functions we are considering. If, for example, we exclude the logarithm function, then it would be impossible to find an antiderivative of $1/x$.

We will include the basic operations of addition, subtraction, multiplication, division, raising to a power, taking of logarithms, exponentiation, and composition of functions. So, for example, we include such things as

$$x^{10}, \quad x^{-2}, \quad \ln x, \quad \ln(x^2 + 1/x), \quad e^{2x+3}, \quad \sqrt{x}.$$

We also include the trigonometric functions and their inverses. We make the restriction that our functions must have a finite number of terms. Thus,

$$1 + x + x^2 + x^3 + \dots, \quad \ln(\ln(\dots \ln(x) \dots)), \quad x^{x^{\cdot^{\cdot}}}$$

and similar expressions are not allowed. Any function that we can write down using a finite number of pencils will be allowed. The set of functions that are allowed are known as the elementary functions.

There are no impossible derivatives. That is, given any elementary function, we can find its derivative and the derivative is another elementary function, as we now discuss. If we look at the product, quotient, and power rules for differentiation and note that taking derivatives is a linear operation (so that if f and g are functions and c is a constant, then $(f + g)' = f' + g'$ and $(cf)' = cf'$) we see that we can account for addition, subtraction, multiplication, division, and raising functions to a power. Considering exponentials, logarithms, and the inverse trigonometric functions, one need only examine the derivatives of such functions to see that they are indeed elementary functions. To take care of the trigonometric functions, we need only to observe that the derivative of a trigonometric function is still a trigonometric function, e. g. the derivative of $\sin x$ is $\cos x$. And to complete the discussion we simply examine the chain rule to verify it for the composition of functions.

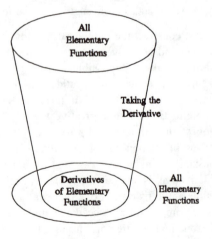

Figure 2

Now that we have defined exactly what functions we allow ourselves to use, we must ask ourselves the question that we asked before: given a function, say f, what does it mean to say that f has an antiderivative? We can state this in a much more mathematical manner since we have defined all the terms in the question. We will restate it: given an elementary function f, does there

exist an elementary function g such that $g' = f$? If such a g does exist then we will say that f is antidifferentiable, and if not then we say that f is not antidifferentiable.

Up to this point you may never have encountered an antiderivative that is impossible (there may have been some that seemed impossible, but turned out only to be very hard), so you may ask if there are any. To think how this may be possible, look at Figure 2. The functions inside the area labeled "Derivatives of Elementary Functions" are those that have antiderivatives, while those outside do not.

We now demonstrate a function that does not have an antiderivative to place this in a more concrete setting where we are more concerned with examples rather than merely the existence of such examples.

The function that we look at is e^{x^2}. To display its nonantidifferentiability we assume that it is possible to antidifferentiate it and obtain a contradiction. In this case the contradiction is that something not zero is equal to zero.

We attack the problem of showing that e^{x^2} does not have an antiderivative by breaking the problem into two steps. In the first step we determine the form of the antiderivative, $g(x)$, of e^{x^2} if it exists. This step consists of a plausibility argument demonstrating the form the antiderivative must have if it exists (this step can be proved using higher mathematics). Then in step two we take the antiderivative, in the form known from step one, differentiate it, set it equal to e^{x^2}, and proceed to the contradiction. The second step is completely justified and is a bit tedious at times.

For step one, if we assume that e^{x^2} has an antiderivative then there must exist a g such that $g'(x) = e^{x^2}$. We now ask ourselves, if such a g exists, then what form will it have? First, we know that since the derivative of e^{x^2} is $2xe^{x^2}$ it seems intuitive that g contains (that is, has as one of its terms) an e^{x^2}. However, since we do get $2xe^{x^2}$ when we differentiate the e^{x^2} it is most likely that the e^{x^2} is multiplied by some other elementary function, say h_1. So we now have a guess that g contains a $h_1(x)e^{x^2}$. This leads to another question: does g contain $h_2(x)\left(e^{x^2}\right)^2$, or a term containing e^{x^2} to any other power? If it does, then since the derivative of $h_2(x)\left(e^{x^2}\right)^2$ is $\left(h_2'(x) + 4x\right)\left(e^{x^2}\right)^2$ we see that it cannot help to cancel out any other powers of $\left(e^{x^2}\right)^2$ other than that of its own power so it seems unlikely that g contains $h_2(x)\left(e^{x^2}\right)^2$, i. e. it does except that $h_2(x) = 0$. We can extend this to all other powers of e^{x^2} so we see that as far as terms containing e^{x^2} our intuitive guess would be that the only one is $h_1(x)e^{x^2}$. However, we would further guess that $g(x)$ may contain some other elementary function, say h_0. So, to sum up our intuitive guess as to the form of g, we guess that $g(x) = h_1(x)e^{x^2} + h_0(x)$, where both h_0 and h_1 are elementary functions.

Using higher mathematics it can be shown, though we will not do it here, that our intuitive guess is correct. Furthermore, we can say something about the form of h_0 and h_1; it can be shown that (once again, we will not do it here) they are polynomials, that is have the form

$$c_n x^n + c_{n-1} x^{n-1} + \dots + c_0,$$

where n is a positive integer and each of the c_i's is a rational number. Note that h_0 and h_1 do not contain any exponential terms.

Now we come to step two: we know what g *must* look like, if it exists, and we use the fact the $g'(x) = e^{x^2}$ to obtain our contradiction.

Since we have that if g exists, then

$$g(x) = h_1(x)e^{x^2} + h_0(x),$$

we have that

$$g'(x) = h_1'(x)e^{x^2} + 2xh_1(x)e^{x^2} + h_0'(x) =$$

$$(h_1'(x) + 2xh_1(x))e^{x^2} + h_0'(x) = e^{x^2}.$$

From the previous equation we arrive at the following two equations by equating the coefficients:

$$h_0'(x) = 0, \quad h_1'(x) + 2xh_1(x) = 1.$$

From the first equation we see that $h_0(x)$ is a constant. The second equation is slightly more complicated. To solve it we will need to use the fact that

$$h_1(x) = c_n x^n + c_{n-1} x^{n-1} + \dots + c_1 x + c_0,$$

where $n \geq 0$ and c_n is not equal to zero.

By substituting the expression for $h_1(x)$ and the associated expression for $h_1'(x)$ into the second equation, we get

$$nc_n x^{n-1} + \dots + 2c_2 x + c_1 + 2c_n x^{n+1} +$$

$$2c_{n-1} x^n + \dots + 2c_1 x^2 + 2c_0 x = 1.$$

Now we combine the coefficients of like terms to get

$$2c_n x^{n+1} + 2c_{n-1} x^n + (nc_n + 2c_{n-2})x^{n-1} + \dots$$

$$+ (3c_3 + 2c_1)x^2 + (2c_2 + 2c_0)x + c_1 = 1.$$

By equating the coefficients we get the following set of equations

$$2c_n = 0$$
$$2c_{n-1} = 0$$
$$(nc_n + 2c_{n-2}) = 0$$
$$\cdot$$
$$\cdot$$
$$\cdot$$
$$(3c_3 + 2c_1) = 0$$
$$(2c_2 + 2c_0) = 0$$
$$c_1 = 1.$$

However, from the first equation we see that c_n must equal zero. This contradicts our assumption that c_n is not zero. Therefore, we have the desired conclusion that something nonzero is equal to zero. Hence, our original assumption must be incorrect. This completes the proof that e^{x^2} is impossible to antidifferentiate using elementary functions.

EXERCISES AND QUESTIONS

1. (a) Give an example of a fifth-degree polynomial equation all of whose roots can be written down exactly.

(b) Are all of the roots in (a) different? If not, give an example where they are.

2. Show how to carry out a straightedge-and-compass construction that will start with a line segment of length one and result in a line segment of length (a) $\sqrt[4]{2}$ (b) $\sqrt{4 + \sqrt{7}}$.

3. Find those values of k for which $g'(x) = x^k e^{x^2}$ is possible, where

$$g(x) = h_1(x)e^{x^2} + h_0(x),$$

and h_0 and h_1 are polynomials.

4. (a) Someone comes to you and says, "Look, look! These 150 pages of calculations show that I've found two even integers whose sum is odd." Suppose that you answer, "I don't have to look. You've made a mistake somewhere." The person then says, "What do you mean I've made a mistake? You haven't even looked at my work. How can you say that I've made a mistake? Here, you've got to read it, every page." What do you say then?

(b) Would your answer be any different if the person was claiming to have trisected the angle instead?

INFINITY: LIMITS AND INTEGRATION

by David L. Hull

Mathematics has depths, infinite depths. We will never get to the bottom of it and there will always be new things to do. Many people seem to have the idea that the subject is all finished and everything is nicely wrapped up in textbooks, but that is not the case at all. The subject is in a constant state of ferment, boiling and bubbling, heaving and seething, going off in many new directions.

When students have mastered the idea of the Riemann integral, they may think, "Well, that finishes that: what's next?" But the integral is not finished. Besides the Riemann integral, there is the Stieltjes integral, the Perron integral, the Denjoy integral. What could they possibly be like? Isn't the integral the integral? How can you find the area under a curve any other way? Students may not realize that the integral has depths, perhaps infinite.

Infinity itself has infinite depths, or heights if you want to look up instead of down. There are different sizes of infinity, infinitely many of them, each infinitely larger than all of the rest. What a triumph of the finite human mind, confined in a finite universe, to be able to deal with infinities of infinities! The universe may be finite, but mathematics is infinite.

The following selection explains how an integral can be different from the Riemann integral, and how there can be more than one size of infinity.

"A great people, that cannot be numbered nor counted for multitude." King Solomon (1 Kings, 3:8)

"How do I love thee? Let me count the ways." E. B. Browning

What is infinity? To the lay person, infinity might represent some largest number, perhaps the cosmos, or simply all there is. Kings, artists, poets, philosophers have pondered infinity during most, if not all, of recorded history. In calculus, you have encountered infinity and its inverse, the infinitesimal. The two main problems of calculus are the tangent problem and the area problem. How does one find the tangent to a given curve at a given point? For a positive function, how does one find the area under the graph of the function between two points on the x-axis? The solution to these problems involves infinity and infinitesimals.

The student of calculus has seen algebra, geometry, and trigonometry in his/her preparation. Calculus is in some ways more of the same. Thus, one might ask, "What else is there?". This essay gives an intuitive approach to some topics from advanced mathematics. Would you believe there is more than one type of infinity?

The creation of mathematics is a human activity which takes place within the context of a culture. The integral you study in calculus is called the Riemann integral which is named for Bernhard Riemann (1826-1866). He gave his formal definition of the integral in 1854. As you have seen, the Riemann integral provides a solution to the area problem for continuous functions. It was smooth functions which were of major interest to mathematicians in the late 1800s. By this period in time mathematicians were doing delta-epsilon proofs, and they had come to view a function as a pointwise correspondence or a rule for associating elements in one set with elements in another set. Most new functions were invented for the purpose of solving practical problems.

Younger mathematicians began to contrive functions as counterexamples to theorems which were thought to be generally true. The mathematical establishment regarded such functions as pathological because it took strange conditions to violate theorems which the mathematical community regarded as valid. These young mathematicians had begun to upset the apple cart of conventional mathematics. (See Chapter 28 of Boyer and Merzbach for a discussion of mathematical activity and comments and attitudes of established senior mathematicians of the late nineteenth century.)

The late nineteenth century was an era in which mathematical ideas could arouse fervor in the intellectual community. The power structure was vehemently concerned that younger mathematicians were working on the wrong problems. Two important young mathematicians were Georg Cantor and Henri Lebesgue, and their ideas were fiercely attacked and severely criticized. If you reflect a moment on human nature, you might

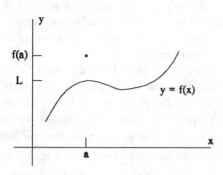

Figure 1

begin to imagine the strong feelings of anxiety, doubt, and insecurity that were felt by Cantor and Lebesgue. Nevertheless, they persevered, and they were able to lay the foundation for some of the most general theories of twentieth century mathematics. Cantor came up with a rigorous definition of infinity, and Lebesgue invented a new type of integral. The following paragraphs of this essay give an introduction to some of their contributions to mathematics.

The notion of the limit of a function f as x approaches a, has to do with infinitesimals. We are concerned with the behavior of $f(x)$ when x is near to a, but x is not equal to a. Intuitively, $\lim_{x \to a} f(x) = L$ means that when x is very close to a, f has a value which is close to L. As Figure 1 illustrates, $f(a)$ need not be equal to L. In the case where $\lim_{x \to a} f(x) = f(a)$, we say f is continuous at $x = a$. Continuity of a function f on an interval $a < x < b$ has to do with smoothness of a function over the interval. If a function f is sufficiently smooth over the interval $a \leq x \leq b$, then the area under its curve can be defined as the Riemann integral of f over the interval from a to b.

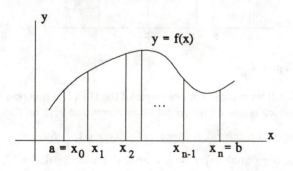

Figure 2

One way of defining the Riemann integral has to do

with upper and lower Riemann sums. Let f be nonnegative and continuous on $a \leq x \leq b$ with $a < b$. We partition the interval $[a, b]$ by a set $P_n = \{x_0, x_1, ..., x_n\}$ with $x_i < x_{i+1}$ for $i = 0, 1, ..., n - 1$, as in Figure 2. U_n, the upper sum corresponding to P_n, is defined as the sum of the area of the largest rectangles.

$$U_n = \sum_{k=1}^{n} f(x_k^*)(x_k - x_{k-1}),$$

where x_k^* is chosen so that $f(x_k^*)$ is the maximum of f for $x_{k-1} \leq x \leq x_k$. Similarly, there is a lower sum L_n corresponding to P_n,

$$L_n = \sum_{k=1}^{n} f(x_k')(x_k - x_{k-1}),$$

where x_k' is chosen so that $f(x_k')$ is a minimum of $f(x)$ for $x_{k-1} \leq x \leq x_k$.

Next we define the mesh of the partition P_n as the maximum length of a subinterval $x_k - x_{k-1}$, where k ranges from 1 to n. If the mesh of P_n goes to zero as n goes to infinity and $\lim L_n = \lim U_n$, where the limit is taken over all possible partitions, we define this limit to be the Riemann integral of f over the interval $[a, b]$.

Notice that we are trying to add up infinitely many infinitesimals in a meaningful way. Perhaps you have seen the work of the artist M. C. Escher. His "Smaller and Smaller" is an example of how he perceived the process of adding infinitely many infinitesimals.

Let us consider a bounded nonnegative function which does not have a Riemann integral. Perhaps you have worked with this function in your calculus class. It is called the Dirichlet function and is defined by

$$f(x) = \begin{cases} 0 & \text{if } x \text{ is rational and } 0 < x < 1 \\ 1 & \text{if } x \text{ is irrational and } 0 < x < 1. \end{cases}$$

A little reflection tells us that between every two rational numbers there is an irrational number, and between every two irrational numbers there is a rational number. We cannot actually draw a graph of this function since the width of a dot made by our pencil would cause us to get two straight lines.

Notice that for any partition P_n, there is always a rational number and an irrational number in any subinterval. Thus the upper sum $U_n = 1$ and the lower sum $L_n = 0$ for every partition P_n, and $\lim L_n = 0$ while $\lim U_n = 1$.

Why does this function not have a Riemann integral? Because it is wildly discontinuous. We introduce another definition of continuity called the sequential definition. The function $f(x)$ is continuous at $x = a$

provided that for every sequence of points $x_1, x_2, \ldots, x_n,$ ... such that $\{x_n\}$ converges to a, $\lim_{n \to \infty} f(x_n) = f(a)$.

To show that f is not continuous at any a in the closed interval [0, 1], first let a be a rational number, and let $\{x_n\}$ be a sequence of irrationals which converges to a. Picking $x_n = a + \pi/2n$ would generate such a sequence. Now $f(a) = 0$ but $f(x_n) = 1$ for each n, so $\lim_{n \to \infty} f(x_n) = 1$. Therefore, f is not continuous at $x = a$. A similar argument shows that f is not continuous at a if a is an irrational number in the interval [0, 1].

You might step back and think "Gee, this function takes on only two values, perhaps there is a way to define its integral." If we knew the mass of the irrational numbers and the mass of the rational numbers, we might let m denote a mass function and define

$$\int_0^1 f(x)\,dm(x) = 0 \cdot m(\text{rationals}) + 1 \cdot m(\text{irrationals}).$$

The above guess has to do with measure and Lebesgue integration.

The notion of a measure is a generalization of the length of a line. For example, if $a < b$, the measure of the open interval (a, b) could be defined as $b - a$. Or if $A = \{x \mid 1 < x < 2 \text{ or } 3 < x < 3.5\}$ we could define the measure of A to be $2 - 1$ plus $3.5 - 3$ or $m(A) = 1.5$. What properties do we want from a measure?

1) $m(\emptyset) = 0$. The measure of the empty set is zero.
2) $m(A) \geq 0$, for a set A.
3) If $A \cap B = \emptyset$, $m(A \cup B) = m(A) + m(B)$, finite additivity.

4) $m\left(\bigcup_{n=1}^{\infty} A_n\right) = \sum_{n=1}^{\infty} m(A_n)$ if $A_i \cap A_j = \emptyset$ for $i \neq j$.

This property is called countable additivity.

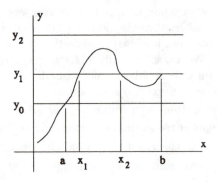

Figure 3

One can regard a measure as a nonnegative, countably additive set function. A measure is defined on a

family of sets. It is possible to define some strange sets of real numbers, and exactly what type of sets are measurable is best left to a course in real variables. For Lebesgue measure, we want the measure of an open interval to be its length.

Henri Lebesgue (1875-1941) discovered a new type of integral. His idea was to partition the y-axis rather than the x-axis. For example (see Figure 3), we might start with a partition of the y-axis into two subintervals $y_0 \leq y < y_1$, and $y_1 \leq y < y_2$. and then project functional values onto the x-axis. Let $I_1 = \{x \mid y_0 \leq f(x) < y_1\}$ and $I_2 = \{x \mid y_1 \leq f(x) < y_2\}$. Notice I_1 is the union of the disjoint intervals $[a, x_1)$ and $[x_2, b)$, while I_2 is the interval $[x_1, x_2)$.

Next form the step function (see Figure 4)

$$h(x) = \begin{cases} y_0 & \text{if } x \text{ is in } I_1 \\ y_1 & \text{if } x \text{ is in } I_2. \end{cases}$$

The integral of h is defined to be

$$\int_a^b h(x)\,dm(x) = y_0 \cdot m(I_1) + y_1 \cdot m(I_2),$$

which is the sum of the area of the rectangles 1, 2, and 3.

Lebesgue's idea was to successively refine the partitions of the y-axis, letting the lengths of the subintervals go to zero. If the areas under the corresponding step functions approached a limit, this limit was defined to be the Lebesgue integral.

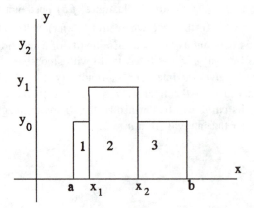

Figure 4

If we return to our example of the Dirichlet function (which is one on the irrational numbers and zero on the rational numbers), we see that for $0 \leq x \leq 1$, $\int_0^1 f(x)\,dm(x)$ $= 0 \cdot m(\text{rationals}) + 1 \cdot m(\text{irrationals}) = m(\text{irrationals})$.

Now we need to find the measure of the irrational numbers in the unit interval. To do this we first find the measure of the rational numbers and subtract the result from 1. Actually, we need to find out how many rational numbers there are. Of course, there are infinitely many rational numbers, but Georg Cantor (1845-1919) came up with the first operational definition of infinity. He said a set was countably infinite if it could be put into one-to-one correspondence with the positive integers. Cantor discovered a proof that the positive rational numbers are countable. His proof involves a diagonal process and a diagram (Figure 5). If one follows along the above path and crosses out duplicates, one will establish a one-to-one correspondence with the positive integers.

How to measure the rationals in [0, 1]? First, enumerate them, say r_1, r_2, r_3, ... The rational numbers are distinct, so there is no overlap. Let $R = \{r_1\} \cup \{r_2\} \cup$ ... be the set of rationals in [0, 1]. Now, by countable additivity,

$$m(R) = m\left(\bigcup_{n=1}^{\infty} \{r_n\}\right) = \sum_{n=1}^{\infty} m(\{r_n\}).$$

Each rational number each should weigh the same amount. If we assign any positive value to the measure of a singleton rational, the Archimedean property of the

$$
\begin{array}{llll}
1/2 & 2/2 \rightarrow 3/2 & 4/2 \rightarrow 5/2... \\
\downarrow & \nearrow \swarrow & \nearrow \swarrow \\
1/3 & 2/3 & 3/3 & 4/3 & 5/3... \\
 & \swarrow & \nearrow \swarrow \\
1/4 & 2/4 & 3/4 & 4/4 & 5/4... \\
\downarrow & \nearrow \swarrow \\
1/5 & 2/5 & 3/5 & 4/5 & 5/5...
\end{array}
$$

Figure 5

real numbers implies that $m(R) = \infty$. Thus, we are forced to define the measure of each individual rational number to be zero. Countable additivity of a measure implies that $m(R) = \sum_{n=1}^{\infty} 0 = 0$. Thus, the rationals carry no mass in the interval [0, 1]. What about the irrational numbers? One might be tempted to reason as follows. Since there is a rational number between every two irrational numbers, the number of rationals must be the same as the number of irrationals. Such intuition does not produce correct results.

We see that the measure of the irrational numbers in [0, 1] must be one, since $1 = m([0, 1]) = m(\text{rationals}) + m(\text{irrationals})$. We also see that the irrational numbers are uncountable. That means the irrational numbers in [0, 1] cannot be put into one-to-one correspondence with the positive integers. Why? If the irrationals were countable, then countable additivity of Lebesgue measure would mean $m(\text{irrationals}) = 0$, and $1 = m([0,1]) = m(\text{rationals}) + m(\text{irrationals}) = 0 + 0 = 0$, a contradiction.

Cantor gave a different proof for the uncountability of the irrational numbers in the unit interval. Cantor's

ideas led to cardinal numbers. Two sets have the same cardinality if the sets can be put into one-to-one correspondence. The first two cardinal numbers are aleph-null and c, where c represents the cardinality of the continuum and aleph-null represents the cardinality of the counting numbers (positive integers). There are many cardinal numbers. For example, the set of all real valued functions defined on the unit interval has a cardinal number which is greater than c.

The work by Cantor and Lebesgue led to the development of modern fields of mathematics such as set theory, functional analysis, point set topology, and probability. These areas of mathematics are actively studied by research mathematicians, and new results and theorems emerge on a regular basis.

The author encourages you to do further browsing and reading. The references present a variety of sources, and some of them are fun and easy to read. The mathematics history book by Maor, and the one by Boyer are interesting and accessible. Maor has reproductions of the artwork of M. C. Escher. If you want to read more about probability, the books by Feller, Parzen, Ross, Gnedenko get progressively more difficult. Two books on real analysis which are worth a browse are the one by Hewitt and the one by Royden.

References

Carl Boyer, Uta Merzbach, *A History of Mathematics*, 2nd Edition, Wiley, 1989.

William Feller, *An Introduction to Probability Theory and Its Applications*, Volume I, 2nd Edition, Wiley, 1957.

B. V. Gnedenko, A. N. Kolmogorov, *Limit Distributions for Sums of Independent Random Variables*, Addison-Wesley, 1968.

Edwin Hewitt, Karl Stromberg, *Real and Abstract Analysis*, Springer-Verlag, 1965.

Eli Maor, *To Infinity and Beyond*, Birkhäuser, 1987.

Emanuel Parzen, *Modern Probability Theory and Its Applications*, Wiley, 1960.

Sheldon Ross, *Introduction to Probability Models*, 4th Edition, Academic Press, 1989.

H. L. Royden, *Real Analysis*, 3rd Edition, Macmillan, 1988.

EXERCISE

1. "We also see that the irrational numbers are uncountable." Are they? Suppose that I choose to count all of the numbers from 0 to 1 as follows:

$$
\begin{aligned}
1 &-- .1 \\
2 &-- .2 \\
&\ldots \\
9 &-- .9 \\
10 &-- .01 \\
11 &-- .02 \\
&\ldots \\
109 &-- .99 \\
110 &-- .001 \\
&\ldots \\
1110 &-- .999 \\
1111 &-- .0001 \\
&\ldots
\end{aligned}
$$

Won't all the numbers get listed?

FALLACIES

Fallacies are arguments that look correct but lead to an absurd conclusion. They can be used to illustrate some important mathematical points, such as division by zero is not allowed and square roots of numbers are never negative. They can also be used to torment people who cannot find where the error in the argument is. This is harmless fun, as long as you sooner or later end the torment and point out where the error occurs. The only danger in fallacies is that someone will forget about the error but remember the conclusion and go through life with the vague impression that someone once proved that $1 = 2$, or that all triangles are isosceles. Since this is obvious nonsense, it may generate confusion about what mathematics can do and how it does it. However, we cannot let the fact that there are people with cloudy minds who will sometimes misinterpret what we say stop us from saying things.

The division by zero fallacy is usually easy to spot. In its simplest form, it goes like this: let

$$x = 1.$$

Multiply both sides of the equation by x:

$$x^2 = x.$$

Subtract:

$$x^2 - x = 0.$$

Factor:

$$x(x - 1) = 0.$$

Divide by $x - 1$:

$$x = 0.$$

So, the conclusion is that $1 = 0$.

The error comes when you divide by $x - 1$. You can't divide by zero, and at the start we let $x = 1$. The reason you can't divide by zero is that if you could, you could prove, just as above, that $1 = 0$. If $1 = 0$, then all of mathematics collapses, so we do not divide by zero. The division-by-zero fallacy not always as obvious as in this example, but it cannot be completely disguised. Whenever you see "Now divide by ... " that is your cue to check and see if the divisor is zero. If it is, nonsense will result.

The square-root fallacy will get by most people, probably because they have not been sufficiently impressed with the fact that, by definition, square roots of positive numbers are positive. Here is a simple form of the square-root fallacy, proving that if the sum of two numbers is 12, both numbers have to equal 6.

Let

$$a + b = 12.$$

Multiply by $a - b$:

$$(a - b)(a + b) = 12(a - b),$$

$$a^2 - b^2 = 12a - 12b.$$

Transpose terms:

$$a^2 - 12a = b^2 - 12b.$$

Add 36:

$$a^2 - 12a + 36 = b^2 - 12b + 36.$$

Perfect squares!

$$(a - 6)^2 = (b - 6)^2.$$

Thus

$$a - 6 = b - 6$$

so

$$a = b.$$

Since $a + b = 12$, it follows that $a = b = 6$.

The error is that the square root of x^2 is *not x*, but $|x|$. Thus, what follows from

$$(a - 6)^2 = (b - 6)^2.$$

is

$$|a - 6| = |b - 6|.$$

That is true, but it does not allow you to conclude that $a = b$. What you can conclude is either $a = b$ or $a + b = 12$. That is also true. This fallacy is also easily spotted if you have the good habit of asking yourself, whenever a square root is taken, if your square root is nonnegative, as it must be.

Fallacies are not the same thing as *howlers*, which are examples of illegal operations giving correct results. For example, you do not solve

$$\ln 3 + \ln x = \ln \frac{9}{2}$$

for x by "cancelling" the logarithms, but if you did you would get

$$3 + x = \frac{9}{2},$$

and so

$$x = \frac{3}{2},$$

which is correct:

$$\ln 3 \quad\;\; = 1.098612$$
$$\ln(3/2) = 0.405465$$

$$\ln 3 + \ln(3/2) = 1.504077 = \ln(9/2).$$

Nor do you simplify 16/64 by cancelling the 6, but if you do you get the correct answer.

Those are howlers, because they make teachers howl. A more complicated example, one that actually occurred, is this "solution" of

$$x^2 + (x + 4)^2 = (x + 36)^2.$$

Here we go:

$$x^2 + x^2 + 4^2 = x^2 + 36^2 \qquad \text{(No, it doesn't.)}$$

$$x^2 + x^2 + 16 = x^2 + 336 \qquad (36^2 = 3(6^2).)$$

$$x^2 + x^2 - x^2 = 336 - 16$$

$$x^4 = 320 \qquad \text{(I don't understand that either.)}$$

$$x = 80 \qquad \text{(Fourth roots are easier that way.)}$$

And there we are, with an answer that checks:

$$80^2 + 84^2 = 6400 + 7056 = 116^2 = (80 + 36)^2,$$

correct.

Here is another howler. This one probably never happened, but was made up by a clever author. The problem is to solve the system

$$\frac{y - 1}{x - 2} = \frac{3}{5},$$

$$\frac{y + 1}{x - 5} = \frac{5}{2}.$$

The first thing we do is forget about the second equation—who needs it? So, just using the first, we have

$$\frac{y}{x} - \frac{1}{2} = \frac{3}{5},$$

so

$$\frac{y}{x} = \frac{1}{2} + \frac{3}{5} = \frac{4}{7},$$

from which it immediately follows that $y = 4$ and $x = 7$. See, I told you that the second equation wasn't needed. Of course, good students always check their work, so let us substitute:

$$\frac{4 - 1}{7 - 2} = \frac{3}{5} \quad \text{and} \quad \frac{4 + 1}{7 - 5} = \frac{5}{2}.$$

Right! Some fussy pedants might say that we can't just ignore the second equation, so to satisfy them, we had better solve it too:

$$\frac{y}{x} + \frac{1}{-5} = \frac{5}{2},$$

$$\frac{y}{x} = -\frac{1}{-5} + \frac{5}{2} = \frac{-1}{5} + \frac{5}{2} = \frac{4}{7},$$

so, just as before, $y = 4$ and $x = 7$, proof that the first equation has been solved correctly. Clever indeed!

Howlers can be amusing, but they do not illustrate mathematical points the way that fallacies do. They can provide some mathematical activity, though—it can be a challenge to find out under what circumstances their erroneous procedures can give correct results. For which a and b can

$$\ln a + \ln x = \ln b$$

be solved by "cancelling" the $\ln$s? Are there any other two-digit numbers like

$$\frac{16}{64} = \frac{1}{4}$$

where "cancelling" the 6s gives a correct result? It would be nice if

$$\frac{d}{dx} uv = \frac{du}{dx} \frac{dv}{dx},$$

were always true, but it isn't: for which u and v is it correct? The "identity"

$$\sqrt{a + b} = \sqrt{a} + \sqrt{b}$$

has been applied innumerable times: is it ever true for non-zero a and b? Such questions are worth looking at.

Calculus fallacies are not as common as algebraic or geometric ones, but there are some good ones. Here is how to show that $1 = 2$ using derivatives. Obviously,

$$3^2 = 3 + 3 + 3, \quad 4^2 = 4 + 4 + 4 + 4,$$

and so on. Thus,

$$x^2 = x + x + \ldots + x,$$

where the sum has x terms. Differentiate both sides:

$$2x = 1 + 1 + \ldots + 1.$$

Since the right-hand side is the sum of x 1s, its value is x, so

$$2x = x,$$

and so $2 = 1$. It is easy to see where the flaw in that argument is, namely that x^2 is a sum of x xs only when x is an integer, and the mathematical principle involved, that being discrete is different from being continuous, is not difficult to grasp.

The next fallacy, $0 = 1$ again, is better educationally. Someone who is unable to explain what is wrong with the "proof" hasn't fully grasped the idea of anti-differentiation. The conclusion follows from an application of integration by parts:

$$\int u \, dv = uv - \int v \, du.$$

Suppose that we do not know that the antiderivative of $1/x$ is $\ln x$, and that we want to use integration by parts to find

$$I = \int \frac{1}{x} dx.$$

Let

$$u = \frac{1}{x}, \qquad dv = dx$$

so

$$du = -\frac{1}{x^2} dx, \qquad v = x$$

and thus

$$I = \left(\frac{1}{x}\right) x - \int x \left(-\frac{1}{x^2}\right) dx = 1 + \int \frac{1}{x} dx = 1 + I.$$

Subtract I from both sides and you get $0 = 1$. To explain the fallacy, do I need to give the hint that antiderivatives always come with "+ C" attached? This fallacy shows why the C cannot be left off.

Another simple fallacy with a similar educational point is the following one. Let us differentiate $1/(1 - x)$ by writing it as $(1 - x)^{-1}$ and applying the power and chain rules:

$$\frac{d}{dx}(1 - x)^{-1} = (-1)(1 - x)^{-2}(-1) = \frac{1}{(1 - x)^2}.$$

Now let us differentiate $x/(1 - x)$ by applying the quotient rule:

$$\frac{d}{dx} \frac{x}{1 - x} = \frac{(1 - x)(1) - x(-1)}{(1 - x)^2} = \frac{1}{(1 - x)^2}.$$

The same derivative! But the functions we started with were different:

$$\frac{1}{1 - x} \qquad \text{and} \qquad \frac{x}{1 - x}$$

are certainly not equal. For instance, the first one is never zero, while the second one is zero when $x = 0$. In fact, the second one is x times the first one, and when you multiply a function of x by x, you get a new function. Something must be wrong! Different functions can't have the same derivative, can they? Maybe one of our methods of differentiation has a flaw in it.

Of course, neither method has a flaw. The reason that the derivatives are the same is that the functions really are the same, almost:

$$\frac{1}{1 - x} - \frac{x}{1 - x} = \frac{1 - x}{1 - x} = 1$$

so

$$\frac{1}{1 - x} = \frac{x}{1 - x} + 1,$$

and functions that differ by a constant have the same derivatives. Understanding this fallacy drives that point home.

Most of the preceding material appears in *Fallacies in Mathematics* by E. A. Maxwell. Before going to another source, here is one more from Maxwell's book, also illustrating a mathematical point. Everyone knows that

$$\int_0^2 f(x) \, dx = \int_0^1 f(x) \, dx + \int_1^2 f(x) \, dx.$$

Let us now make a change of variable in the first integral: let

$$x = 2y, \qquad dx = 2 \, dy,$$

so, after remembering to change the limits of integration,

$$\int_0^2 f(x) \, dx = 2 \int_0^1 f(2y) \, dy.$$

It does not matter what we call the dummy variable of integration, so changing it back to x gives

$$2 \int_0^1 f(2x) \, dx = \int_0^1 f(x) \, dx + \int_1^2 f(x) \, dx.$$

Now suppose that f has the property that $2f(2x) = f(x)$ for all x. Then the integral on the left is the same as the first integral on the right, and we get

$$0 = \int_1^2 f(x) \, dx.$$

If this were a pretentious textbook, this could be put in

the form of a

Theorem: If f is an integrable function such that

$$f(2x) = \frac{1}{2}f(x) \quad \text{for all } x, \text{ then} \int_1^2 f(x)\, dx = 0.$$

Readers would look at it, think, "That's nice," and pass on. However, it is always a good idea to look at specific cases of general results. $f(x) = 1/x$ is a function that has the property that $f(2x) = (1/2)f(x)$ for all x. Thus, the theorem says that

$$\int_1^2 \frac{1}{x}\, dx = 0.$$

But

$$\int_1^2 \frac{1}{x}\, dx = \ln 2 - \ln 1 = \ln 2.$$

Thus, ln 2 = 0. But my calculator says that ln 2 = .69314718... and, as a check, my computer algebra system tells me that

ln 2 = .69314718055994530941712321214581765...

and neither of those looks like zero. So, the theorem is no good. True theorems cannot give false consequences. The question is why the theorem is no good for this function, and the answer is that $\int_0^1 \frac{1}{x}\, dx$ is an improper integral that does not converge. That integral was used in the proof, and if it does not exist it cannot be used, so there is no proof. The mathematical lesson to be drawn from this fallacy is to be careful with improper integrals and, before using one, be sure that it exists. A good exercise is to go through the proof of the theorem with proper integrals (start with a lower limit of ε instead of 0) and see that the fallacy disappears.

There follow some more fallacies, more or less complicated but each with a mathematical lesson, drawn from *Riddles in Mathematics* by Eugene P. Northrop. First we use l'Hôpital's Rule to show that 1 is infinitely large. Everyone who knows about geometric series knows that

$$\frac{1}{1 - x^3} = 1 + x^3 + x^6 + x^9 + \ldots$$

and so

$$\frac{x^2}{1 - x^3} = x^2 + x^5 + x^8 + \ldots .$$

Hence

$$\frac{1 - x^2}{1 - x^3} = 1 - x^2 + x^3 - x^5 + x^8 - x^9 + \ldots .$$

Now let us take the limit of both sides of that equation as x goes to 0. The right-hand side approaches 1. Let us use l'Hôpital's Rule on the left-hand side:

$$\lim_{x \to 0} \frac{1 - x^2}{1 - x^3} = \lim_{x \to 0} \frac{2x}{3x^2} = \lim_{x \to 0} \frac{2}{3x} = \infty.$$

And so $1 = \infty$.

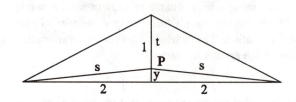

Figure 1

Now that you have learned not to trust l'Hôpital's Rule, we can go on to try to convince you that the usual process for finding maximums and minimums—set the derivative equal to zero—does not work either. The problem is to find the point P on the altitude of the isosceles triangle (Figure 1) so that the sum of the distances from the three vertices is a minimum. That is, we want to minimize

$$D = 2s + t.$$

As always in maximum-minimum problems, we want to get everything in terms of one variable. Since

$$2^2 + y^2 = s^2$$

we have

$$s = \sqrt{4 + y^2}, \quad t = 1 - y,$$

and so

$$D = 2\sqrt{4 + y^2} + (1 - y).$$

Now we differentiate:

$$\frac{dD}{dy} = 2 \cdot \frac{1}{2} \cdot \frac{1}{\sqrt{4 + y^2}} \cdot 2y - 1.$$

Now we find when the derivative is zero:

$$\frac{2y}{\sqrt{4 + y^2}} = 1, \qquad 2y = \sqrt{4 + y^2},$$

$$4 + y^2 = 4y^2, \quad 3y^2 = 4, \quad y^2 = \frac{4}{3}, \quad y = \frac{2}{\sqrt{3}}.$$

Whoops! That would put the point P up *above* the vertex at the top, hardly a likely place for a minimum of the sum of the distances from P to the three vertices. It doesn't look like a maximum, either. Nor does taking y to be negative and putting P below the base of the triangle seem to give the minimum that we are after. Something odd is going on.

Speaking of something odd, everyone knows that the series expansion for sin x has only odd powers of x in it:

$$\sin x = x - \frac{x^3}{3!} + \frac{x^5}{5!} - \frac{x^7}{7!} + \ldots$$

However, everyone also knows that

$$\sin^2 x = 1 - \cos^2 x,$$

so

$$\sin x = (1 - \cos^2 x)^{1/2}.$$

Now use the binomial theorem to expand that as

$$\sin x = 1 - \frac{1}{2}\cos^2 x - \frac{1}{8}\cos^4 x - \frac{1}{16}\cos^6 x - \ldots$$

But

$$\cos x = 1 - \frac{x^2}{2!} + \frac{x^4}{4!} - \frac{x^6}{6!} + \ldots.$$

When we substitute that into the series for sin x in terms of cos x, do you see what we will get? Raise cos x to any even power and all the powers of x in the series expansion will be even. Add together any combination of even powers of cos x and the sum will have only even powers of x in it. So, the series expansion for sin x has nothing but even powers of x in it. But we know that it has only odd powers of x in it. That is also odd.

EXERCISES AND QUESTIONS

1. Find for which a and b the equation
$$\ln a + \ln x = \ln b$$
can be solved correctly by cancelling the *ln*s.

2. What's wrong with the application of l'Hôpital's Rule that showed that 1 and infinity were the same?

3. Why did the maximum-minimum problem come up with a wrong answer? Where *is* the minimum sum?

4. Does the series expansion for sin x have all even powers of x in it? If so, what are their coefficients? If not, why not?

5. The point in favor of fallacies is that by understanding one, a person understands better the mathematical principle whose violation made the fallacy possible. A point against them is that many people will not be able to understand why they are fallacious, become confused, and tend to think that mathematics is also confused and can contain error. Which point is more important? Should fallacies be kept from people until it is certain that they will be able to understand them?

6. It is a truth of logic that from a false statement any statement will follow. For example, "If the moon is made of green cheese, then pigs have wings" is a true theorem. Show that from $1 = 0$ it follows that $2 = 1$, that $e^2 = \pi$, and that $x^2 = x$ for all x. Can you prove that if $0 = 1$ then pigs have wings?

MASTERING THE MYSTERIES OF MOVEMENT

by David Bergamini and the editors of *Life*

Every now and then, someone decides that calculus is not just for students who need it for something but should be for everybody. Not the details or the calculations, of course; what people ought to appreciate, someone thinks, are the ideas behind the subject, what it is for, and what it can do. Quite right, that someone is. Calculus is an important part of our science and technology, hence part of our culture, and every educated person should know a little about everything that is important. Someone then writes a popular account of calculus and puts it before the public.

Popular calculus, and more generally popular mathematics, is difficult to write. One reason for that is that mathematics is in part a language and it is not easy to tell someone who knows nothing about a language anything significant about it. How do you explain to someone who speaks only English, a language almost without cases, about the fifteen separate and distinct cases in Finnish? You have to give *examples*, but your reader will make nothing of them, so you can't. You work under great difficulties. It is the same with mathematics. You have to give examples, but you can't. Another reason why it is hard to popularize mathematics is that there is usually nothing to point to. The other sciences have things easier. In physics, you can point at the newest giant particle accelerator, thousands of feet in diameter, or thousands of feet long, hurling protons or something at targets at 99.9% of the speed of light. Zap! Pow! It is impressive. In astronomy, there are new telescopes looking yet more billions of light-years into deepest space and bringing back amazing news about neutron stars, supernovas, and black holes. Gosh! Wow! It is impressive. In medicine, there are new devices to look inside of people and new machines that use lasers and ultrasound to make wrong things right. They too are impressive. Even geology has impressive new ways of getting at oil. But pity poor mathematics! It has no gadgets, nothing to show and tell about, no things that are impressive. Advances in mathematics are new ideas, and ideas are invisible. Big news in mathematics is made when a big problem is solved, but what the problem was about is nothing that can be seen or touched, so the solution does not mean very much to a member of the general public. Thus, the only mathematical news that you are likely to see in a newspaper is the news that some computer has found a new gigantic prime, or that some other computer has computed π to another few million or billion decimal places. That is all very well—news is news—but it is not news about real mathematics. It is very, very difficult to write about mathematics for a general audience.

The following selection, whose author had the help of the *Life* magazine organization, is an attempt to popularize calculus. It is well done. I should not have said that, since after reading it there is a possibility that you will not agree, and thus feel disappointed or let down, but people familiar with other attempts to write about calculus without actually using any calculus (and including only one equation) agree that it is hard to imagine anyone, or any team, doing better. If you are not impressed, it is not because of a deficiency in you but instead it is yet one more confirmation of the unfortunate reality that it is difficult to make calculus popular.

Nothing in the world is immune from change. The hardest rock on the driest desert expands or contracts in the shifting sunlight. The steel gauge blocks at the National Bureau of Standards, even though they are stored in temperature-controlled subterranean vaults, are subject to seasonal fluctuations in length thought to be caused by radiation from the surrounding walls. Everything grows or shrinks, warms up or cools down, changes its position, its color, its composition—perhaps even its spots.

Inescapable as the process of change is, and vital as it is to understanding the laws of nature, it is difficult to analyze. Being continuous, it offers no easy point at which the mind can catch it and pin it down. For centuries it baffled mathematicians. Some starts, to be sure, were made toward a mathematics of movement. The Greeks did so when they thought of curves as tracings made by moving points, and when they analyzed curving lines, instant by instant, through the techniques of slicing them into infinitesimally fine segments. Descartes did so when he conceived of the items in an equation as functions between variables, and most of all when he supplied a way to draw graph-pictures of fluid situations and relationships. But by and large the world of mathematics was populated by waxworks—shapes and numbers that stood stock-still.

Then, in 1665 and 1666, England's incomparable Isaac Newton produced a prodigious brain child, now called calculus, which for the first time permitted the mathematical analysis of all movement and change. In calculus Newton combined the fine-slicing technique of the Greeks and the graph system of Descartes to devise a marvelously automatic mental tool for operating on an equation in order to get at infinitesimals. So quickly did calculus prove its effectiveness that in a few years its creator used it to work out the laws of motion and gravitation—the fundamental laws of physics which explain why the solar system acts as it does, or why any moving object reacts as it does to outside forces like gravity, the tension of a spring or the push of a man's hand. By its ability to probe the fleeting mysteries of movement, calculus today has become the principal pipeline between practical science and the reservoir of mathematical thought. Every airplane, every television set, every bridge, every bomb, every spacecraft owes it a tithe of indebtedness.

The different kinds of change which calculus can analyze are as diverse as a queen's wardrobe. If the factors involved in any fluid situation can be put in terms of an equation, then calculus can get at them and uncover the laws they obey. The change under scrutiny may be as dramatic as the gathering speed of a missile lifting from its pad or as quiet as the varying grade of a mountain road. It may be as visible as the pounds added around a once-svelte waistline or as invisible as the ebb and flow of current in a power line. It may be as audible an the crescendo of a Beethoven concerto or as silent as the build-up of flood force behind a dam.

Calculus analyzes all these situations by invoking two new mathematical processes—the first fundamental operations to be added to the canon of mathematics since the laws of addition, subtraction, multiplication, division and finding roots were laid down some 4,000 years ago. These new operations are called *differentiation* and *integration*, and they are the reverse of each other in much the same way that subtraction is the reverse of addition, or division of multiplication. Differentiation is a way of computing the rate at which one variable in a situation changes in relation to another at any point in a process—at a given instant in time, for example, or at a given point in space. The actual method employed in differentiation is to divide a small change in one variable by a small change in another; to let these changes both shrink until they approach zero; then—and this is the key—to find the value which the ratio between them approaches as the changes become indefinitely small. This value is what mathematicians call a "limit," and it is the answer they are seeking, the end result of differentiation—the rate of change at a given instant or point. Integration works back the other way from differentiation; it takes an equation in terms of rate of change and converts it into an equation in terms of the variables that do the changing.

Through differentiation, a mathematician can probe deep into a fluid situation until he finds some unchanging factor that reflects the action of a constant law of nature. In this fashion Newton and later theorists made a discovery which is still not easy for laymen to absorb. This discovery was that the constant factor in many processes of nature is the rate at which a rate of change changes. Deciphering this seeming double talk may appear hopeless. But anyone who drives a car is familiar with the rate of change of a rate of change whether he realizes it or not. The speed of the car—so many miles per hour—is a rate of change of distance with respect to time. In speeding up or slowing down, the car's speed itself changes, and changes at a rate—acceleration or deceleration—which is the rate of change of the rate of change.

In nature, gravity acts to make a falling object move at a rate which increases at a constant rate. For processes involving actual physical movement, Newton defined this rate of a rate as the *acceleration*. And he called the gravity causing it a *force*. He defined force in general as something which causes an object to accelerate. As applied through calculus, this definition—laid down three centuries ago—has enabled scientists to do no less than identify the three fundamental forces of the cosmos; the force of gravitation; the force of magnetism, or electric charge; and the force that binds together the atomic nucleus.

In contrast to the spectacular role that calculus has played in unlocking the secrets of the universe, the nomenclature surrounding differentiation and integration is woefully prosaic. The rate of change of a *y* or *x*, found by differentiation, is called a *derivative*—a derivative of *y* with respect to *x*, written *dy/dx*, or of *z* with respect to *y*, written *dz/dy*. A derivative's counterpart, found by integration, is called an *integral*, and is symbolized by ∫ an old-fashioned letter S, which was short, originally, for "sum" or "summation." When integration is performed on an equation written in terms of derivatives, it converts the equation back into one in which the *x* and *y* have doffed their rate-of-change disguises and resumed normal algebraic appearance.

A definition for dieters

The handles and hieroglyphs attached to the techniques of calculus may look alien, but the ideas behind them are easily recognized. Being a rate of change, a derivative means, simply, the speed of a process; so many miles per hour or feet per second if it refers to change in position; so many pounds per week if it refers to a triumph in dieting; so many geniuses per childbirth if it refers to I. Q. statistics; so much cornstalk per candle power of sunshine if it refers to the growth of a corn crop. The integral corresponding to each of these derivatives would be the miles traveled, pounds lost, geniuses gained or length of cornstalk grown.

When used abstractly in an equation, a derivative can most readily be thought of in terms of the curve which represents this equation on a graph. At any point, the curve is rising or falling at a rate of so many *y*-units per *x*-unit. This slope up or down is the exact geometric equivalent of the rate of change—the derivative—of *y* with respect to *x*. Engineers often express the grade of a hill, the pitch of a roof or the steepness of an airplane's climb in identical terms; as so much altitude gained per unit of horizontal distance traversed. But in these applications the slope is normally conceived of as being measured over some definite span of distance. In calculus, on the other hand, the derivative is thought of as an instantaneous slope at a single point on a curve.

That this elusive concept of instantaneous slope is no figment of the mathematical imagination can be seen in an artillery shell as it arcs toward target. At any single moment the shell is moving in a definite direc-

tion. This direction is an instantaneous slope with respect to the ground, a rate of change in the shell's altitude with respect to its horizontal position. In terms of a graph, the speed of the shell, moving up or down, can also be considered as an instantaneous slope on a curve—a rate of change in the shell's altitude with respect to the time that has been elapsing since the shell was fired. A mathematician would normally write such a derivative—the velocity of climb or fall, or the rate of change of vertical distance—as *dy/dt*, in which the *t* stands for time.

The counterpart of a derivative, an integral, can also be visualized in terms of a graph. Suppose that *y* equals some expression of *x* and that this equation is plotted as a curve. Then the integral of *y* is the area between the curve and the horizontal line, or axis, running along below it. Why this is so can be seen by imagining that the area under the curve is filled by a picket fence with a scalloped top. As the fence is being built, each new picket adds to the area of the fence. In fact, the height of each added picket is a measure of the rate at which the area of the fence is growing; a six-foot picket, for instance, adds twice as much area as a three-foot picket. The integral of the rate of change, therefore, must be the actual factor in the situation that is changing; namely, the area of the fence itself. The geometric equivalent of each picket is simply the height of a curve—the vertical, or *y* coordinate of each point on a curve. Integrating *y* must, therefore, give the total area under the curve.

Many of the most practical applications of calculus stem from integration's ability to sum up *y*-length pickets and determine areas. Through it a mathematician can determine the volume of all manner of irregular shapes, such as airplane fuselages or oil-storage tanks; he can also find the areas of curved surfaces—the amount of sheet metal in a molded car body of the lifting surface on the wings of a jet.

There is one major difficulty in the process of integration—a difficulty so enormous and so recurrent that most of our largest computers today have been built specifically to cope with it. This is the problem of so-called "boundary conditions." When the area of a picket fence is measured, the boundary conditions are established by the two pickets that mark the end of the fence. But there are no ends to many of the curves that represent equations. The area beneath this type of curve may be indefinitely large. To give it boundaries, the mathematician erects the equivalent of end posts to mark off the particular part of the area he is interested in. He then integrates the equation represented by the curve between these two verticals. Often, however, in the case of equations which are arrived at experimentally, the proper way to interpret such an equation cannot be found except by integrating it, and the boundary conditions necessary for integrating it cannot be found except by understanding how to interpret it. To avoid this impasse, the scientist, in effect, chooses arbitrary picket posts and lets a computer run off dozens of laborious solutions, which give him insight into the equation and the process of change which it symbolizes.

To work out the rules of calculus, Newton visualized what would happen if one point on a graph-curve slid down into a point nearby. As the slide begins, the average slope of the curve between the two points is the number of y-units separating them vertically, divided by the number of x-units separating them horizontally. As the slide continues, both distances in this fraction diminish toward zero and finally vanish when the two points merge. But this does not mean that the fraction itself vanishes. A ratio of 1:2, for instance, need not suddenly become zero just because its numerator and denominator become indefinitely small. When last heard from, as they disappear arm in arm into the fastnesses of infinity, the numerator may be one zillionth and the denominator two zillionths, but the ratio between them is still 1:2.

Finding the value which a fraction approaches as its numerator and denominator diminish toward nothingness is called "taking a limit." If the numerator equals half the denominator, the limit is one half. If the numerator equals 10 times the denominator, the limit is 10. As two points on a curve slide together, the vertical and horizontal distances between them remained coupled, even as they fade away, by the relationship of y to x expressed in the original equation of the curve. As they merge, therefore, the ratio of their distances approaches a definite limit which can be evaluated in terms of y and x. This limit, 1/2 or 10 or whatever it may be, is the slope of the curve at the precise spot where the two points merge—the rate of change of vertical y with respect to horizontal x or, put another way, the derivative of y with respect to x.

The subtle train of reasoning which enabled Newton to differentiate equations and find the derivative or limiting value of the ratio—written dy/dx or dy/dt—is the fundamental process of calculus. It can be roughly paraphrased as follows: in a developing situation, the difference between the state of affairs at one moment and the state of affairs at the next moment is an indication of how the situation is shaping up; and if the ratio of the net changes that take place between the two moments is evaluated as a limit—a limit approached when the interval between the two moments is imagined as diminishing toward zero—then that limit shows how fast developments are taking place. The logic of

calculus can be applied to moments of time, points on a curve, temperatures in a gas or any state of affairs which can be related by equations; the same rules of differentiation apply to all of them.

A simple gift from Galileo

The way these rules work, and the reason for their enormous usefulness, can best be illustrated by applying them to the classically simple equation $y = 16t^2$, first expressed in a much simpler form by the physicist Galileo Galilei. This brief, unpretentious expression is one of the most useful in all of physics because it shows how gravity acts on a freely falling object—an elevator run amuck, a hailstone or a jumper descending to ground. Since almost all movements and changes on earth are heavily influenced by gravity, the equation of free fall indirectly plays a part in innumerable human actions—from taking a step or lobbing a tennis ball to lifting a steel girder or launching an astronaut into orbit.

Timing an object as it falls from a given height is the most straightforward method of gauging the effects of gravity. It was this technique which Galileo used, about 1585, to arrive at his free-fall equation. According to legend, Galileo dropped small cannon balls from the colonnades of the leaning tower of Pisa. According to his own account, he used the less fanciful means of timing bronze balls as they rolled down a ramp. The results of Galileo's experiments subsequently led to the equation for free fall, $y = 16t^2$, with y representing the distance fallen in feet and t the elapsed time in seconds after the start of the fall.

By differentiating this equation twice—so as to shave away successive layers of change and inconstancy—Newton uncovered the essential nature of gravitation. Differentiating the equation once, he found that the speed with which a jumper is falling at any moment equals 32 times the number of seconds which he has been falling. Differentiating the equation a second time, he found that the jumper's acceleration—the rate of increase in his speed—is always 32 feet per second, every second.

The fact that in the free-fall equation acceleration equals a constant number, 32, indicates the end of the trail. This 32 need not be differentiated further; it does not change, and its rate of change is zero. It represents a law of nature: that every free-falling object falls to earth with a constant acceleration of 32 feet per second, every second.

Having ascertained this fact by calculus, Newton was able to set his mathematical sights far beyond the earth and to deduce the law of universal gravitation—one of the most important results ever to be achieved by mathematics. It is the law which governs the movements of all celestial bodies—from human beings in orbit to entire systems of stars.

Looking back with awe on what a little deduction could accomplish in the mind of Isaac Newton, later thinkers have ranked him as the greatest physicist and one of the greatest mathematicians the known. Albert Einstein wrote: "Nature to him was an open book, whose letters he could read without effort." Newton himself said: "I do not know what I may appear to the world; but to myself I seem to have been only like a boy playing on the seashore, and diverting myself in now and then finding a smoother pebble or a prettier shell than ordinary, whilst the great ocean of truth lay all undiscovered before me."

Newton began to use his astounding inventiveness while still a child, to build toys for himself, including a wooden water clock that actually kept time and a flour mill worked by a mouse. His brilliance did not really catch fire, however, until he read Euclid at the late age of 15. The story goes that he rushed impatiently on to Descartes' relatively abstruse *La Géométrie*. Thereafter his progress was meteoric. Five years later, while still a graduate student at Cambridge, he had already worked out the basic operations of calculus—the rules of integration and differentiation, which he called the laws of "fluxions and fluents."

Newton put together his great invention and applied it in a preliminary way to the problems of motion and gravitation in a two-year burst of creativity, while rusticating during the epidemic of plague which swept England in 1665 and 1666. In retrospect it seems as if the whole framework of modern science arose from his mind as miraculously as a jinni from a bottle. But as Newton himself said, he "stood on the shoulders of giants." Many men had wrestled with the same problems; it was his genius to fuse their separate inspirations. The twin processes of differentiation and integration in calculus, for instance, were rooted in two classic questions of Greek antiquity: how to construct a tangent line (a line that just touches a curve at a given point), and how to calculate an area which is bounded on one side by a curve. The problem of the tangent, or "touching", line, was equivalent to the problem of finding the slope of a curve at any point and therefore of finding the derivative of an equation. The area problem was equivalent to the problem of integrating the equation that gives the rate of growth of an area.

A wine keg of infinitesimals

By viewing any curve as a succession of infinitely short segments, or any area as an accumulation of infinitely fine slices, the Greeks—particularly Archimedes—had solved a number of specific problems concerning rates of change. Mathematicians of the 16th and 17th Centuries also used infinitesimal methods, though seldom with rigorous Greek proofs. Kepler, for instance, had employed infinitesimals to give vintners a formula for gauging the volume of wine kegs. In Descartes' time and in the 15 years after his death, his compatriot, Pierre de Fermat, and the Englishman, John Wallis, had begun to cast infinitesimals in the helpful analytic molds of equations. Then, in about 1665, Newton's professor at Cambridge, Isaac Barrow, became the first man to realize that the tangent problem and the area problem are two sides of the same coin—in effect, that integration is the reverse of differentiation.

When Newton first began to unite all these preliminary insights in the single well-knit structure of calculus, he showed Barrow some of his early results. Barrow was so enthusiastic that he generously let it be known about Cambridge that Newton had done what he himself had failed to do. A few years later, in 1669, when he was retiring, he was instrumental in getting Newton, then 26, appointed as his successor to the Lucasian professorship of mathematics at the university—one of the most desirable chairs of mathematical scholarship in the academic world. Thereafter, honors and inspirations came to Newton in a steady stream. Over the next four decades he formulated the law of gravitation and used it to explain the movements of the planets, moon and tides; analyzed the color spectrum of light; constructed the first modern reflecting telescopes; performed innumerable alchemistic experiments; tried to reconcile with Scripture the date of 4004 B. C., which was currently accepted as the time of Adam's creation; served as a member of parliament; was appointed warden and then master of the British mint; was knighted by Queen Anne in 1705 and was repeatedly elected president of Britain's select scientific club, the Royal Society, from 1703 until his death in 1727.

Strangely enough, Newton revealed his monumental discoveries to only a few of his scientific cronies. Many explanations have been given for his inordinate secretiveness. It has been said that he was always too busy with new ideas to find time to write up old ones, and that he had a passionate distaste for the wrangles and criticism which inevitably raged around scientific pronouncements in those days. Then, too, he was just not much of a talker. While he was in parliament, his only recorded utterance was a request to open the window. On one occasion the astronomer Edmund Halley came to him, after a discussion with England's most eminent scientists, to ask if he knew what path a planet would take around the sun if the only force influencing it was a force that diminishes according to the square of its distance from the sun. Newton immediately gave the answer: the path would be elliptical. When asked how he knew, he explained casually that he had worked out the problem years before as a graduate student. In other words, he had worked out one of the fundamental laws of the universe and told nobody about it. Encouraged by Halley to re-create his original calculations, he went on to produce his masterwork, the *Principia*.

Newton's *Principia* is generally recognized as the most influential, conclusive and revolutionary scientific work ever to appear in print. In it, he not only explained why the solar system works the way it does but also laid down the laws of dynamics which are still the chief ingredients of practical engineering physics—of missile shoots or thruway construction. Most of these laws Newton had worked out through calculus, but like Archimedes before him, he chose to present his finished work in universally understood mathematics—as a lengthy Greek proof, couched almost entirely in the terms of classical geometry.

Not even the skillful coaxings of Halley could convince Newton to publish his calculus—not, that is, until another mathematician, the German Gottfried Wilhelm von Leibniz, had independently re-created the entire mental machinery. Leibniz invented calculus 10 years after Newton, in 1675, and in 1684 published his account of it, 20 years before Newton was to give the first published explanation of his own version.

Like Newton, Leibniz was as successful and practical as the mathematics he originated. The son of a well-to-do university professor, he learned Greek and Latin by the age of 12, attended university, took a law degree, and went on to become the counsel to kings and princes in an illustrious career that sometimes verged on the shady. He traveled all over Europe tracing dubious lineages to establish the dubious rights of princelings to vacant thrones. He formulated many of our modern principles of international power politics—including the phrase "balance of power." During trips to Paris, he studied algebra and analytic geometry under the great optical physicist Christian Huygens. And while jogging along in coaches on diplomatic missions, he created new mathematics simply for pleasure, including his own version of calculus.

Although Newton accomplished far more with calculus than Leibniz did, Leibniz had a superior notation for it—one he polished so carefully that we still use it. It was Leibniz who first wrote derivatives as *dy/dx* or *dx/dy*—forms that suggest the fractional rate-of-change measurements to which they apply. (Newton wrote the derivative of *y* as $\dot{y}$ and the deriva-

tive of *x* as $\dot{x}$. The dots in Newton's symbolism led rebellious 19th Century Cambridge students to protest against the "dotage" of English notation and to advocate the "pure d-ism" of continental notation.)

Unfortunately, Newton and Leibniz in their later years became embroiled in a chauvinistic dispute as to who had invented what first. The result was that scholars on the continent, helped by Leibniz' notation, went on to develop calculus further, while English mathematicians, hampered by the less felicitous notation devised by Newton, foundered in a morass of perplexities.

Salute to a lion's paw

The supremacy of the continental approach did not emerge, however, while Newton was still alive. At least twice after the rivalry had broken out, Leibniz and his followers posed problems with which they hoped to stump Newton. Each time Newton tossed off the answers in a single evening after coming home from his work at the mint. One of these problems was a particularly devilish one: to find the shape of the curve down which a bead will slide under the influence of gravity so as to move from a higher point to a lower point in the shortest possible time. The problem was important as an early example of "maximizing-minimizing" questions which confront mathematicians today—maximizing industrial productivity or minimizing the amount of fuel required to get to the moon. Newton solved the problem overnight and transmitted his solution anonymously the next morning through the channels of the Royal Society. Upon its receipt, Johann Bernoulli, the disciple of Leibniz who had posed the problem, is reported to have said, "Tanquam ex ungue leonem," which, freely translated, means "I recognize the lion by his paw."

Logicians of the next generation sharply criticized both Newton and Leibniz for having used the equivalents of infinitesimals—for having added up nothings to create the somethings of areas, and for having shaved down rates of change to instantaneous slopes measured in no time at all. The Irish metaphysician Bishop Berkeley, in an essay entitled "The Analyst," examined the logic of Newton's "fluxions" and concluded, "They are neither finite quantities, nor quantities infinitely small, nor yet nothing. May we not call them the ghosts of departed quantities ...?" Mathematicians of the 19th Century were to satisfy such critics by invoking new standards of rigor for calculus. But meanwhile it met the test of success—it worked. Scientific problems capitulated to it as the walls of Jericho to Joshua's trumpets. Indeed the chief danger was one of self-satisfaction. Using calculus, scientists explained every process in nature as a sequence of actions and reactions, of causes and effects. ...

EXERCISES AND QUESTIONS

1. "It $[y = 16t^2]$ shows how gravity acts on a freely falling object—an elevator run amuck, a hailstone or a jumper descending to ground." The first and last maybe, but let us investigate the hailstone. Suppose that it falls from 19,600 feet: $y = 16t^2$ tells us how long it takes to get to the ground. How long? $v = 32t$ tells us how fast it is going. How fast when it gets to the ground? Wow! Faster than the speed of sound! Don't go outside when hail is falling! What did the author miss and why was it missed?

2. "The nomenclature surrounding differentiation and integration is woefully prosaic." How true! "Differential" for example—how dull! What we need are zippier terms, with more punch and whoosh. Can you suggest some? Can you think of examples of non-prosaic nomenclature in other sciences?

3. Or, or second thought, is it true after all? Why shouldn't mathematics be prosaic? Why, for that matter, should mathematics be popularized? Why shouldn't the public be let alone and not pestered with mathematics?

THE EDUCATION OF T. C. MITS

by Lillian R. and Hugh G. Lieber

Not all calculus instruction goes on in classrooms, nor does everyone study it only out of compulsion. The truth of that statement can be deduced from the existence of books of *popular mathematics*—books that are not textbooks that try to explain mathematics and mathematical ideas, including calculus, to a general audience. Such books exist and new ones continue to appear, and that implies that someone must be buying them. In fact, enough people must buy them so that their publishers find enough profit in them (or sufficiently small losses) to make their publication worthwhile. That they are bought implies that they are read, and that shows that the first sentence of this paragraph is true.

Authors of books of popular mathematics usually are not concerned only with teaching some topic of mathematics, even though the temptation to tell someone else about something that you have mastered and find delightful is hard to resist. They have wider purposes as well. The following excerpt is from *The Education of T. C. Mits*, one of a series of books by Lillian and Hugh Lieber, and contains both kinds of popularization: explanation of mathematics, and ideas about mathematics. "T. C. Mits" stands for "The Celebrated Man In The Street", showing that the authors meant their book to be readable by anyone. You can have your own opinion about how successful the authors are in explaining what a derivative is, and you can argue about the correctness of their general ideas, but the point of putting the excerpt here is to acquaint you with a part of mathematical literature that you may not have known about, one that you may want to explore and profit from. Another point is to show you that all writing does not have to in blocks of text, divided now and then into paragraphs. The Liebers wrote all of their books in the same style as *T. C. Mits*, and you can also have your own opinion about how successful *that* is.

X. THE OFFSPRING

We just want to indicate briefly here
one major idea of Newton's Calculus:

Suppose you are taking a trip
in an automobile
and traveling at a steady rate of
40 miles an hour.
How far can you go in 2 hours?
Obviously the simple formula
(2) $d = rt$
(distance = rate × time)
will give you a quick answer.
But suppose that
your rate is not constant;
you can easily see that
this formula will no longer work.
And since we often have need
for formulas which will apply
to motions in which
the rate is not constant,

let us see how this can be done.

To do this easily,
let is first
plot the graph of equation (2)
for the case when $r = 40$,
namely
(3) $d = 40t$.
We first make a table,
giving t any values we please,
and calculating from (3)
the corresponding values of d:

t	d
0	0
1	40
2	80
3	120
4	160
5	200

and then plot these points (Figure 1).

then stop for an hour,
and then continue for 3 hours at

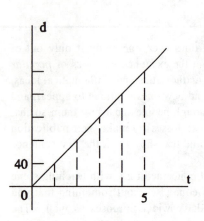

Figure 1

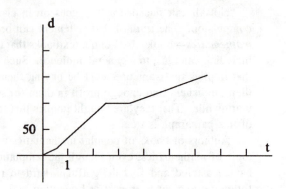

Figure 2

Now since equation (2)
may be written

$$r = \frac{d}{t}$$

(to find the rate,
divide distance by time),
we see from the graph that
the rate may be found by
dividing the values of
any dotted line
(which represents distance traveled)
by the corresponding value of *t*.

And so the graph
completely shows
the motion in question,
the time being shown
along the horizontal axis,
the distance along
the vertical axis,
and the rate being their ratio.
And obviously,
a motion having a constant rate
will be represented by
a straight line.

Now,
what about a motion in which
the rate is NOT constant?
Suppose, for example, that
you go at a rate of
20 miles per hour for 1/2 hour,
then increase your speed to
40 m. p. h.,
and keep that up for 2 hours,

the rate of 35 m. p. h.,
what would the graph look like?
Obviously it would look like this (Figure 2):
And similarly,
the following "broken line" graph

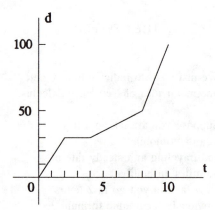

Figure 3

tells what story?
For each straight portion of the line
the rate is uniform.
But at each CHANGE of the slope of the line
the rate changes to
a new value which remains the same
until the next break.
Note that after each break
the change, as shown in these graphs,

is a sudden change (Figure 3),
no allowance being made for the
process of accelerating or
slowing down.

To show this process,
we must have a CURVE as shown
on Figure 4
where x is the time and
y the distance covered.
Here any particular rate is
not kept up for an appreciable time
but is CHANGING ALL the time.

How can we now "catch" a thing
which is so elusive?
This was the problem solved by
the Calculus:
Suppose first that
the motion from A to B were
a uniform motion instead of
an accelerated one.
Then it would be represented by
the straight line AB instead of
the curve AB.
And it would show that
in time AC
the distance BC was covered,
at a constant rate equal to
BC/AC.
Now as you take the point B
nearer and nearer to A,
the straight line AB approaches

at point A.
Thus we may say that
the actual rate at A is
the "limit" of BC/AC.
And whereas
this rate lasts only an instant,
(for as soon as you get away from A
the slope of the tangent line is
obviously different),
still
we can "catch" it and
express it mathematically
(and thus be able to work with it).
Thus,
if we represent AC by Δx
(read "delta x"),
which simply means the
difference in the x-value
from A to B,
and BC by Δy,
then,
as B approaches A,
this ratio $\Delta y/\Delta x$ approaches
a limiting value.
This limiting value of $\Delta y/\Delta x$ is
represented by dy/dx.
And so we have
$$dy/dx = r,$$
the rate AT THE POINT A—
and r of course changes
from point to point.

Now
if we know
the equation of the original curve,
the Calculus gives us
the necessary machinery
(called "Differentiation")
by which we may find
$$dy/dx$$
at any point.
And, vice versa,
if we know the value of dy/dx,
that is, if we know the
"Differential Equation,"
we can find,
by means of the Calculus
(by "Integration")
the equation of the original curve.

Now in most physical problems,
in this ever-changing world,
the idea is

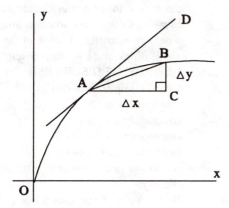

Figure 4

more and more to the line AD
which is tangent to the curve

to set up a differential equation
which represents what is happening
in a small local region,
and from this,
by "Integration,"
to find out, for example,
the entire sweep of
the path of a planet.
We can hardly expect,
from this brief sketch,
that anyone can get
even a slight idea of the
power of the Calculus
as a tool of Science.
Suffice it to say, here,
that
it is a method which
enables us to study
an ever-changing world,
rather than only those things,
like the figures in Geometry,
which very accommodatingly
stand still while
we are measuring them.
It is an instrument for the study of
a swift, dynamic world.
Why then is it not
the last word in Mathematics?
What more is there to be desired?

But wait till you see Part II!

The Moral: Learn to study
          ON THE WING!

XI. A SUMMARY OF PART ONE

We have tried in Part 1 to
give you the following ideas:
(1)   A man trying to think
         without mathematics is
         like a helpless child
         (see Chapters I, II, and III).
(2)   A "practical" man
         working with his hands alone,
         without the aid of theory,
         may be just a fool
         (see Chapters II, V, VI).
(3)   The value of
         Mathematics and Science
         is not limited to

the gadgets which they give us,
but is also in their
philosophy
(see Chapters V and VI).
(4)   Generalization and abstraction
         (two powerful tools of thought)
         are important in
         all thinking.
         You cannot really think
         without them.
         But you must learn to use them
         properly.
         If used carelessly
         they are "dynamite" and
         may blow you up!
(5)   Do not always demand a
         "Yes" or "No" answer.
         For example:
         "Shall we cling to the traditions
         of our great forefathers?
         Yes or no?"
         The history of mathematics shows
         just how much of Euclid
         we must keep
         and how much we must discard.
         You will see this in some detail
         in Part II.
         But outside of Mathematics
         in the social studies,
         you will hear people quoting
         blindly:
         quoting the Constitution,
         quoting Karl Marx,
         quoting Theodore Roosevelt,
         (By the way,
         the men who wrote the Constitution,
         as well as other men so often quoted,
         would be horrified at some of the
         applications made by their disciples.
         BEWARE OF DISCIPLES!)
         with the implication that
         you must either
         completely accept or
         completely reject.
         In Mathematics, however,
         we do not just quote authority.
         We say:
         "In the light of our knowledge today,
         Euclid was right in this and
         wrong in that."
         And this is a wholesome way
         to look at the past.
         It is partly good and partly bad;

we must select,
in the best light of
our knowledge now.

(6) Do not jump at conclusions
(see Chapters I, II, and III).

(7) Do not rule out hunches
because they are sometimes wrong.
Read some of the original
writings of Faraday or
other great scientists—
you will be surprised to find
how much of their work
started as "hunch."

BUT

(8) Do not think all
your hunches are wonderful!
Some of them may be terrible!
Follow them up cautiously!
Encourage them but
watch them!

(9) Try to judge
statements and theories
in the light of
important long-time activities
of the human race—
like Science or
Mathematics or
Art.
They reveal "human nature"
better than anything else.
In them you will see that
Internationalism and Democracy
are very deep in the human spirit.
(see Chapter VI).

(10) And so you see that
Mathematics is not for
the engineer only,
or only for someone who
needs its formulas.
It is a way of thinking,
a way of life,
VERY IMPORTANT FOR EVERYONE.

(11) Most courses in Mathematics
do not leave time
to consider these things.
They are too full of technique.
We MUST stop now and then
from the manipulation of
techniques
to see what
general ideas we can get from them,
which will be useful for
ALL of us.

QUESTIONS AND EXERCISES

1. Plot a graph of your velocity against time, say for today from $t = 0$ at 7 a. m. to $t = 5$ at noon. What does the area under the curve represent?

2. Why should a general reader care about the ideas of calculus? Why should anyone read a book on popular mathematics? There are (I think) no books on popular metallurgy, or, if there are, no one I know reads them; why should mathematics be any different?

3. "Science or Mathematics or Art reveal 'human nature' better than anything else." List at least four ways in which art reveals human nature. List at least three ways in which science reveals human nature. List at least two ways in which mathematics reveals human nature. Does the difficulty in making the lists increase from first to last? Why, or why not?

4. Do you agree with the Liebers' point (11)? If so, how could what they suggest be done in a calculus class? If not, why not?

5. "A man trying to think without Mathematics is like a helpless child (see Chapters I, II, and III)." You have not seen chapters I, II, and III, but what could possibly have been in them to justify such an extreme statement? Surely millions of people are thinking at this very moment, without mathematics, and not like helpless children. Were the Liebers crazed fanatics? *What did they mean?*

THE SOCIOLOGY OF MATHEMATICS

by Lars Gårding

What place does mathematics have in the world? What does it do for people? Here is one answer.

To round off things we shall now give a short analysis of the role of mathematics in society. It starts with biographies of three persons or rather collectives A, B, and C, representing, in order, the public, the users of mathematics, and the professional mathematician.

Let us introduce the notation A for the more or less educated public as a mathematician. We then have to imagine a person who has lived and worked for thousands of years. He will even have different ages at the same time. The reader who is bothered by this phenomenon can conceive of A as sometimes split into several individuals. We shall sketch his development and his achievements in mathematics.

To start with, A is a child who learns to count. We assume that A lives in an environment where the numbers are important and that he has normal contact with grown-up people. Then, at the age of about six, A has a firm control of the first 10-20 integers and their order. He can also add small numbers and count with the aid of his fingers. He gets interested in larger numbers and asks for their names. Once he counts himself to sleep and comes to 100. A breathless moment. During his continued life as an illiterate A enlarges his mathematical territory to all numbers under 1000 and gets a good grasp of those fractions and geometrical figures that his language has names for. If he plays games of cards or dice, or otherwise is forced to make mental calculations, he makes them with accuracy. If A lives in a country where goods are bought and sold, where money is used, or where time is measured by the clock, he uses these talents every day. This A has existed for many thousands of years, and still makes up over half of humanity. His almost daily contact with the integers and perhaps also the fractions makes him feel at home in the number model. He has an intuitive feeling for the kind of practical geometry that is used in, for instance, carpentry. But his knowledge is limited. To be able to handle large numbers and make long calculations, A has to go to school. There he ceases to be illiterate and can start absorbing parts of man's cultural heritage.

As a first task, A has to learn to read and write. On a large scale this did not happen until the nineteenth century. "Reading, writing, and 'rithmetic" were the three pillars of the education of the people. Mathematics came third, but the objective was very clear: to teach effective algorithms for the four arithmetic operations and to acquaint the pupils with the current systems of weight, volume, money, and time. A was first taught the names and signs of the integers and to make simple additions and subtractions. After that came the multiplication table, to be known by heart. Equipped with this indispensable tool, A learned the algorithms of multiplication and division, not without a considerable effort. There was a constant supply of practical applications. The teaching was very concrete with an emphasis on skill and the result was good. After school A had enough practice not to forget what he had learnt. In so far as A pondered what so-called higher mathematics could be about, he imagined a very comprehensive multiplication table or perhaps a fifth arithmetic operation.

Let us now follow A a bit further to high school where he studies Latin, history, languages, and some mathematics. The teaching is now less utilitarian and A meets scientific mathematics in two forms, geometry according to Euclid, and algebra including equations of the first and second degree. A is a good pupil and does all that is required of him but after school he has no use for his knowledge and it deteriorates very soon. As a public servant A now and then remembers the theorem of Pythagoras, but he has forgotten what a hypotenuse is and he has difficulties in computing percentages. As an editor A keeps the pages of his journal free from all kinds of formulas. If mathematics turns up as a subject of conversation, A wonders if there is a connection between musical and mathematical ability. This picture of A, true 100 years ago, is still true in all essentials. The difference depends on the fact that tedious calculations are no longer made by hand. A learns about the same things as before but he gets more explanation and less drill. The new things are: set theory to explain the number system, the binary system to explain computers, and the reading of simple diagrams to make A understand the complex society he lives in. As an adult A has very little reason to occupy himself with mathematics, and his view of the subject can be stated very briefly: it is something that has to do with computers, satellites, and nuclear power. He does not know any details but he believes that the binary system and sets somehow enter the picture. As an editor, A still keeps his pages free from formulas except for a short period when Einstein's formula $E = mc^2$, at times reproduced as $E = mc2$ or $F = m3^2$, was an accepted incantation. A treats figures produced by a computer with greater respect than those computed by hand. On social occasions A behaves as before.

B. Like A, B appears in many shapes at many times. We let B denote people who use mathematics as a tool without contributing to the subject. B stands between A and the professional mathematician, who is the subject of the next section. It is not so easy to differentiate between B and C, but let us say that the typical B nowadays is an engineer who designs a bridge or writes advanced programs for a computer. This B has had a brilliant career. It extends over thousands of years and we can only sketch it here.

B appeared for the first time 4000 years ago in the fertile valleys of the Nile and Euphrates. He did bookkeeping for princes, recording their stocks of grain, oil, wine, cattle, soldiers, and slaves. He wrote out the laws and built temples and palaces with the sign of civilization: straight lines and right angles. He measured distances, made maps, and registered the movements of the celestial bodies, and sometimes won the admiration of the crowd by predicting an eclipse of the sun. He ran a school where smart young boys learned arithmetic and bookkeeping. His life did not change much for 3000 years. In Italy, during the Renaissance he still did bookkeeping, but now for a rich merchant. After the invention of gunpowder B devoted himself to the art of ballistics. After this first success on the battlefield, B started to build modern civilization.

He constructed, built, improved old processes and invented new ones. Among his achievements are the steam engine, big ships, railways, steel, long range guns, new explosives, cars, airplanes, telecommunication, synthetic materials, satellites, nuclear power, etc. In all his work, B uses mathematics as a tool, mostly just to do simple computations, but now and then he employs complicated mathematical models, for instance, when B discovered the planet Uranus or constructed optical lenses. As a rule, B got his models from C but sometimes B had to invent the model and its theory all by himself. This happened when B was Newton and thought about the movements of the planets. He then found the solution of the problem in the law of gravita-

tion and invented the mathematical instrument, infinitesimal calculus, which made it possible for him to compute the orbits of the planets from this law. After his time as Newton, when B worked as a physicist, mathematics was absolutely necessary for him in order to formulate the fundamental laws that he discovered, e. g., the laws of hydrodynamics and electricity and the principles of atomic physics. Physics is a very difficult task for B and he often looks for new mathematical models. As a rule he no longer has sufficient force to construct them himself and it has happened that C has anticipated him. When B was Einstein and invented general relativity he needed a kind of geometry that Riemann had already put on the market. B had the same experience when he founded quantum mechanics. The mathematical tools, among others, group theory, were already available in stock. Today, when B works as a theoretical physicist, he is trying out lots of things in this stock, although infinitesimal calculus remains the main tool.

Otherwise, B is more active than ever. To his traditional fields of activity, engineering and physics, B has added numerical analysis on a large scale, data processing, and branches of the social sciences, biology, and medicine. Sometimes his pursuits are rather modest, for instance, when B is a doctor and wants to prove something with statistics. He then has to solicit the advice of another B, a specialist on the applications of mathematical statistics.

B's views on mathematics as a subject and his opinion of C vary between uncritical admiration and arrogant superiority. As a professor of physics or engineering B usually thinks that the teaching of mathematics should be done by physicists (engineers) and not be left to his colleague C, a hopeless theoretician. Lately, there have been signs of better relations between the two parties. B has gotten a better mathematical understanding and C has adjusted himself to a growing motley crew of students who want to use mathematics for very diverse purposes.

C. The mathematician C, the man who studies the subject for its own sake, appeared together with B for the first time 4000 years ago in the fertile valleys of the Nile and Euphrates. It is likely that B and C were the same person for some time and that B became C when he got more interested in the tool than in its use. The metamorphosis took place when B had plenty of time and found it to his pleasure to sit and think. C had his first great period in Greece from 500 to 0 B. C. B had then a time of stagnation after some initial successes, but C was busy thinking about mathematical problems, discussing them with others, and writing down what he found. His results are in Euclid's *Elements* and in the works of Archimedes and Apollonius. It was a brilliant debut and the *Elements* was a fantastic success in spite of its pedantic style and abstract reasoning. After this show of force C kept quiet for over 1000 years. His second appearance, with the solutions of third and fourth degree equations, was all right but did not measure up to his debut. But in the seventeenth century C created a sensation by inventing infinitesimal calculus, and he has worked hard and been rather successful ever since. He is also still respected by A, the general public. But his traditional prestige depends to a large extent on A's inability to distinguish between B and C. A is likely to give C undue credit for some of B's achievements, for instance, electronic computers.

We know already how A and B look at C and his subject and we shall now say a few words about the position of C in our society and of himself and others. Most of the time, C is a university teacher but he can also be employed by industry or some research institute. He is a specialist in some branch of mathematics and he has written two or three articles that are known all over the world—if only in a small circle. At times he works very hard and sleeps badly. This is when he is trying to prove a new theorem. He is often unsuccessful but sometimes everything works out wonderfully, and then he is deeply satisfied. If he does not write good articles himself, there are others who do. C is proud of his collective self that has written and still writes so many wonderful things. He knows that his subject has unlimited possibilities and that it will always attract gifted young people. In his relations to A and B he is a realist in that he does not pretend to communicate on levels where it is not possible. But he sometimes finds B irritating, and it has happened that he thinks that A learns the wrong kind of mathematics and then decides to do something about it. We shall come back to his decision later on in this chapter.

QUESTIONS

1. Is the classification complete? Have any groups of people who have anything to do with mathematics been left out? If so, classify them as $D, E, \ldots$.

2. What are the approximate sizes the groups? Of the 5 billion people in the world, how many are As, how many are Bs and how many are Cs? If there are Ds, Es, ..., how many of them are there?

3. Are the sizes appropriate? Should any of the groups be larger or smaller, and why?

4. Do you think that the future bring any changes in the nature or composition of the groups?

5. "B became C when he got more interested in the tool than in its use." Can you think of examples other than mathematics where people have become so enamored or fascinated with a tool that it became more important than what it was supposed to be used for?

6. "A's inability to distinguish between B and C." Here is a chance for sociological research. Take a survey: ask a random selection of As to name as many mathematicians as they can and—if they can name any—determine how many responses refer to Cs and how many to Bs (and how many refer to As), if need be by asking your teacher to make the classification. Or, to test the hypothesis that the author is correct, prepare a list Bs and Cs in random order and ask As to classify them correctly.

CALCULUS AND OPINIONS

by David L. Hull

Decisions are hard to make (where shall I go to college?) and decisions that have to be made in the face of uncertainty (what if I don't like it there?) are harder still. How can we make good decisions? Some people go with their intuitions. This may turn out well, but remember that intuition tells us that the world is flat and the sun goes around it. Other people trust to luck and flip a coin. This may also turn out well, but the casinos of Las Vegas pay their bills and earn their considerable profits from people trusting their luck. Astrology and other forms of fortune-telling are popular and always have been, as have appeals for supernatural intervention. People are always looking for help, from anywhere, when it comes to making decisions.

Mathematics and calculus can help, sometimes, in making some decisions. There are no equations that can be solved to tell you where to go to college, or what job to take, or what person to marry, but for some smaller decisions mathematics can remove some of the uncertainty. How effective is this new drug? Seventeen of thirty-two patients showed great improvement; how sure can I be that more than half of the potential market of 300,000 would benefit more from it than from their present treatment? Is it worth investing $1.2 million in development and testing? What's the chance that the drug would make a profit? What's the chance it would make a *big* profit? Questions like that can be answered, partially, using ideas from probability and statistics, and the ideas of probability and statistics depend, with no uncertainty at all, on the ideas of calculus.

Opinion polls are used, a lot, in making decisions. What you get, or do not get, on television depends on the opinion of a sample of the television-watching public. Which politician gets to run for high office depends, sometimes, on what polls show. Opinion polls are important.

Anyone who has watched election night on television knows that sample opinion surveys affect our political and social world. It is getting increasingly expensive for a person to run for political office, and those with money do not want to back losers. Thus, sample surveys are used to determine the popularity of a candidate, and ultimately political polls determine who gets to run for political office. Opinion polls affect our daily lives in other ways too. Sample surveys are used to establish the taste appeal of new foods, and more generally, opinion polls are used by the marketing departments of corporations to investigate the potential of new products and services.

Modern opinion polling gives us a way to communicate with our government. For example, your home town might need to pass an operating levy for its school system. A community could perceive a need for a new library, park, municipal building or the like. Leaders in the community would want to take the pulse of the voters. Would the voting public support a project by enacting an additional tax? If not, what might convince the voters to support such a project?

Let us consider a specific example. In a recent election in Franklin County, Ohio (Columbus), the Central Ohio Transit Authority (COTA) had a 0.25 percent sales tax levy on the ballot. The Columbus metropolitan area views itself as one of the few growth areas in the midwest, but cities are not considered major if they have no public transportation. Without the added revenue from the sales tax it was likely that COTA would have to go out of business. Civic and political leaders viewed passage of the tax levy as vital to the continued growth and prosperity of the Columbus area.

About a month before the election, the area's major newspaper, the Columbus *Dispatch*, ran an article with the headline "COTA Sales Tax Appears Headed for Approval." In the body of the article, the *Dispatch* reported that a mail poll of 1,206 county residents showed 60% of the voters favoring the tax levy. An estimated margin of error of plus or minus 3.5 percentage points was also reported.

Although the *Dispatch* did not talk about reliability, what they had done was to give a 99% confidence interval for the true proportion of voters favoring the COTA levy. We should give credit to the *Dispatch* for reporting an error bound, because some newspapers give estimates without any notion of error.

Calculus comes into play when we begin to talk about error bounds and reliability, or confidence. The *Dispatch* poll is trying to estimate a parameter, p, the true proportion of registered voters who favor the COTA levy. They use an estimator which is a rule or function of the values obtained in a random sample. In this case the function $\hat{p}$ is defined by

$$\hat{p} = \frac{\text{number favoring the levy}}{\text{number of voters surveyed}}.$$

The function $\hat{p}$ is called a random variable or chance phenomenon. We all know that a function is a fixed rule and not a variable. In fact, $\hat{p}$ is a good rule, and the rule is constant. What varies is the random sample. If one chooses a different random sample, one would get a different $\hat{p}$, since different samples will give different numbers of voters favoring the levy.

There is a theorem from probability theory called the Central Limit Theorem, which says that probabilities for $\hat{p}$ correspond to areas under the normal curve. The normal curve is a bell-shaped curve with formula

$$f(x) = \frac{1}{\sqrt{2\pi}} e^{-x^2/2}, \quad -\infty < x < \infty.$$

More precisely, if n is the sample size and p is the true proportion, the Central Limit Theorem says for $a > 0$, the probability that

$$-a \le \frac{\hat{p} - p}{\sqrt{\dfrac{p(1-p)}{n}}} \le a$$

is nearly equal to

$$\int_{-a}^{a} \frac{1}{\sqrt{2\pi}} e^{-x^2/2} \, dx.$$

We could say the above in terms of a formula

$$P\left\{-a \le \frac{\hat{p} - p}{\sqrt{\dfrac{p(1-p)}{n}}} \le a\right\} = \int_{-a}^{a} \frac{1}{\sqrt{2\pi}} e^{-x^2/2} \, dx.$$

Your calculus background tells you that finding formulas for

$$\int e^{x^2} \, dx \quad \text{or} \quad \int e^{-x^2} \, dx$$

is not an easy thing to do. Fortunately, the area under the normal curve between nay limits can be computed by numerical integration methods, and the results are tabulated. For example, if

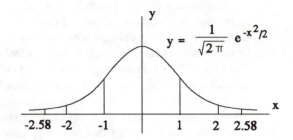

Figure 1

$$z = \frac{\hat{p} - p}{\sqrt{\dfrac{p(1-p)}{n}}},$$

then

$$P\{-1 \le z \le 1\} = .65, \quad P\{-2 \le z \le 2\} = .95,$$

and

$$P\{-2.58 \le z \le 2.58\} = .99.$$

See Figure 1.

Before sampling takes place, the expression

$$P\left\{-a \le \frac{\hat{p} - p}{\sqrt{\dfrac{p(1-p)}{n}}} \le a\right\}$$

can be manipulated to obtain

$$P\left\{\hat{p} - a\sqrt{\dfrac{p(1-p)}{n}} \le p \le \hat{p} + a\sqrt{\dfrac{p(1-p)}{n}}\right\}.$$

There are some theorems about convergence which allow us to replace

$$\sqrt{\dfrac{p(1-p)}{n}} \quad \text{by} \quad \sqrt{\dfrac{\hat{p}(1-\hat{p})}{n}}$$

to obtain the expression

$$P\left\{\hat{p} - a\sqrt{\frac{\hat{p}(1 - \hat{p})}{n}} \leq p \leq \hat{p} + a\sqrt{\frac{\hat{p}(1 - \hat{p})}{n}}\right\}.$$

The expression

$$\hat{p} \pm a\sqrt{\frac{\hat{p}(1 - \hat{p})}{n}}$$

is called a random interval for p. Recall that for $a = 2$, the probability that the interval

$$\hat{p} \pm 2\sqrt{\frac{\hat{p}(1 - \hat{p})}{n}}$$

contains the true value of p is approximately .95. Also for $a = 2.58$, the probability that the interval

$$\hat{p} \pm 2.58\sqrt{\frac{\hat{p}(1 - \hat{p})}{n}}$$

contains p is .99.

After we sample, the probability is gone. Mother Nature knows whether or not the true value of p is in the interval

$$\hat{p} \pm a\sqrt{\frac{\hat{p}(1 - \hat{p})}{n}}$$

but we do not know whether p is in the interval. Since probability vanishes after the sample is taken, we say we have confidence in our result. If $a = 2$, we would have .95 × 100% = 95% confidence. If $a = 2.58$, we would have .99 × 100% = 99% confidence. Basically, if one took one hundred random samples and computed a 95% confidence interval for each sample, about 95 of the intervals would contain the true proportion p.

Let us return to the *Dispatch* poll of the COTA levy. The random sample of size $n = 1206$ yielded 724 voters favoring the levy. Thus

$$\hat{p} = \frac{724}{1206} \approx .60.$$

If we choose $a = 2.58$ in the formula

$$\hat{p} \pm a\sqrt{\frac{\hat{p}(1 - \hat{p})}{n}}$$

we obtain

$$.60 \pm 2.58\sqrt{\frac{.6 \cdot .4}{1206}} \approx .60 \pm .036.$$

So we see that the *Dispatch* rounded .036 to .035 and reported 60% plus or minus 3.5 percentage points. We also know that the interval 56.3% to 63.5% is a 99% confidence interval.

The above theorems and calculations are based on a single random sample. That means that all registered voters in Franklin County were equally likely to be chosen. The *Dispatch* bought a mailing list of all registered voters in Franklin County. Then the *Dispatch* poll used a random number generator to select its sample.

On election day, the levy passed with about 76% in favor. The voting took place a month after the survey was taken, and during the ensuing period, community leaders continued to work for the issue's passage. Evidently their efforts paid off.

There is another way that calculus is used in proportion sampling. It is a simple maximum-minimum problem to determine the sample size needed to estimate a proportion within a given error bound and with a given confidence level. We will answer the question "How large a sample should I take in order to be within three percentage points of the true proportion p with 95% confidence?"

The Central Limit Theorem tells us that a 95% confidence interval will be given by

$$\hat{p} - 2\sqrt{\frac{p(1 - p)}{n}} \leq p \leq \hat{p} + 2\sqrt{\frac{p(1 - p)}{n}}.$$

If we let α denote the error, then

$$\alpha \leq 2\sqrt{\frac{p(1 - p)}{n}}.$$

The worst error would occur if p were at one of the endpoints of the confidence interval. So, we choose

$$\alpha = 2\sqrt{\frac{p(1 - p)}{n}}.$$

To answer the question of being within three percentage points of the true p we let $\alpha = .03$. Now $0 < p < 1$, and we let (see Figure 2) $f(p) = p(1 - p)$. Setting $f'(p) = 0$, $f(p)$ is a maximum at $p = 1/2$. Let $p = 1/2$ in the formula

$$.03 = 2\sqrt{\frac{p(1 - p)}{n}}.$$

and solve for n to get $n = 1111.11$. A random sample of size $n = 1112$ will give the required accuracy.

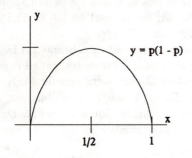

Figure 2

A spectacular failure in political polling occurred in the *Literary Digest* magazine's prediction on the 1936 presidential election. The *Digest* was a prestigious magazine and about 2.4 million people responded to its mailing of 10 million questionnaires. The *Digest* predicted that Kansas governor Alfred Landon would beat incumbent president Franklin D. Roosevelt by 57% to 43%. However, President Roosevelt won the 1936 election by a 62% to 38% landslide, and the *Literary Digest* went bankrupt shortly thereafter. (For a more complete discussion and further references, see David Freedman et al, *Statistics*, second edition, W. W. Norton, 1991.)

What went wrong with the poll? The selection process for the recipients of the poll was biased. The names and addresses of the ten million people came from sources like telephone books and club membership lists. In 1936, there was high unemployment and many families did not have telephones. Thus, the sampling process was biased against the poor, and the poor voted overwhelmingly for Roosevelt. In addition, the poll had a large number of individuals who did not return the polling questionnaire. Statisticians call this phenomenon nonresponse bias, as nonrespondents may be very different from respondents.

Current professional polling organizations such as Gallup or Harris are able to take a sample of 3,000 voters and predict the results of a national election within one percentage point with 95% confidence. Such accuracy means that sample surveys have been given a solid foundation, and mathematics has been the primary tool in the development of the statistical fields of sampling theory and the design of sample surveys. Sampling is also an active research area, and people are using sophisticated mathematics to write doctoral dissertations on sampling theory and methodology.

Modern polling technology has come a long way since 1936, but consumers of sample surveys need to be wary. What do you think of your favorite radio station's call-in poll? Statisticians think such polls are quite like the *Literary Digest* poll of 1936.

QUESTIONS AND EXERCISES

1. What would the newspaper headline have been if only 624 of the sample of 1206 had said that they were in favor of the new tax? Would the headline have been different if there were 6240 in favor in a sample of 12060?

2. The difference between the percentage of people in favor of the new tax in the poll and in the election—between 60% and 76%—is quite large. The author explains it by the continued campaigning for the tax in the month between the poll and the election. That would mean that the campaigning cut the proportion of those opposed nearly in half, from 40% to 24%. It seems to me unlikely that advertising, especially advertising for a tax to support busses, could have had that much effect in so short a time. Can you think of other explanations for the difference?

3. $P\{-2 \leq z \leq 2\} = .95$ and $P\{-2.58 \leq z \leq 2.58\} = .99$; what is the number inside the parentheses if you want .975 on the right? This is work that is barely worth doing by humans: see if your computer algebra system can find the number for you. After it has, find what that .99 *really* is. That is, get it to a few more than two significant figures

4. Does anyone on the *Dispatch* know calculus? I doubt it. I bet that whoever ran the poll looked up numbers in tables to report that the margin for error was plus or minus 3.5%. That is what bankers do when they need to find how much a monthly payment is needed to repay a loan: they look up numbers in tables or do the equivalent, enter numbers on a computer keyboard. The inference is that mathematics is less important than running banks and newspapers. As long as we have a few technicians to prepare tables or computer programs, we are all right. The inference from *that* is that we should be teaching how to run banks and newspapers in schools instead of calculus. Do you agree?

MATHEMATICS AS A SOCIAL FILTER

by Philip J. Davis and Reuben Hersh

In the United States, at this very moment, hundreds of thousands of people are studying calculus. Perhaps that needs to be restated: at this very moment, hundreds of thousands of people are enrolled in calculus courses. Either way, the number is absolutely astounding. Never in the history of the known universe has so large a proportion of the population of 18-year-olds in any nation been engaged in studying calculus. We are in the midst of a huge social experiment, never tried before, whose outcome cannot be predicted. For all we know, historians a few hundred years down the road will look back with a sneer at us turn-of-the-21st-century Americans, condescendingly pointing out with the unerring vision of hindsight how misguided and doomed to failure the experiment was, and implying that they, and also their readers, are ever so much smarter than their simpleminded ancestors were in 2000. Or, they may do research into the origins of the social movement that had such immense consequences and did so much to shape the modern world. But it is more likely that the calculus explosion of the second half of the 1900s will go unnoticed, or rate only a footnote or two in a specialized history or two. Whatever the verdict of history is, the explosion is a fact and it is worth considering.

One question that should be asked is, *why*? Calculus and mathematics have not always been part of everyone's education. Until the middle of the nineteenth century in this country, it was possible to have what everyone would agree was a fine education without ever having worried about algebra or geometry. Daniel Webster was never taught arithmetic, since it was then thought to be a subject not very dignified, suitable only for clerks, shopkeepers, and like people, but not for a person of quality. For some reason, mathematics began seeping into the common curriculum and by around 1870 it was firmly entrenched. It has been advancing ever since, with calculus moving from something to be taken up in the junior or senior year of college to the sophomore year, then to the freshman year, and then into the high schools, where it is still spreading with no signs of a retreat. This multiplication of mathematics has yet to be completely explained. The following selection offers one reason: calculus has become a hurdle that must be jumped in order to enter a variety of desirable occupations. Thus, the more calculus is studied, and the earlier, the easier it will be to get into them. Thus, the more mathematics is studied, and the earlier, the easier it will be to jump the hurdle of calculus. Thus, everyone studies mathematics.

If the authors are correct when they argue that for many people calculus is only a hurdle and is not at all necessary for the desirable occupations, then there are social implications. People who cannot jump the hurdle will not be able to try to do some things that they might be able to do very well. This is not fair to those people. If there are groups of people who, for one reason or another, tend not to do very well at calculus, then this is not fair to those groups. They have been filtered out with the wrong filter. Society loses: people who could have done a good job at something have not had the chance, and we need all the people who can do jobs well that we can get. Those filtered out lose. Does anyone win? Is calculus actually *harmful*?

In the life and work of the teacher of mathematics, there is a strange contradiction. He or she studied mathematics by choice. He loves to lose himself in this ideal world of clarity and precision. There is nothing he would like better than to invite others to join him. For someone who loves math, teaching math should be a ball.

Sad to say, it isn't quite that way. Many of the students in mathematics classes are there by compulsion. Often they have little taste for mathematics, and many have great difficulty learning even a very little bit of it. What mathematics teacher can forget the first time he showed a class something especially elegant, something beyond the basic facts and problems in the text book? When the presentation is complete, a hand goes up. "Will we be responsible for that on the final?"

Certain complaints are heard in the corridor, year after year. The differential equations students are poorly prepared. They haven't learned what they should in calculus. The Calculus III students are poorly prepared. They haven't learned what they should in Calculus II. And so on back to Calculus I, and to College Algebra, all the way down to Intermediate Algebra.

We change textbooks, change curricula, but the complaints don't change.

The conclusion is to blame the high schools. Many college math teachers are convinced that most high school teachers do a terrible job. The college teachers whom I have heard express this view never visit high schools and never talk to high school math teachers. The high school teachers I've met are intelligent and care sincerely how well their students do. The problem, they explain, is that the students coming into high school don't know what they need to in order to do high school work. The trouble is in the middle schools. And of course if you look in the middle school, you'll be told that the trouble is in the elementary school. In the elementary school you'll be told that the trouble is in the home and the family. Thus everyone is to blame, and no one is to blame.

The situation I have just described from the vantage of a college math teacher is just the back side of the "crisis" often described in headlines in the nation's better newspapers. "Crisis of scientific illiteracy," "rising flood of mediocrity," "falling far behind Japan and even the Soviet Union."

From the viewpoint of international educational competition, the situation is indeed a crisis and calls for immediate emergency measures involving non-trivial amounts of money. Strangely enough, however, the actions that are taken seem to be rather modest. I would like to reconsider the situation, again from a teacher's point of view, and see if there is another way of thinking about the problem. My conclusions may turn out to violate the calls of our educational competition with Japan and the Soviet Union. If so, I can only hope that this will not be judged seditious. What is worse, they may turn out to violate the perceived self interest of my profession. For this, I will surely be declared seditious. But let us see.

The first question that would be asked by an innocent observer is this: "Why are so many people studying a subject for which they have, it seems, so little interest, affection, or aptitude?"

The answer, of course, is simple: "It's required."

Yes, but required by what? By whom? For what reason?

The requirement of calculus and differential equa-tions for engineers is beyond challenge. You can't get through many an advanced engineering course without knowing differential equations. The same applies in physics and chemistry. The nature of the material that must be learned in those fields makes a mathematics prerequisite inescapable. But what about the require-ment of the University's Business School? I have occasionally asked business students whether they ever used the calculus they had learned in Math 180. "Never," they reply. I suppose I could go beyond students and ask working business persons, executives, entrepreneurs and so on if they ever use calculus, but the very idea seems ridiculous. Of course they don't.

Why is calculus required by the University's Business School? I could go over there and ask around, but it doesn't seem really necessary. I do know two relevant facts: (1) Most other business schools require calculus. (2) The business school has many more applicants than it is able to accept.

These two facts suggest two motivations for our university business school's math requirement:

1. Lack of a math requirement could mean lower prestige for the school; conversely, instituting a math requirement means higher prestige in the world of University Schools of Business Administration. (This is the world that is real to a professor of Business Administration, just as the world of mathematics as a profession is the real world in the eyes of a college math teacher.)

2. The requirement of passing calculus cuts down the number of applicants, making it easier to decide whom to admit every year.

As to point 2), a remarkably frank statement was made by John Kenneth Galbraith in his book *Economics, Peace, and Laughter*. Commenting on the models of mathematical economics, he says this:

> Moreover, the models so constructed, though of no practical value, serve a useful academic function. The oldest problem in economic education is how to exclude the incompetent. The requirement that there be an ability to master difficult models, including ones for which mathematical compe-tence is required, is a highly useful screening device.

Galbraith adds a dour footnote:

> There can be no question, however, that prolonged commitment to mathe-

matical exercises in economics can be damaging. It leads to the atrophy of judgement and intuition ...

Granting that these two reasons are in effect, a spokesman of a university school of business could still argue that the calculus requirement has educational validity. Perhaps in some of their courses some of the teachers sometimes may wish to take a derivative or to find an integral of something. It would remain a matter of dispute whether the calculus requirement is really necessary. I think it isn't, but I don't claim I can prove it. The point that interests me is point 2). Mathematics is serving as a filter, a way to sort out those who will and those who won't be allowed to get a business degree.

It is being used in the same way, of course, with respect to engineering students, but there the role of mathematics is more deeply entrenched and its usefulness is more apparent. In the case of business, we have a different story. There seems to be no necessity to make math a requirement. There *is* a practical necessity to make a selection among the students who want to go to the business school. The business professors decide to use math for that purpose. Is that OK? How should we (math teachers) feel about it? First of all, there is nothing inevitable about the choice of math as a filter. Some other filters that could be used, or that have been used are: family connections, political connections, income, ability in sports, personal charm, brutality and aggressiveness, trickiness and sneakiness, devotion to public welfare, etc. The first five have been relevant criteria of admission of students to U. S. institutions of higher learning; the last three are suggested, somewhat in jest, for particular relevance to a school of business. Each of the criteria can be taken in more than one way. Take family connections, for instance: in American schools it has sometimes been helpful to be the son or daughter of an alumnus or alumna. In England, at one time, it would have been helpful to have aristocratic family connections. In Mao's China, on the other hand, it was helpful to be from a proletarian family and harmful to have bourgeois parents.

It must be admitted that consideration of the alternatives to math makes the use of math as a filter seem more reasonable. Compared to the other possibilities, a math requirement has an appearance of impartiality, of objectivity, and of rationality. How much easier for the faculty of the school of business to require calculus and rely on the math department to do the grading and sorting than to impose some other criterion, such as the last on our list, and have to interview hundreds of students and rate them on this criterion.

But what price do mathematicians pay for the privilege of being a gatekeeper, an obstacle which must be overcome in the race to that good job, that professional career, to which young Miss and Mister America know they are entitled?

In order to evaluate the use of the mathematics filter, we should consider it from three points of view. That of the business school (or the economists, or any other group that is using mathematics as a criterion of admission), that of the mathematicians (who have to administer the courses and tests constituting the filter) and that of society as a whole, especially the candidates and applicants who are being filtered.

From the viewpoint of a college of business administration, I suppose the whole thing works tolerably well. The pool of applicants is cut down without any effort or expense on their part, and without any blame or resentment accruing to the business school. After all, flunking the math prerequisite is something that takes place outside of the business school.

Next, we have to evaluate the math filter from the point of view of the mathematicians. How should we feel about a university's school of business having a calculus requirement? There is an easy, almost automatic, answer to that question. We should feel great! More math requirements means more math students. More math students means more jobs for math teachers. Our math department gets one or two new positions. Great!

This is a normal and logical response. One might call it the material or the physical response (as opposed to the spiritual response).

But there is another response mathematicians could make, which I would call the principled response. It would go like this: We already have too many students in our classes who don't want to be there. To have classes which are interesting, we need to have students who are interested. As for the students who are in our classes only by compulsion, let them go free! We do not desire their dormant bodies while their minds are elsewhere!

Well, this spiritual or idealistic mathematical response will most likely bring a smile to the reader's face. It just isn't practical! But, practical or not, it is part of the felt response (spoken or unspoken) of the mathematicians to proposals for increasing math requirements. We really don't want to teach unwilling students in required courses. We want to teach interested students in voluntary courses. Even if there is no possibility of getting what we want, the way we feel affects the situation and must be taken into account. It follows then, that, as to the mathematics filter, the mathematicians are ambivalent.

Finally, there is the general population, society as a whole, especially the candidates or applicants who must be passed or rejected by the math filter.

One effect of the math filter is to introduce a serious bias against women, Blacks, Hispanics and Native Americans. This bias has, of course, been noticed by the advocates and spokespersons of these groups. The general response has been to institute campaigns to raise the mathematical competence of these groups and to improve the mathematical education available to them. I think that significant progress has been made in these endeavors. At the same time, the general situation in which the math filter excludes disproportionate numbers of women, Blacks, Hispanics and Native Americans continues to prevail and is unlikely to change much in the foreseeable future.

There is no doubt that whatever improvements in math education will be achieved as a response to the math filter are a benefit to society as a whole and to the young people in question. But why is the filter itself sacrosanct? Why not challenge its justification in terms of the specific school or job for which it is imposed? There is no law of nature, God, or government that everybody must know the quadratic formula. Mathematics is interesting and important, but so are art, religion, literature, and many other things.

In a just, more rational world, mathematics would be used as a filter only for posts for which it is demonstrably required. Such a change would be gladly accepted by the mathematics professor. We do not really want to be the gate keepers and agents of exclusion.

QUESTIONS

1. During World War II, the U. S. Navy determined that those of its officers who had at some time passed a course in calculus made better officers than those who had not. Let us assume, without going into how the Navy measured goodness, that this is so. We may also assume, since it is true, that naval officers in World War II never had to use calculus in the performance of their duties. How, then, can the Navy's finding be explained?

2. It is well known that people can rise to occasions. There is no need to give a list of people who, while very young, have distinguished themselves in a wide variety of jobs and vocations. Why, then, do there have to be filters like calculus?

3. Can you think of reasons why it is that mathematics is so widely used as a filter? Can you think of other filters that would be as good or better? Why are

they not used instead?

4. Notice that the authors made it a point not to ask teachers at the business school about calculus, nor did they interview any practitioners of business. This may be another example of the know-it-allness of mathematicians, but that is beside the point. I once said to a graduate of my school, a mathematics major whose job was running a television station in Knoxville, Tennessee, that he wasn't using his mathematical training. "Oh, no," he said, "I use it every day." Surprised, I enquired further: what he meant was that he used, every day, the mathematical modes of thought that he had acquired in his study of calculus and like things. He had no reason to lie to me, so I am convinced that he was convinced that he was using in his business, every day, his mathematical training. Could that be, or was he deluding himself? If it is so, does this justify the calculus requirement of business schools?

SOLVING EQUATIONS IS NOT SOLVING PROBLEMS

by Jerome Spanier

When college students refer to the "real world" they mean the world outside of college. This is an error, since the world inside a college is just as real as the one outside. The reason for the error is of course that what goes on in a college, both inside and outside of classrooms, may seem artificial and inconsequential: things don't *count* the way they will later. Maybe so; in any event, how college graduates spend their time is usually quite different from how they spent their time as students. If they like to think that they inhabit a more real world after they graduate it does no one any harm.

Part of the business of colleges is to prepare students for the so-called real world. Thus, part of the business of calculus courses should be to prepare students for the job of solving mathematical problems that occur there. How to do that? The following selection suggests an answer, and also says quite a bit about what applied mathematicians do and how they do it.

In 1955 I was a Ph. D. candidate at the University of Chicago completing a dissertation in Topology under Shiing-Shen Chern. Although I had been trained exclusively in pure mathematics, jobs in industry were beginning to become available to mathematicians and I was interested. Some of my teachers expressed disapproval of such notions, but Professor Chern did not. He told me he believed that working on physical problems was interesting and difficult, and he encouraged me to keep an open mind. I found that I was curious to learn more about the applications of the mathematics I had studied, so upon graduation I took an industrial rather than a teaching position.

The first six months were very difficult for me. I was anxious to begin the transition from pure to applied mathematician, but at a loss to know how. My supervision ranged from minimal to nonexistent (Ph. D. mathematicians were evidently expected to know what to do with themselves) so I was pretty much on my own. I read considerably and gained some satisfaction from what I was learning, but I seemed to be moving no closer to an encounter with "real problems."

I had nearly decided to give up and seek an academic appointment when I tried a different approach. My position was in a laboratory whose research was concerned with the then-blossoming field of atomic energy. I made a point of speaking to physicists, engineers, chemists—as many different scientists as possible—about the nuclear reactor problems they were working on. Gradually I began to learn enough about reactor physics to appreciate some of these problems. My reading turned to physics and engineering more than to mathematics *per se.*

Finally, after more than a year, I felt ready (with the help of non-mathematician co-workers) to formulate and tackle some of the problems with which I had become somewhat familiar. This painful introduction to mathematical modeling was absolutely essential to my progress as an applied mathematician. My first job lasted nearly twelve years and the work I began then has continued as a research theme throughout my career.

Of the 25 years which have passed since 1955, I spent the first sixteen working as an industrial applied mathematician and the last nine teaching applied mathematics. I have given considerable thought to doing and teaching applied mathematics. Although as a student I felt uncertain about the relation between mathematics and its applications, I am now absolutely convinced that the use of mathematics to solve real problems is an interesting and challenging process which leads to genuinely new mathematics every bit as valid as work in abstract mathematics. After all, ancient geometry was created, in part, to answer questions raised in measuring land areas, and information theory arose out of consideration of modern communication systems.

I am also convinced that if we want to teach students to apply mathematics effectively, we must do much more than just put techniques for solving various equations in their hands. In this essay I shall advocate that a liberal dose of applying mathematics ought to be an integral part of the education of every applied mathematician.

The nature of applied mathematics

In thinking about criteria for a proper education in

applied mathematics, it seems appropriate to compare preparation for a more traditional mathematical career with what appears to be necessary training for work in the applications. Abstract mathematics proceeds from a system of axioms, or assumptions, to a collection of theorems (truths) and counterexamples (falsehoods) which can be established within such a system. Further work proceeds by extending and sharpening existing results, or by specializing to fit particular realizations. The teaching of techniques of proof plays an important role in training persons for work in abstract mathematics.

By contrast, work in applied mathematics begins with the statement of a problem which has arisen in a discipline outside of mathematics. The applied mathematician's first task is to create a mathematical image—or model—of the physical process which incorporates *realistic* assumptions and constraints. From this model, mathematical evidence is gathered which must then be compared with physical evidence typically gained through experiments or observation. This comparison determines the worth of the mathematical model. If the model is found wanting it must be altered and new mathematical evidence must be gathered. Quite clearly the original problem is of paramount importance to the effectiveness of the modeling process.

If one accepts these contrasting requirements, it seems evident that the teaching of applied mathematics calls for new approaches. Traditional courses in mathematics are technique-oriented: students are taught how to deal with a problem once it has been formulated mathematically. Thus, students are told how to solve algebraic equations, differential equations, integral equations—all kinds of equations—and how to formulate and prove theorems about such systems; but they are not taught where or how these equations arise. And yet the applied mathematician in industry or government must be prepared to help translate the engineer's or the economists's description of a problem into mathematical language *before* he can apply any of the ideas he has learned in traditional courses. The critical steps required to transform a concrete problem into mathematical symbols and equations, as well as the subtler techniques needed to evaluate the worth of the resulting mathematical model, are simply not treated in traditional courses.

Mathematical modeling

Partly to remedy this defect, it has become quite fashionable in the past decade to offer courses in mathematical modeling. A great variety of such courses has been introduced; the feature these seem to share is

an emphasis on problem-orientation, rather than on technique. Students in these courses are exposed to "solved" real problems, ranging from those that are simply stated and accessible to undergraduates to complex problems arising in modern technology that require the most sophisticated mathematical treatment. Although such modeling courses may be excellent, they fail to give students the experience of tackling significant *unsolved* problems on their own except perhaps through projects undertaken as part of the course requirements. My contention is that when students are forced to develop their own approaches to unsolved problems they benefit much more than from exposure to the mathematics alone.

Although modeling courses can be excellent vehicles for acquainting students with some aspects of the work of an applied mathematician, they cannot really prepare students adequately for industrial problem-solving. The reasons are, first, that industrial problem-solving almost always involves communication with non-mathematicians; second, it usually involves teamwork and shared responsibilities; and third, industrial work of a substantial nature requires a considerable depth of penetration into the discipline in which the problem arises. In many other respects as well, courses in modeling offer rather pale imitations of actual industrial problems.

A superior educational device in many respects is an applied mathematics practicum. This might range from an internship in which a student is introduced (normally individually) into a working environment, to a full-scale duplication of many aspects of the working environment in a classroom setting.

Internship vs. educational project activity

An internship might seem to provide an ideal way to introduce students to industrial work. However, I believe that although there may be some value in sending students to the working environment, there is much more to be gained in bringing the working environment into the classroom.

When a student accepts an internship, he may (and often is) given an assignment to assist an individual who acts as the student's supervisor. Rarely does the student become involved in substantial independent activity and almost never does he become part of a working team. He may be asked to do a fair amount of reading on his own and he will often be given a computer programming assignment. In a small number of instances he may be performing more nearly the work of a research assistant. While each of these activities is useful, the overall situation is less than ideal and the student is often bored and/or frustrated. Usually, no team activity is encountered, and (often) no real applied mathematics is learned. Total control of the educational experience is turned over to the sponsoring firm, and this does not usually serve the best interests of the student. ...

Requirements for training applied mathematicians

In my analysis of requirements for training students for work in the applications of mathematics, I do *not* advocate wholesale replacement of fundamental mathematics courses by modeling and practicum work. Frankly, my own view is that the successful applied mathematician must know as much mathematics as possible and must acquire some experience, of the sort I have described, in dealing with open-ended real problems. The traditional and the new educational elements need to be related skillfully for maximum benefit.

A graduate-level modeling course and a unique practicum in which students grapple with real problems lie at the heart of Claremont's new program in applied mathematics. The program has been effective in attracting students and in placing them after graduation. Experience gained in developing and testing this novel program suggests that the following constitute very important requirements in training applied mathematicians and prove to be valuable even more generally:

(1) a focus on problem-solving;
(2) experience in communications, both oral and written;
(3) familiarity with cognate disciplines;
(4) exposure to at least one paradigm of applied mathematics research;
(5) confidence acquired in open-ended problem solving.

While many of the specifics of the program I have described may not be easily duplicated elsewhere, some features of it deserve strong consideration by any institution that would train students to use mathematics effectively.

QUESTIONS

1. "In this essay I shall advocate that a liberal dose of applying mathematics ought to be an integral part of the education of every applied mathematician." Note the adjective: *applied* mathematician, not every mathematician. Note also the noun: part of the education of every applied *mathematician*, not the education of everyone who studies calculus. Do you think that the program that the author advocates should be part of the education of every mathematics major, or of every student of calculus? If so, why? And if not, why not?

2. Look at the five things that the author says are "valuable even more generally." How many of them could be part of a calculus course? If one could be a part, in what way could it? *Should* they be part of a calculus course? What are the costs, what are the benefits, and do the benefits justify the costs?

3. The vast majority of applied mathematicians, past and present, never had the benefit of the author's program. Nevertheless, they were and are able to solve problems well enough so that their employers found and find it worth their while to keep them on their payrolls. Therefore, it can be argued, no change in the present system is needed. Would you argue this way? Why, or why not?

APPLIED MATHEMATICS IN ENGINEERING

By Brockway McMillan

Around 1960, the eminent physicist Eugene P. Wigner wrote a much-reprinted paper called "The unreasonable effectiveness of mathematics in the natural sciences". Wigner was surprised that mathematics should prove so useful in physics, chemistry, geology, in all of the natural sciences. The natural sciences are in the business of investigating the world and finding out what makes it work, and Wigner found it wonderful that mathematics, which has nothing to do with nature, should be able to solve problems that scientists have. It is not only wonderful, he said, it is miraculous:

> The miracle of the appropriateness of the language of mathematics for the formulation of the laws of physics is a wonderful gift which we neither understand nor deserve.

Is it a miracle, wonderful, and unreasonable, or is it reasonable and only to be expected?

It is a question that the average person, not knowing much about mathematics or science, would probably have no trouble answering: "Sure it's reasonable," the average person would most likely say. The reason is that in the popular mind science, mathematics, and engineering are all mixed together in an undistinguished mass of technical stuff. People who don't have to think about science, mathematics, or engineering can think that Thomas Edison was a great scientist (he was great, but no scientist) and Albert Einstein was a great mathematician (also great, but no mathematician), so it is no surprise that they find it unsurprising that mathematics applies to science and engineering.

With a little more knowledge of science and mathematics the surprise can develop. Science is interested in *things*: astronomy is always gazing at the stars, geology is always chipping at rocks, and physics is always trying to find smaller and smaller things that can be put together in various ways to make up all the things that are. Mathematics does not deal in things. No one has ever seen a derivative go by outside the window, touched a prime number, or smelt a triangle. High-school geometry, if taught correctly, is presented as a body of theorems that are deduced from postulates using the laws of logic. Nothing is said about the "truth" of the postulates. Truth or falsity is irrelevant: here are the postulates and here is what follows from them; what is in the world is irrelevant, the *world* is irrelevant, the theorems of geometry would be true if the world did not exist. However, the theorems of geometry come in handy when you want to apply fertilizer to a field or order cement for a dam. How can that be? How can something independent of the world be so useful in it? It seems unreasonable.

Looked at that way it *is* surprising. Sciences use induction to get at truth: they observe that part of reality they are concerned with and make guesses about why reality behaves as it does. Why did the apple bump Newton on the head? Maybe, physics said, there's a force that pulls masses together. Why did people get smallpox? Maybe, medicine said, little germs pass from one person to another. Science looks at the world, while mathematics does not. Science induces, mathematics deduces. Every mathematical result has the same form: "If A then B." If A is true, then B is true. Is A true? Mathematics does not know, and mathematics does not care. And yet mathematics *applies*. Here is a sonnet, by a mathematician, C. R. Wylie, Jr., that expresses the unreasonableness of the effectiveness of mathematics:

> Not truth, nor certainty. These I foreswore
> In my novitiate, as young men called
> To holy orders must abjure the world.
> "If ... , then ... ," this only I assert;
> And my successes are but pretty chains

Linking twin doubts, for it is vain to ask
If what I postulate be justified,
Or what I prove possess the stamp of fact.

Yet bridges stand, and men no longer crawl
In two dimensions. And such triumphs stem
In no small measure from the power this game,
Played with the thrice-attenuated shades
Of things, has over their originals.
How frail the wand, but how profound the spell!

Actually the effectiveness of mathematics is not all that unreasonable. Mathematics is that subject that considers quantities. It is about numbers. It answers questions about how much, how many, how far, how fast. So, it should be no surprise that when a science has a question that involves quantities, its answer will involve mathematics. *Of course* it will! If geology wants to do more than classify rocks into types, if it wants to predict earthquakes, then it will have to measure and count and use mathematics. If biology wants to do more than sort species into phyla, if it wants to determine how they evolve, then it will have to calculate and compute and use mathematics. Number and quantity—the world is full of them and millions of scientific questions involve them, so it is not seem unreasonable that mathematics should help answer them.

The next selection describes how mathematics helped answer the question "How can we send a rocket to the moon?" and it also helps answer "How is mathematics effective?"

It is the purpose of this paper to illustrate and defend the following assertions:

Mathematical solutions can be found only for mathematical problems. It is the first job of the applied mathematician to state the central problems in explicit mathematical form. It is then his job to solve these problems. Computational algorithms and numerical solutions are neither necessary nor sufficient for either task.

It is my conviction that any lesser concept of applied mathematics than is implied by these assertions is not merely incorrect, but harmful. I shall set forth the illustration and defense of the assertions, and the defense of the stated conviction, in elliptical terms: they are offered in a form that some might call a parable and others a case study in sociology.

It is a practice in the softer sciences, such as sociology, for an author to begin his paper with a brief autobiography or other statement of personal background. This alerts the reader to possible bias on the author's part in matters involving judgment. The case study I present will, of course, be fully factual and objective. Nevertheless, a statement of the author's bias is a good precaution, lest by inadvertence some value judgment find expression.

Two schools of thought exist relative to applied mathematics: there are the formalists and the spiritualists. The formalists hold the view that there is some unique body of knowledge and subject matter which one must command in order to qualify as an applied mathematician. The spiritualists, on the other hand, believe that applied mathematics is defined not by subject matter, but by the spirit or attitude of the mathematician who engages in it.

I am without reservation a spiritualist. In my view, applied mathematics has no mathematical boundaries which distinguish it in any way from the domain of mathematics as a whole. Furthermore, I deplore any effort by definition or decree to erect such boundaries. There is no question that some branches of mathematics have found more widespread application that some others. This is a fact of historical interest, and not without technical interest. But the fact is not itself a valid basis for boundaries that restrict inquiry or interfere with education.

Thornton Fry was a spiritualist when, in 1941, he defined mathematics as "what mathematicians do." In the context of Fry's paper, I think it is fair to conclude that the modifier "applied" was understood, in reference both to "mathematics" and "mathematicians." Fry elaborated his definition by characterizing mathematicians: he referred to their demand for logical precision, their skill at abstraction and idealization ("model building"), their passion for economy of thought, their desire for completeness (from these latter two comes their predilection for generalization), and their respect for craftsmanship. I commend Fry's account to every reader.

Suffice it here to say, as a caveat for what follows, that I am unreservedly a spiritualist.

Our case study centers upon a segment of the aerospace industry, and recounts the development of a new domain of engineering. Our prime focus is on the role of the mathematician in this development, and upon related cultural phenomena. Similar cases, I am sure, could be studied in other subcultures—e. g., in the communications industry. Technical material which illustrates the features of most interest to our study lies closer at hand, however, in the dynamics of flight vehicles than in the analogous dynamics of electrical circuits.

Our history begins at the time that a certain society decided, for reasons which are not a part of the present study, to build a rocket to go to the moon. For many years this society had supported a large and occasionally prosperous aircraft industry. It seemed natural to turn to that industry for the development of the related, but new, technology of rockets.

As we shall see, the development of this technology was greatly assisted by a certain mathematician of that society. From our, and his, point of view, the development took place in five phases: first, improvisation; second, simulation; third, conceptualization; fourth, assimilation; fifth, recrimination.

Rockets are not airplanes, but aircraft engineers knew something about them. It was obvious, for example, that to go to the moon would require more fuel, per pound of payload, than to loft fireworks over Expo 67. But nobody was sure exactly how much more fuel. Indeed, an important question from the very start was the question of size: how big must a rocket be, to go to the moon? Long before they could begin to lay out the detailed design of such a rocket, or even to make a forward-looking plan, the engineers needed some estimates of size to orient their thinking.

They set about in a straightforward way to get their answers by improvisation. They built a rocket, and tried it. It flew a short distance and fell into the sea.

They thought about this. It looked as if the thrust hadn't lasted long enough. Maybe they should make the hole in the nozzle a little smaller, to reduce the rate of ejection.

They tried that, too. They built a second rocket, with a smaller nozzle. This one burned a long time. But it just sat on the pad without going anywhere at all.

It is possible to imagine that such a process of trial and error, if continued long enough, could ultimately build up (perhaps even at great profit to the industry involved) an empirical body of doctrine and design data

by which one could finally determine the characteristics of a rocket to go to the moon.

Actually, things didn't go that way. Someone found a mathematician who was willing to work on the project. He may have been a space cadet, or just disappointed in his plea for a grant from NSF, but at any rate he was willing to work.

Despite the careful schooling by which he had been trained, this mathematician had somewhere heard that force equals mass times acceleration and that action and reaction were equal and opposite. He thought that by using these principles, he could write some differential equations and by this process construct a mathematical model of a rocket. This in fact he did. His equations related the vertical acceleration of the rocket to the rate of burning of the fuel, to the velocity of ejection of the burned gases, and to the mass of the rocket and its fuel.

This is of course a modern history, so we will postulate the existence of a computer. Equally modern, our mathematician put his equations on the computer. And thus began phase 2: simulation.

To check the model, they put into the computer the mass, burning rate, and ejection rate that characterized their first experimental rocket. The computer correctly concluded that this rocket would fall into the sea. Then they tested the model against rocket number two, and again got the correct conclusion: the computer found that number two would sit on the stand and go nowhere.

By this time, the engineers were delighted, though perhaps their bosses and stockholders were not, because now they could conduct simulated tests at a cost of a few hundred dollars per trial, rather than at costs measured in the millions. And the leaders of the society were delighted also, for now their engineers could do in hours what had previously taken them months to accomplish. For the first time, their hopes rose that they could reach the moon before their rivals.

Now any resemblance to a value judgment at this juncture is purely coincidental. But I submit that today there are some domains of engineering which are still essentially in phase 1, improvisation, or are no farther than phase 2, in which improvisation is facilitated by simulation. Not many of these domains have been subjected to social analysis as penetrating as the one we are here entered into. Consequently, their state of development may not be as starkly evident as in the case under study here. But what we are seeing in our case is typical. As engineers extend the base of their knowledge and doctrine, they go through phases 1 and 2, rapidly or slowly, and sometimes stop at this point.

In our case, the engineers were fortunate in their mathematical colleague. In the first place, he was not an empiricist. He was not satisfied by results without understanding. Furthermore he believed, as I do, that mathematics is the art of avoiding computation. Therefore, while the engineers were grinding away at their simulations—and indeed, while some of them were designing a graphical display console so they could get data out of their computer faster—our mathematician was staring at his differential equations.

There is in fact a secret that he told no one but me: he even made a simpler model. He wrote down a one-dimensional model, with one degree of freedom, replacing the six-degree-of-freedom system being used in the simulation. He was able to solve the resulting second-order differential equation. With this solution, he discovered momentum; and he discovered the conservation of momentum; and he discovered kinetic energy; and he discovered the gravitational potential, and recognized that this measured a second form of energy; and he discovered the conservation of energy; and he defined specific impulse and escape velocity. He wrote two memoranda, one about integral invariants of a certain twelfth-order system of differential equations, and another containing some universal inequalities relating specific impulse and mass fraction, on the one hand, to terminal velocity and escape velocity, on the other.

These two memoranda initiated phase 3: conceptualization. Please note that it began without our mathematician using one result of a simulation, and indeed without his examining one numerical solution of a differential equation. It began because he sought to understand the system of equations with which he was dealing, and to explore the properties of their solutions.

As will appear later in this history, there were engineers who didn't recognize that the mathematician had solved any problem. After all, he hadn't told them how big a rocket must be, to put a pound of payload on the moon. But let us look at what he had done.

Conceptualization was, in fact, less a phase than an epoch. For conceptualization initiated phase 4: assimilation. With an explicit mathematical model at hand, the differential equations of dynamics, and with the right fundamental concepts to use in discussing this model, what started out as mathematics suddenly became rocket engineering. The engineers now had a structure within which to define their problems and to put them into explicit mathematical form, and a language with which to discuss relations and solutions. And they had limiting solutions and inequalities on which to base judgments as to how well they were doing, or as to how well it was reasonable to try to do.

It's true that the mathematician didn't tell them how big to make their rocket, because that turned out not to be a mathematical problem. It was he who showed them that this problem reduced to two other engineering problems: that of estimating the mass fraction which could be achieved in the design of the rocket's structure, and that of estimating the specific impulse which could be expected from the propellant. The concepts that mathematics provided told the engineers how to break their design problem into logically defined components, gave them a basis for setting objectives to be accomplished by the component designs, and told them how sensitive their final result was to success or failure in the components.

At this point, our case history has demonstrated the assertion with which we began. Mathematics made rocket engineering possible, just as it has made possible all other branches of engineering, not by providing numerical solutions or computational algorithms, but by providing an explicitly mathematical structure within which to formulate problems for solution. Mathematics gave the engineers something to talk about, and provided the language used in the discussion. Whether or not the mathematician also supplied numerical solutions, or techniques for getting them, is merely coincidental.

In the present case, a couple of nineteenth-century chemists named Adams and Bashforth had long ago provided step-by-step methods for solving differential equations. Prior to the opening of our history, these had been brought up to date and fitted to the high-speed computer, so that our mathematician was little concerned with this part of the problem.

Those acquainted with the current history of rocketry, however, will recognize that this lack of concern is also merely coincidental. For example, when one must pursue an orbiting body through many orbits and must predict its ephemeris far into the future in a real environment, numerical methods other than the step-by-step are required to avoid the accumulation of artifacts. Here, many mathematicians have contributed.

To return to our case history, we observe that the onset of phase 4, assimilation, imposed a choice upon our mathematician. For rocket engineering now existed as a named branch of engineering. Mathematical models were interpreted as engineering fact and mathematical concepts were endowed with physical reality. The reference and debt to mathematics were forgotten. Our mathematician had to choose between continuing his work on rockets, and being called a rocket engineer, or picking up his tools and moving on to some new field, where for a few years he would again be called an applied mathematician. He ultimately made the latter choice, but not until he had experienced the beginning of phase 5: recrimination.

Phase 5 involves two subcultures: that of the rocket engineers, and that of a group of scholars in our society who carefully call the domain of their discourse "pure mathematics." These are brilliant, imaginative people with intense curiosity about mathematical phenomena. Some of them read the memorandum about the integral invariants of a certain twelfth-order system of differential equations. They discovered that there are other systems of equations which have integral invariants. Other such scholars found that integral invariants of systems of ordinary differential equations are simply special cases of more general invariants of more general mathematical systems. These people belittled their colleagues who insisted on studying systems of ordinary differential equations, and both groups scorned those outside their subculture who found that thinking about the motion of rockets was helpful to the understanding of integral invariants. And this scorn was carefully planted and nurtured in the young students in the subculture.

Disdain of people who think in terms different from those in current vogue in one's own subculture is a social phenomenon of universal occurrence. It appears in the arts as well as in science. It is important to phase 5 because of its essentially destructive effect. It is not, of course, so destructive that it threatens our society—it could not be and at the same time be universal. But it is harmful in that it rigidifies the structure of society and acts against growth or change. It limits the opportunities for thought and for learning, and it discourages intellectual adventure. In our particular case, it imposed social barriers, meaningless but effective, between those studying material of potential relevance to rocket engineering and the knowledge of that relevance.

More immediately destructive phenomena meanwhile became evident within the community of rocket engineers itself. In fact, the difficulty was appearing even during phase 2 of simulation. For it was then that one of the engineering supervisors said, in reference to our mathematician: "Look, we got to get rid of that guy. All he does is sit there and write memoranda. What we need is results."

I cite this quotation to describe the phenomenon of recrimination, not to illustrate its impact. For in fact the harmful effect of this particular engineer's attitude was soon negated. The mathematician's successes rapidly passed him by. His lack of understanding was so explicit that he soon was transferred to Contract Administration, where the technical demands better matched his capacity.

But recrimination can be evidence of a lack of understanding that is harmful. After assimilation has taken place and the whole fabric and structure of the engineering domain has been defined in terms of mathematical concepts and operations, there are still

those who ask "Well, Mr. Mathematician, what have you solved for us lately?"

What is harmful here is not, of course, the question but the attitude that gives it rise—the belief that mathematicians serve by solving problems which engineers pose. When the engineer says, "Don't think; calculate. I know what my problems are," he alienates the professional mathematician by consigning him to the servant class. Worse, he closes the door to understanding by the mathematician of the engineer's true problems, and therefore denies himself the creative thought—the formulation of concepts and the isolation of problems—that is the essence of the mathematician's potential contribution.

Let us not forget, either, that hard problems may indeed be intrinsically hard. Anyone, be he mathematician or engineer, who believes that mathematics is a magic key which opens all doors is certain to be disappointed and frustrated. Such emotions, felt on either side, do not foster creative working relations between mathematicians and engineers.

Indeed, if one looks at rocketry today he sees a good sample of what mathematics has done, and has not done, in its service. In fact, of course, it took over a quarter of a millennium to progress from mass-times-acceleration to specific impulse and escape velocity. Today with all this, with the sophisticated development in dynamics and celestial mechanics of the last hundred years, and with modern control theory at hand, engineers still have only a set of guiding principles for the design of a rocket to the moon. The design itself results from thousands of man-years of detailed effort. Flight is possible only after many tests of elements, components, and the whole system. To select trajectories and launch times, hundreds to thousands of trial trajectories may have to be computed.

All these detailed efforts use the tools and concepts provided by mathematics, but the departures—caused by real materials, by the actual gravitational field of the solar system, by the qualities of gyroscopes, and by the laws of fluid flow—from the mathematical idealizations which were employed to define the concepts and set up the original mathematical models are too great to be ignored in a final design. Mathematics points the road to the top of the mountain, but it cannot make the mountain any lower. To believe otherwise is to invite disappointment.

This paper must close with two caveats. Its elliptical discourse has in fact exceeded escape velocity and opened out into hyperbole. I have flagrantly distorted the intellectual history of rocketry. But I submit that the distortion presents a fair picture in symbolic terms of the intellectual growth of any new domain of engineering.

I may also have implied that all developments in engineering are made by mathematicians. Being a spiritualist, I must insist that indeed many such developments are. I cannot claim, however, that when a mathematical advance was made, it was made by someone who realized at the time that he was a mathematician.

QUESTIONS

1. "Disdain of people who think in terms different from those in current vogue in one's own subculture is a social phenomenon of universal occurrence." A house is on fire. The problem is to put the fire out. The problem is given to an engineer, an applied mathematician, and a pure mathematician. The engineer says, "Call the fire department." The applied mathematician says, "Apply 120,000 gallons of water." The pure mathematician says, "A solution exists."

There is a genre of such jokes. Here is another one. The engineer, the applied mathematician, and the pure mathematician are travelling through Australia and pass a field of sheep, half of whom have black wool. The engineer says, "I see that half of the sheep in Australia are black." The applied mathematician says, "No, no—all you can say is that half of the sheep in that field are black." The pure mathematician says, "Half of the sheep in that field are black on at least one side."

What do the jokes say about the three subcultures? Which subculture do you think made them up?

2. "Applied mathematics is bad mathematics"—a title intended to shock, a little—by P. R. Halmos is a paper well worth reading (it can be found in *Mathematics Tomorrow*, edited by Lynn A. Steen, Springer-Verlag, New York, 1981). Among other things it contains

> You can usually (but not always) tell an applied mathematician from a pure one just by observing the temperature of his attitude toward the same-different debate. If he feels strongly and maintains that pure and applied are and must be the same, that they are both mathematics and the distinction is meaningless, then he is probably an applied mathematician. About this particular subject most pure mathematicians feel less heat and speak less polemically: they don't really think pure and applied are the same, but

they don't care all that much. I think what I have just described is a fact, but I confess I can't help wondering why it's so.

Do you have any conjectures as to why it's so?

3. "Mathematics is the art of avoiding computation." What? *Avoiding* computation? Isn't mathematics the art *of* computation? Do you think that the assertion is true? Do you think that people in general think it is true? How would you try to convince someone who thought that it was false that it was true, at least in a sense?

4. "He may have been ... disappointed in his plea for a grant from NSF." "NSF" stands for "National Science Foundation," a branch of the Federal government that spends hundreds of millions of dollars each year on various scientific endeavors, much of it in grants to private investigators who have made application for support. In particular, the NSF supports many mathematical research projects, and many of them fall into the category of "pure" mathematics. Pure mathematics is mathematics for its own sake, with no thought of any possible application. Moreover, much of pure mathematics cannot be applied to any practical problem whatsoever and never will be. Why should the government spend its taxpayers' money in this way? Should it stop? What would happen if it did stop?

FAMILIAR QUOTATIONS

First edition edited by John Bartlett

There are no new deep truths about human existence. The big thoughts have been thought. The history of humanity comprises thousands of generations and billions of people and the collective wisdom that the race has accumulated is not going to be revolutionized by this generation, or by the next. Of course, to each new person the world is new, and each new person makes discoveries that seem new and astonishing ("Some politicians will actually *tell lies* to get elected!"; "The purpose of education is to *keep things the way they are!*") but they have come before, many times, to other people who were once new, in generations that have long passed away. Let us never underestimate the past.

Bartlett's *Familiar Quotations* is a collection of part of the accumulated wisdom of the race. "Familiar" is the key word: the first edition was meant to include quotations that everyone knew, more or less vaguely. If everyone knows something then it is part of our culture, one of the things that has made us what we are. It will be generally accepted, and part of what everyone agrees with. There is even a good chance that it will be true. To know a people, read their quotations. Open Bartlett's at random and you will find familiar things like

> Beauty's but skin deep.
> (John Davies of Hereford, c. 1565-1618)

> A place for everything and everything in its place.
> (Isabella Mary Beeton, 1836-1865)

and other items, not as familiar but worth reading

> The difference between a moral man and a man of honor is that the latter regrets
> a discreditable act, even when it has worked and he has not been caught.
> (H. L. Mencken, 1880-1956)

It is a fine book to browse through.

What does Bartlett's say about calculus? Nothing at all, which says something about the place of calculus in human existence. However, it does say some things about mathematics. They appear below, and so they represent, more or less, the collected wisdom of all the ages on the subject.

John Adams (1735-1826)

I must study politics and war that my sons may have liberty to study mathematics and philosophy. My sons ought to study mathematics and philosophy, geography, natural history, naval architecture, navigation, commerce, and agriculture, in order to give their children a right to study painting, poetry, music, architecture, statuary, tapestry, and porcelain.
(Letter to Abigail Adams (May 12, 1780))

Francis Bacon (1561-1626)

Historie make men wise; poets, witty; the mathe-matics, subtile; natural philosophy, deep; moral, grave; logic and rhetoric, able to contend.
(*Essays* (1625), "Of Studies")

Roger Bacon (c. 1214-c. 1294)

If in other sciences we should arrive at certainty without doubt and truth without error, it behooves us to place the foundations of knowledge in mathematics.
(*Opus Majus*, Robert Burke translation, bk. I, ch. 4)

Havelock Ellis (1859-1939)

The mathematician has reached the highest rung on the ladder of human thought.
(*The Dance of Life* (1923), ch. 3)

Galileo Galilei (1564-1642)

Philosophy is written in this grand book—I mean the universe—which stands continually open to our gaze, but it cannot be understood unless one first learns to comprehend the language and interpret the characters in which it is written. It is written in the language of mathematics, and its characters are triangles, circles, and other geometrical figures, without which it is humanly impossible to understand a single word of it; without these, one is wandering about in a dark labyrinth.
(*Il Saggiatore* (1623), Stillman Drake and C. D. O'Malley translation (1960))

Carl Friedrich Gauss (1777-1855)

It may be true that people who are *merely* mathematicians have certain specific shortcomings; however, that is true of every exclusive occupation.
(Letter to H. C. Schumacher (1845))

Mathematics is the queen of the sciences.
(From Sartorius von Waltershausen, *Gauss zum Gedachtness* (1856))

Johann Friedrich Herbart (1776-1841)

Psychology cannot experiment with men, and there is no apparatus for this purpose. So much the more carefully must we make use of mathematics.
(*Lehrbuch zur Psychologie* (1816), Margaret K. Smith translation (1891))

Oliver Wendell Holmes, Jr. (1841-1935)

The law embodies the story of a nation's development through many centuries, and it cannot be dealt with as if it contained only the axioms and corollaries of a book of mathematics.
(*The Common Law* (1881))

William James (1842-1910)

The union of the mathematician with the poet, fervor with measure, passion with correctness, this surely is the ideal.
(*Collected Essays and Reviews* (1920), ch. 11, Clifford's "Lectures and Essays" (1879))

Sir James Hopwood Jeans (1877-1946)

All the pictures which science now draws of nature and which alone seem capable of according with observational fact are mathematical pictures. ... From the intrinsic evidence of his creation, the Great Architect of the Universe now begins to appear as a pure mathematician.
(*The Mysterious Universe* (1930))

James Clerk Maxwell (1831-1879)

All the mathematical sciences are founded on relations between physical laws and laws of numbers, so that the aim of exact science is to reduce the problems of nature to the determination of quantities by operations with numbers.
(*On Faraday's Lines of Force* (1856))

Benjamin Peirce (1809-1880)

Mathematics is the science which draws necessary conclusions.
(*Linear Associative Algebra* (1870), first sentence)

Plato (c. 428-348 B. C.)

I have hardly ever known a mathematician who was capable of reasoning.
(*The Republic* (Jowett translation) VII, 531-E)

Michael Polanyi (1891-1976)

An art which has fallen into disuse for the period of a generation is altogether lost. There are hundreds of examples of this to which the process of mechanization is constantly adding new ones. These losses are usually irretrievable. It is pathetic to watch the endless efforts—equipped with microscopy and chemistry, with mathematics and electronics—to reproduce a single violin of the kind the half-literate Stradivarius turned out as a matter of routine more than two hundred years ago.
(*Personal Knowledge* (1958))

Bertrand Russell (1872-1970)

Thus mathematics may be defined as the subject in which we never know what we are talking about, nor whether what we are saying is true.
("Recent work on the principles of mathematics" (1901), in *International Monthly*, vol. 4, p. 84)

Mathematics, rightly viewed, possesses not only truth, but supreme beauty—a beauty cold and austere, like that of sculpture, without appeal to any part of our weaker nature, without the gorgeous trappings of painting and music, yet sublimely pure, and capable of a stern perfection such as only the greatest art can show.
(*The Study of Mathematics* (1900))

Mathematics takes us still further from what is human, into the region of absolute necessity, to which not only the actual world, but every possible world,

must conform.
(Ib.)

Sir D'Arcy Wentworth Thompson (1860-1948)

Numerical precision is the very soul of science.
(*On Growth and Form* (1917), ch. 1)

The harmony of the world is made manifest in Form and Number, and the heart and soul and all the poetry of Natural Philosophy are embodied in the concept of mathematical beauty.
(Ib. 10)

The perfection of mathematical beauty is such ... that whatsoever is most beautiful and regular is also found to be most useful and excellent.
(Ib.)

Leonardo da Vinci (1452-1519)

Mechanics is the paradise of the mathematical sciences because by means of it one comes to the fruits of mathematics.
(*The Notebooks* (1508-1518), vol. I, ch. 20)

Izaak Walton (1593-1683)

Angling may be said to be so like the mathematics that it can never be fully learnt.
(*The Compleat Angler* (1653-1655), Epistle to the Reader)

Alfred North Whitehead (1861-1947)

The study of mathematics is apt to commence in disappointment. ... We are told that by its aid the stars are weighed and the billions of molecules in a drop of water are counted. Yet, like the ghost of Hamlet's father, this great science eludes the efforts of our mental weapons to grasp it.
(*An Introduction to Mathematics* (1911), ch. 1)

The science of pure mathematics, in its modern developments, may claim to be the most original development of the human spirit.
(*Science and the Modern World* (1925), ch. 1)

QUESTIONS

1. Bertrand Russell: "Mathematics takes us still further from what is human ... ." Alfred North Whitehead: "The science of pure mathematics ... may claim to be the most original development of the human spirit." Is there a contradiction here? Is mathematics human, or is it not?

2. Havelock Ellis puts mathematicians on the highest rung of human thought, which is nice for mathematicians. If there is a rung, there must be a ladder. Who is on the other rungs, and in what order? If there is a highest rung, there must be a lowest one too: what unfortunate person stands there?

3. Oliver Wendell Holmes says that mathematics is too limited for the law and Michael Polanyi says that mathematics is too limited to duplicate a Stradivarius. Does that mean that mathematics does not occupy the highest rung of human thought?

4. The quote by John Adams contains a hierarchy, but in what direction it going? Up, in order of increasing goodness from the mundane to the worthwhile, or down, in order of increasing badness from the useful to the frivolous?

5. Does Francis Bacon have the causes and effects in the right order? Does the study of morality make people grave, or is it that only people who are naturally grave study morality?

MEMORABILIA MATHEMATICA

edited by Robert Edouard Moritz

What good is studying calculus, or mathematics in general? Everyone knows the answers that we give today, but not everyone knows what answers were given yesterday or the day before that. You may be surprised at some of the ones that appear in the collection that follows. Many of the quotations in it were fifty years old, or older, when it appeared in 1914.

You may be surprised, and about some of them think, "What nonsense! How could anybody write such dumb things?" Don't think that. Don't fall into the common error of patronizing the past and thinking that we, right now, have all the right answers and that our ancestors, though they no doubt did the best that they could considering the handicaps they had to work under, were not quite as smart, well-informed, and wise as we are. Our turn will come, and people at some time in the future will look back at what we write and ask how could those turn-of-the-21st-century Americans have been so stupid, short-sighted, misguided, or prejudiced to think as they did. They may even laugh, uproariously, at some very serious things that were written by some very solemn people. Ideas go in and out of fashion even as clothes do, and old ideas can be revived (though always slightly changed, so that they can be called new) just as old styles of dressing come back (though always slightly changed).

---

117. Everything that the greatest minds of all times have accomplished toward the *comprehension of forms* by means of concepts is gathered into one great science, *mathematics.*—J. F. Herbart.

202. There is no study in the world which brings into more harmonious action all the faculties of mind than the one [mathematics] of which I stand here as the humble representative and advocate. There is none other which prepares so many agreeable surprises for its followers, more wonderful than the transformation scene of a pantomime, or, like this, seems to raise them, by successive steps of initiation to higher and higher states of conscious intellectual being.—J. J. Sylvester.

216. In most sciences one generation tears down what another has built and what one has established another undoes. In Mathematics alone each generation builds a new story to the old structure—Hermann Hankel.

231. [In mathematics] we behold the conscious logical activity of the human mind in its purest and most perfect form. Here we learn to realize the laborious nature of the process, the great care with which it must proceed, the accuracy which is necessary to determine the exact extent of the general propositions arrived at, the difficulty of forming and comprehending abstract concepts; but here we learn also to place confidence in the certainty, scope and fruitfulness of such intellectual activity.—H. Helmholtz.

265. There is probably no other science which presents such different appearances to one who cultivates it and to one who does not, as mathematics. To this person it is ancient, venerable, and complete; a body of dry, irrefutable, unambiguous reasoning. To the mathematician, on the other hand, his subject is yet in the purple bloom of vigorous youth, everywhere stretching out after the "attainable but unattained" and full of the excitement of nascent thoughts; its logic is beset with ambiguities, and its analytic processes, like Bunyan's road, have a quagmire on one side and a deep ditch on the other and branch off into innumerable by-paths that end in a wilderness.—C. H. Chapman.

275. What is physical is subject to the laws of mathematics, and what is spiritual to the laws of God, and the laws of mathematics are but the expression of the thoughts of God.—Thomas Hill.

302. It may well be doubted whether, in all the range of Science, there is any field so fascinating to the explorer—so rich in hidden treasures—so fruitful in delightful surprises—as that of Pure Mathematics. The charm lies chiefly in the absolute *certainty* of its results: for that is what, beyond all mental treasures, the human intellect craves for. Let us only be sure of *something*! More light, more light! "And if our fate be death, give light and let us die!" This is the cry that, through all the ages, is going up from perplexed Humanity, and Science has little else to offer, that will really meet the demands of its votaries, than the conclusions of Pure Mathematics.—C. L. Dodgson.

306. He who knows not mathematics dies without knowing *truth*.—C. H. Schellbach.

310. Mathematics is the gate and key of the sciences. Neglect of mathematics works injury to all knowledge, since he who is ignorant of it cannot know the other sciences or the things of this world. And what is worse, men who are thus ignorant are unable to perceive their own ignorance and so do not seek a remedy.—Roger Bacon.

313. The advancement and perfection of mathematics are intimately connected with the prosperity of the State.—Napoleon.

320. As the sun eclipses the stars by his brilliancy, so the man of knowledge will eclipse the fame of others in assemblies of the people if he proposes algebraic problems, and still more if he solves them.—Brahmagupta.

324. What science can there be more noble, more excellent, more useful for men, more admirably high and demonstrative, than this of the mathematics?—Benjamin Franklin.

329. Mathematics is the life supreme. The life of the gods is mathematics. All divine messengers are mathematicians. Pure mathematics is religion. Its attainment requires a theophany.—Novalis.

401. [Mathematics] engages, it fructifies, it quickens, compels attention, is as circumspect as inventive, induces courage and self-confidence as well as modesty and submission to truth. It yields the essence and kernel of all things, is brief in form and overflows with its wealth of content. It discloses the depth and breadth of the law and spiritual element behind the surface of phenomena; it impels from point to point and carries within itself the incentive toward progress; it stimulates the artistic perception, good taste in judgment and execution, as well as the scientific comprehension of things. Mathematics, therefore, above all other subjects, makes the student lust after knowledge, fills him, as it were, with a longing to fathom the cause of things and to employ his own powers independently; it collects his mental forces and concentrates them on a single point and thus awakens the spirit of individual inquiry, self-confidence and the joy of doing; it fascinates because of the view-points which it offers and creates certainty and assurance, owing to the universal validity of its methods. Thus, both what he receives and what he himself contributes toward the proper conception and solution of a problem, combine to mature the student and to make him skillful, to lead him away from the surface of things and to exercise him in the perception of their essence. A student thus prepared thirsts after knowledge and is ready for the university and its sciences. Thus it appears, that higher mathematics is the best guide to philosophy and to the philosophic conception of the world and of one's own being.—E. Dillman.

405. Probably among all the pursuits of the University, mathematics pre-eminently demand self-denial, patience, and perseverance from youth, precisely at that period when they have liberty to act for themselves, and when on account of obvious temptations, habits of restraint and application are peculiarly valuable.—Isaac Todhunter.

410. In mathematics I can report no deficience, except it be that men do not sufficiently understand the excellent use of the Pure Mathematics, in that they do remedy and cure many defects in the wit and faculties intellectual. For if the wit be too dull, they sharpen it; if too wandering, they fix it; if too inherent in the senses, they abstract it. So that as tennis is a game of no use in itself, but of great use in respect it maketh a quick eye and a body ready to put itself into all positions; so in the Mathematics, that use which is collateral and intervenient is no less worthy than that which is principal and intended.—Lord Bacon.

415. Another great and special excellence of mathematics is that it demands earnest voluntary exertion. It is simply impossible for a person to become a good mathematician by the happy accident of having been sent to a good school.—Isaac Todhunter.

417. He who gives a portion of his time and talent to the investigation of mathematical truth will come to all other questions with a decided advantage over his opponents. He will be in argument what the ancient Romans were in the field: to them the day of battle was a day of comparative recreation, because they were ever accustomed to exercise with arms much heavier than they fought; and reviews differed from a real battle in two respects: they encountered more fatigue, but the victory was bloodless.—C. C. Colton.

418. Mathematics is the study which forms the foundation of the course [at the West Point Military Academy]. This is necessary, both to impart to the mind that combined strength and versatility, the peculiar vigor and rapidity of comparison necessary for military action, and to pave the way for progress in the higher military sciences.—Congressional Committee on Military Affairs.

420. Most readers will agree that a prime requisite for healthful experience in public speaking is that the attention of the speaker and hearers alike be drawn wholly away from the speaker and concentrated upon the thought. In perhaps no other classroom is this so easy as in the mathematical, where the close reasoning, the rigorous demonstration, the tracing of necessary conclusions from given hypotheses, commands and secures the entire mental power of the student who is

explaining, and of his classmates. In what other circumstances do students feel so instinctively that manner counts for so little and mind for so much? In what other circumstances, therefore, is a simple, unaffected, easy, graceful manner so naturally and so beautifully cultivated. ...

One would almost wish that our institutions of the science and art of public speaking would put over their doors the motto that Plato had over the entrance of his school of philosophy: "Let no one who is unacquainted with geometry enter here."—W. F. White.

423. Would you have a man reason well, you must use him to it betimes; exercise his mind in observing the connection between ideas, and following them in train. Nothing does this better than mathematics, which therefore, I think should be taught to all who have the time and opportunity, not so much to make them mathematicians, as to make them reasonable creatures.—John Locke.

428. It hath been an old remark, that Geometry is an excellent Logic. And it must be owned that when the definitions are clear; when the postulata cannot be refused, nor the axioms denied; when from the distinct contemplation and comparison of figures, their properties are derived, by a perpetual well-connected chain of consequences, the objects being still kept in view, and the attention ever fixed upon them; there is acquired a habit of reasoning, close and exact and methodical; which habit strengthens and sharpens the mind, and being transferred to other subjects is of general use in the inquiry after truth.—George Berkeley

429. Our future lawyers, clergy, and statesmen are expected at the University to learn a good deal about curves, and angles, and numbers and proportions; not because these subjects have the smallest relation to the needs of their lives, but because in the very act of learning them they are likely to acquire that habit of steadfast and accurate thinking, which is indispensable to success in all the pursuits of life.—J. C. Fitch.

431. Instruction should aim gradually to combine knowing and doing. Among all sciences mathematics seems to be the only one of a kind to satisfy this aim most completely.—Immanuel Kant.

432. I consider mathematics the chief subject for the common school. No more highly honored exercise for the mind can be found; the buoyancy which it produces is even greater than that produced by the ancient languages, while its utility is unquestioned.—J. F. Herbart.

446. Mathematics, while giving no quick remuneration, like the art of stenography or the craft of bricklaying, does furnish the power for deliberate thought and accurate statement, and to speak the truth is one of the most social qualities a person can possess. Gossip, flattery, slander, deceit, all spring from a slovenly mind that has not been trained in the power of truthful statement, which is one of the highest utilities.—S. T. Dutton.

450. Mathematical knowledge adds vigor to the mind, frees it from prejudice, credulity, and superstition.—John Arbuthnot.

453. Those that can readily master the difficulties of Mathematics find a considerable charm in that study, sometimes amounting to fascination. This is far from universal; but the subject contains elements of strong interest of a kind that constitutes the pleasures of knowledge. The marvelous devices for solving problems elate the mind with the feeling of intellectual power; and the innumerable constructions of the science leave us lost in wonder.—Alexander Bain.

458. In destroying the predisposition to anger, science of all kind is useful; but the mathematics possesses this property in the most eminent degree.—Dr. Rush.

459. The mathematics are the friends to religion, inasmuch as they charm the passions, restrain the impetuosity of the imagination, and purge the mind from error and prejudice. Vice is error, confusion and false reasoning; and all truth is more or less opposite to it. Besides, mathematical truth may serve for a pleasant entertainment for those hours which young men are apt to throw away on their vices; the delightfulness of them being such as to make solitude not only easy but desirable.—John Arbuthnot.

460. There is no prophet which preaches the superpersonal God more plainly than mathematics.—Paul Carus.

515. Mathematics is no more the art of reckoning and computation than architecture is the art of making bricks or hewing wood, no more than painting is the art of mixing colors on a palette, no more than the science of geology is the art of breaking rocks, or the science of anatomy the art of butchering—C. J. Keyser

805. Mathematicians practice perfect freedom.—Henry Adams.

828. As there is no study which may be so advantageously entered upon with a less stock of preparatory knowledge than mathematics, so there is none in which a greater number of uneducated men have raised themselves, by their own exertions, to distinction and eminence. Many of the intellectual defects which, in such cases, are commonly placed to the account of educational studies, ought to be ascribed to the want of a liberal education in early youth.—Dugald Stewart.

835. It is only in mathematics, and to some extent in poetry, that originality may be attained at an early

age, but even then it is very rare (Newton and Keats are examples), and it is not notable until adolescence is completed.—Havelock Ellis.

839. Leibniz lived to the age of 70; Euler to 76; Lagrange to 77; Laplace to 78; Gauss to 78; Plato, the supposed inventor of the conic sections, who made mathematics his study and delight, who called them the handles or aids to philosophy, and is said never to have let a day go by without inventing some new theorems, lived to 82; Newton, the crown and glory of his race, to 85; Archimedes, the nearest akin, probably, to Newton in genius, was 75, and might have lived on to be 100, for aught we can guess to the contrary, when he was slain by the impatient and ill-mannered sergeant, sent to bring him before the Roman general, in the full vigor of his faculties, and in the very act of working out a problem; Pythagoras in whose school, I believe, the word mathematician (used, however, in a somewhat wider than its present sense) originated, the second founder of geometry, the inventor of the matchless theorem which goes by his name, the precognizer of the undoubtedly mis-called Copernican theory, the discoverer of the regular solids and the musical canon who stands at the very apex of this pyramid of fame (if we may count the tradition) after spending 22 years studying in Egypt, and 12 in Babylon, opened school when 56 or 57 years old in Magna Graecia, married a young wife when past 60, and died, carrying on his work with energy unspent to the last, at the age of 99. The mathematician lives long and lives young; the wings of his soul do not early drop off, nor do its pores become clogged with the earthy particles blown from the dusty highways of vulgar life.—J. J. Sylvester

1001. When he had a few moments for diversion, Napoleon not infrequently employed them over a book of logarithms, in which he always found recreation.—J. S. C. Abbott.

1111. A peculiar beauty reigns in the realm of mathematics, a beauty which resembles not so much the beauty of art as the beauty of nature and which affects the reflective mind, which has acquired an appreciation of it, very much like the latter.—E. E. Kummer.

1123. The most distinct and beautiful statements of any truth must take at last the mathematical form.—Henry David Thoreau.

1405. Without mathematics one cannot fathom the depths of philosophy; without philosophy one cannot fathom the depths of mathematics; without the two one cannot fathom anything.—Bordas-Demoulins.

1411. He is unworthy of the name of man who is ignorant of the fact that the diagonal of a square is incommensurable with its side.—Plato

1505. It is only through Mathematics that we can

thoroughly understand what true science is. Here alone we can find in the highest degree the simplicity and severity of scientific law, and such abstraction as the human mind van attain. Any scientific education setting forth from any other point, is faulty in its basis.-—Auguste Comte.

1543. The silent work of the great Regiomontanus in his chamber at Nuremberg made possible the discovery of America by Columbus.—F. Rudio.

1568. Mighty are numbers.—Euripides.

1629. Why are *wise* few, *fools* numerous in the excesse?
'Cause, wanting *number*, they are *number lesse.*
—Lovelace.

1845. Then nothing should be more effectually enacted, than that the inhabitants of your fair city should learn geometry. Moreover the science has indirect effects, which are not small.

Of what kind are they? he said

There are the military advantages of which you spoke, I said; and in all departments of study, as experience proves, and one who has studied geometry is infinitely quicker of apprehension.—Plato.

1911. The method of fluxions is probably one of the greatest, most subtle, and sublime discoveries of any age: it opens a new world to our view, and extends our knowledge, as it were, to infinity; carrying us beyond the bounds that seemed to have been prescribed for the human mind, at least infinitely beyond those to which the ancient geometry was confined.—Charles Hutton.

1924. Among all the mathematical disciplines the theory of differential equations is the most important. It furnishes the explanation of all those elementary manifestations of nature which involve time.—Sophus Lie.

EXERCISES AND QUESTIONS

1. Make a list of ten good things mentioned in the quotations that come from the study of mathematics that you think are the most important.

2. Extend the list of analogies in quotation #515 to at least four other areas.

3. Which three of the quotations contain benefits that are the most ridiculous, improbable, or far-fetched? Why do you think that their writers made such claims?

4. Make a survey: ask a random sample of people what important good things come from the study of mathematics, emphasizing that the question is serious and frivolous answers are not acceptable. How much

overlap is there with the things mentioned in the quotations, or in your list from exercise 1?

5. I would bet that many people asked about the benefits of mathematical study would mention what many of the writers of the quotations said, in different ways, namely that it is good for the mind. Do you agree with that? Not everyone does, nor does every mathematician.

ANECDOTES

edited by Clifton Fadiman

There are two classes of anecdotes about mathematicians. One class is made up of anecdotes that circulate within the mathematical community. Though funny or illuminating ("Don't invite [name of famous mathematician] to stay the night—he won't leave for a month.") they are no doubt not very different from the anecdotes that circulate in other subcultures. If anything they will be duller, since mathematicians put most of their brilliance into their work, leaving very little to sparkle outside of it. We will not bother with anecdotes of this class. The second class contains those anecdotes that are not restricted to mathematicians but are part of the things that everybody knows: common cultural coins, circulating generally. It is worthwhile to look at these, because they can tell us something about what society thinks about mathematics and mathematicians. They can also tell us something about society.

The anecdotes that follow have been taken from a large collection of anecdotes about everybody. They were included in the collection just because they were familiar, and ones that "everybody" knows. Their introductions include information that everybody should know about the subjects of the anecdotes.

Archimedes (287-212 BC), Greek mathematician and scientist, who worked in Sicily under the patronage of the tyrant Hiero of Syracuse. He made numerous inventions and discoveries, including the form of water-pump known as an Archimedes screw, still in use in some third-world countries.

1. Hiero believed that an artisan to whom he had given a quantity of gold to shape into a crown had adulterated the metal with silver. He asked Archimedes if there was any way that his suspicions could be proved or disproved. According to the traditional story, the answer occurred to Archimedes while he was taking a bath. He noticed that the deeper he went into the water, the more water overflowed, and that his body seemed to weigh less the more it was submerged. Leaping from the bath, he is said to have run naked through the streets of Syracuse crying, "*Eureka*!" (I have found it!) The concept he had grasped, now known as Archimedes' Principle, is that the apparent loss of weight of a floating body is equal to the weight of water it displaces, and that the weight per volume (density) of a body determines the displacement. Archimedes realized that by immersing first the crown, then the same weights of silver (less dense) and gold (more dense), different volumes would be displaced, and so he was able to demonstrate that the crown was indeed adulterated.

2. His version of the possibilities opened up by the invention of the lever and the pulley led Archimedes to make his famous utterance: "Give me a place on which to stand, and I will move the earth." Hiero challenged him to put his words into action and help the sailors to beach a large ship in the Syracusian fleet. Archimedes arranged a series of pulleys and cogs to such effect that by his own unaided strength he was able to pull the great vessel out of the water and onto the beach.

3. The lack of a suitable surface could not deter Archimedes from drawing mathematical diagrams. After leaving the bath he would anoint himself thoroughly with olive oil, as was the custom of the time, and then trace his calculations with a fingernail on his own oily skin.

4. When the Roman general Marcellus eventually captured Syracuse, he gave special orders that the life of Archimedes should be protected. A Roman soldier, sent to fetch the scientist, found him drawing mathematical symbols in the sand. Engrossed in his work, Archimedes gestured impatiently, indicating that the soldier must wait until he had solved his problem, and murmured, "Don't disturb my circles." The soldier, enraged, drew his sword and killed him.

Bernoulli, Jacques (1654-1705), Swiss mathematician of Flemish descent and one of a notable family of mathematicians. He is known for his work on the calculus of variations and the theory of complex numbers.

1. "[Bernoulli] had a mystical strain which ... cropped out once in an interesting way toward the end of his life. There is a certain spiral (the logarithmic or equiangular) which is reproduced in a similar spiral after each of many geometrical transformations. [Bernoulli] was fascinated by the recurrence of the

spiral, several of whose properties he discovered, and directed that a spiral be engraved on his tombstone with the inscription *Eadem mutata resurgo* (Though changed I shall arise the same.)"

De Moivre, Abraham (1667-1754), British mathematician of French Huguenot descent. His major work was *The Doctrine of Chances*; he also made significant contributions to trigonometry.

1. (In his old age, twenty hours' sleep a day became habitual with De Moivre.) "Shortly before [his death] he declared that it was necessary for him to sleep some ten minutes or a quarter of an hour longer each day than the preceding one. The day after he had reached a total of over twenty-three hours he slept up to the limit of twenty-four hours and died in his sleep."

Descartes, René (1596-1650), French philosopher and mathematician. His *Meditations* (1641) undermined the Aristotelian concept of reality European thought had adhered to for two millennia. His approach was summed up in his famous line: "*Cogito, ergo sum*" (I think, therefore I am). In his youth, Descartes mainly lived quietly in Holland, though he corresponded with scholars all over Europe. Queen Christina of Sweden lured him to Stockholm in 1649 as her philosophy tutor. The Swedish climate and the rigors of court hours killed him.

1. To Queen Christina Descartes tried to explain his mechanistic philosophy: the view that all animals are mechanisms. The queen countered this by remarking that she had never seen a watch give birth to baby watches.

[As a gloss on this anecdote, the General Editor offers his clerihew:

> Said Descartes, "I extol
> Myself because I have a soul
> And beasts do not." Of course
> He had to put Descartes before the horse.]

2. Descartes's coordinate system was one of his main contributions to the development of mathematics. It is said that the idea came to him during a period of idleness in his military service as he lay on his bed watching a fly hovering in the air. He realized that the fly's position at every moment could be described by locating its distance from three intersecting lines (axes). This insight was the basis of Cartesian coordinates.

Diderot, Denis (1713-84), French philosopher. With Voltaire, Diderot was the guiding spirit of the French Enlightenment. After 1750 he had the enor-

mously influential post of editor of the *Encyclopédie*, through which the tenets of the Enlightenment and the triumph of rationalism and science were propagated.

1. In 1773 Diderot spent some months at the court of St. Petersburg at the invitation of the Russian empress, Catherine the Great. He passed much of his time spreading his gospel of atheism and materialism among the courtiers, until it was suggested to the empress that it would be desirable to muzzle her guest. Reluctant to take direct action, Catherine requested the aid of another savant, the Swiss mathematician Leonhard Euler, a devout Christian. As Diderot was almost entirely ignorant of mathematics, a plot was hatched to exploit this weakness. He was informed that a learned mathematician had developed an algebraical demonstration of the existence of God, and was prepared to deliver it before the entire court if Diderot would like to hear it. Diderot could not very well refuse. Euler approached Diderot, bowed, and said very solemnly, "Sir, $(a + b^n)/n = x$, hence God exists. Reply!" Diderot was totally disconcerted, and delighted laughter broke out on all sides at his discomfiture. He asked permission to return to France, and the empress graciously consented.

Dirichlet, Peter Gustav Lejeune (1805-59), German mathematician.

1. Dirichlet was opposed to writing letters; many of his friends had in the course of their entire lives received no communication from him. However, when his first child was born he broke his silence; he wired his father-in-law: "$2 + 1 = 3$."

Euclid (fl. 300 BC), Greek mathematician who lived and worked in Alexandria, Egypt. His principal work, the *Elements*, remained the standard textbook on geometry until the end of the nineteenth century.

1. Euclid was employed as tutor of mathematics in the royal household. King Ptolemy I complained about the difficulty of the theorems that Euclid expected him to grasp, wondering whether there was not an easier way to approach the subject. Euclid gently rebuked him: "Sire, there is no royal road to geometry."

Euler, Leonhard (1707-83), Swiss mathematician. He worked mainly in Russia (1727-41, 1766-83) and in Frederick the Great's Berlin Academy (1741-66). Although blind in his last years, he labored strenuously until his death, making major contributions in geometry, calculus, and number theory. The number *e*, the base of natural logarithms, is sometimes called Euler's number.

1. When Euler first came to Berlin from Russia, Frederick the Great's mother, the dowager queen Sophia Dorothea, took a liking to him and tried to draw him

out on a number of topics. Euler, no courtier, replied in monosyllables. "Why," asked the dowager queen, "do you not wish to speak to me?" Euler replied, "Madame, I come from a country where, if you speak, you are hanged."

2. (Says E. T. Bell:) "After having amused himself one afternoon calculating the laws of ascent of balloons—on his slate, as usual—[Euler] dined with Lexell [a mathematical colleague] and his family. 'Herschel's Planet' (Uranus) was a recent discovery. Euler outlined the calculation of its orbit. A little later he asked that his grandson be brought in. While playing with the child and drinking tea he suffered a stroke. The pipe dropped from his hand, and with the words, 'I die,' Euler ceased to live and calculate."

Fermat, Pierre de (1601-65), French mathematician renowned for his contributions to the theory of numbers.

1. As Fermat engaged in mathematics for his own amusement, many of his most important contributions were recorded in margins of books or in notes to his friends. In about 1637 he scribbled in his copy of Diophantus's *Arithmetic*, "The equation $x^n + y^n = z^n$, where x, y, and z are positive integers, has no solution if n is greater than 2," and added, "I have discovered a most remarkable proof, but this margin is too narrow to contain it." The problem has gone down in mathematical lore as "Fermat's Last Theorem," and generations of mathematical adepts have taxed their ingenuity to reconstitute his proof.

Galois, Évariste (1811-32), French mathematician. He founded the branch of modern mathematics known as group theory. Dogged by tragic ill luck, he died of peritonitis after a duel, leaving his revolutionary mathematical discoveries to be published posthumously.

1. At the hospital to which Galois, fatally wounded, was taken, his younger brother sat weeping by his bedside. Galois tried to comfort him. "Don't cry," he told him, "I need all my courage to die at twenty."

Gauss, Carl Friedrich (1777-1855), German mathematician. He made major contributions in most fields of mathematics and physics, and is regarded as one of the greatest mathematicians of all time.

1. At school, Gauss showed little of his precocious talent until the age of nine, when he was admitted to the arithmetic class. The master had set what appeared to be a complicated problem involving the addition of a series of numbers in arithmetic progression. Although he had never been taught the simple formula for solving such problems, Gauss handed in his slate within seconds. For the next hour the boy sat idly while his

classmates labored. At the end of the lesson there was a pile of slates on top of Gauss's, all with incorrect answers. The master was stunned to find that at the bottom the slate from the youngest member of the class bearing the single correct number. He was so impressed that he bought the best available arithmetic textbook for Gauss and thereafter did the best he could to advance his progress.

2. Someone hurrying to tell Gauss that his wife was dying found the great mathematician deep in an abstruse problem. The messenger blurted out the sad news. "Tell her to wait a minute until I've finished," replied Gauss absently.

Gibbs, Josiah Willard (1839-1903), US mathematical physicist; professor at Yale (1871-1903). He is considered the founder of chemical thermodynamics, which is based mainly upon the Gibbs free-energy function.

1. At a Yale faculty meeting the language departments had been making out a strong case for being given more money. "Mathematics is also a language," remarked Gibbs.

Kepler, Johannes (1571-1630), German astronomer. He was one of the first to support Copernicus's heliocentric theory and published (1609, 1619) the three laws of planetary motion named after him. He inherited the astronomical records of his master Tycho Brahe and from 1610 used a telescope for his own observations.

1. Kepler also believed in the Pythagorean music of the spheres, that each celestial body in its course gave out a characteristic note or notes. The notes sounded by Earth, he said, were *mi*, *fa*, *mi*, indicating *mi*sery (*miseria*), *fa*mine (*fames*), and *mi*sery.

Laplace, Pierre-Simon, Marquis de (1749-1827), French mathematician and astronomer. His major contribution to science was a detailed study of gravitation in the universe; his conclusions were published in his five-volume *Traité de mécanique céleste* (*Celestial Mechanics*, 1798-1827). Neutral in his views, he escaped execution during the Revolution. He served for a brief period as minister of the interior under Napoleon and was created a marquis by Louis XVIII.

1. Laplace presented a copy of an early volume of his *Mécanique céleste* to Napoleon, who studied it very carefully. Sending for Laplace, he said, "You have written a large book about the universe without once mentioning the author of the universe."

"Sire," Laplace replied, "I have no need of that hypothesis." (*Je n'ai pas besoin de cette hypothèse.*)

2. Joseph-Louis Lagrange worked with Laplace on

his *Mécanique céleste* and indeed made an original contribution to the thinking behind it. Laplace failed to acknowledge this contribution, an omission that the generous Lagrange appears not to have resented. When Lagrange heard of Laplace's reply to Napoleon, he is said to have shaken his head at his colleague's skepticism, commenting, "But it is a beautiful hypothesis just the same. It explains so many things."

Newton, Sir Isaac (1642-1727), English physicist and mathematician. He discovered the law of gravitation and went on to formulate the laws of motion that underlie classical mechanics. He became Lucasian Professor of Mathematics at Cambridge (1669), and in this field his major contribution was the discovery of the calculus (an honor contested by Leibniz). The reflecting telescope was a product of his work on optics. His most important publications were *Principia Mathematica* (1686-87) and *Optics* (1704).

1. In an eighteenth-month period during 1665 to 1666 the plague forced Newton to leave Cambridge and live in his mother's house at Woolsthorpe in Lincolnshire (a house that can still be seen and is preserved as a museum). One day he was sitting in the orchard there, pondering the question of the forces that keep the moon in its orbit, when the fall of an apple led him to wonder whether the force that pulled the apple toward the earth might be the same kind of force that held the moon in orbit around the earth. This train of thought led him eventually to the law of gravitation and its application to the motion of the heavenly bodies. (Voltaire, who heard the anecdote from Newton's stepniece Mrs. Conduitt, and the antiquarian William Stukeley are early sources for this story. If not wholly apocryphal, it is probably an embroidery of the truth. It is certainly a fact that during his stay at Woolsthorpe Newton achieved the insights that led to his greatest scientific work.)

3. A woman, hearing that Newton was a famous astrologer, visited him to ask him where she had lost her purse—somewhere between London Bridge and Shooters' Hall, she thought. Newton merely shook his head. But the woman was persistent, making as many as fourteen visits. Finally, to get rid of her, Newton donned an eccentric costume, chalked a circle around himself, and intoned, "Abracadabra! Go to the facade of Greenwich hospital, third window on the south side. On the lawn in front of it I see a dwarfish devil bending over your purse." Away went the woman—and according to the story, that is actually where she found it. (This story is probably apocryphal, but it neatly illustrates the popular reputation of scientists in the seventeenth-century mind.)

4. An admirer asked Newton how he had come to make discoveries in astronomy that went far beyond anything achieved by anyone before him. "By always thinking about them," replied Newton simply.

8. Newton, Cambridge University's representative to parliament in 1689, was not well adapted to life as a parliamentarian. Only on one occasion did he rise to his feet, and the House of Commons hushed in expectation of hearing the great man's maiden speech. Newton observed that there was a window open, which was causing a draft, asked that it be closed, and sat down.

10. In 1696 Jean Bernoulli and G. W. Leibniz concocted two teasing problems they sent to the leading mathematicians in Europe. After the problems had been in circulation for about six months, a friend communicated them to Newton, who, when he had finished his day's work at the Mint, came home and solved both. The next day he submitted his solutions to the Royal Society anonymously, as he did not like to be distracted from the business of the Mint by embroilment in scientific discussions. The anonymity did not, however, deceive Bernoulli. "I recognize the lion by his paw!" he exclaimed.

11. Newton invited a friend to dinner but then forgot the engagement. When the friend arrived, he found the scientist deep in meditation, so he sat down quietly and waited. In due course, dinner was brought up—for one. Newton continued to be abstracted. The friend drew up a chair and, without disturbing his host, consumed the dinner. After he had finished, Newton came out of his reverie, looked with some bewilderment at the empty dishes, and said, "If it weren't for the proof before my eyes, I could have sworn that I have not yet dined."

12. To the very end of his life Newton's scientific curiosity was unquenched. According to one authority his (somewhat improbable) last words were: "I do not know what I may appear to the world. But to myself, I seem to have been only like a boy playing on the seashore, diverting myself in now and then finding a smoother pebble or prettier shell than ordinary, whilst the great ocean of truth lay all undiscovered before me."

Pascal, Blaise (1623-62), French mathematician and writer on religion. From his youth on, Pascal did important work in mathematics and physics and in 1641 made the first calculating machine. In his early thirties he underwent a profound religious experience and became a Jansenist. Some of the fruits of his meditations on religion are contained in his *Pensées* (1669).

1. Pascal's father began his son's education with a course of reading in ancient languages. When the nine-year-old Pascal inquired as to the nature of geometry, he

was told that it was the study of shapes and forms. The boy immediately proceeded to discover for himself the first thirty-two theorems in Euclid—in the correct order. The elder Pascal saw that it was no use attempting to steer his son away from mathematics and allowed him to pursue his studies as he wished. (The story comes from Pascal's sister and borders on the apocryphal.)

Plato (c. 428 - c. 348 BC), Greek philosopher, who founded the Academy at Athens. His writings include *The Apology*, *Phaedo*, and *The Republic*. The great British thinker A. N. Whitehead once commented that all Western philosophy consists of footnotes to Plato.

1. A student, struggling with the abstract concepts of Platonic mathematics, asked Plato, "What practical end do these theorems serve? What is to be gained from them?" Plato turned to an attendant slave and said, "Give this young man an obol [a small coin] that he may feel that he has gained something from my teachings, and then expel him."

2. Plato considered the abstract speculations of pure mathematics to be the highest form of thought of which the human mind was capable. He therefore had written over the entrance to the Academy "Let no one ignorant of mathematics enter here."

Russell, Bertrand Arthur William, 3d Earl (1872-1970), British philosopher. His *Principia Mathematica* (1910-13), written with A. N. Whitehead, explored the relationship between pure mathematics and logic. He campaigned for numerous social, political, and moral causes, suffering imprisonment for pacifism during World War I (1918) and for civil disobedience during the Campaign for Nuclear Disarmament (1961). He won the 1950 Nobel Prize for literature.

2. Russell's friend G. H. Hardy, who became a Professor of pure mathematics at Cambridge in 1931, once told him that if he could find a proof that Russell would die in five minutes' time, he would naturally be sorry to lose him, but the sorrow would be quite outweighed by pleasure in the proof. Russell, wise in the ways of mathematicians, observed, "I entirely sympathized with him and was not at all offended."

QUESTIONS AND EXERCISES

1. In the first anecdote about Gauss, the formula that he discovered for himself was

$$1 + 2 + 3 + ... + n = \frac{n(n + 1)}{2}.$$

He did it by writing the sum twice, once in the reverse order, and then adding:

$$S = 1 + 2 + 3 + ... + (n - 1) + n,$$

$$S = n + (n - 1) + (n - 2) + ... + 2 + 1,$$

$$2S = (n + 1) + (n + 1) + ... + (n + 1) + (n + 1),$$

where the last sum has n terms. Using the same idea (or a different one), get a formula for

$$1 + 3 + 5 + ... + (2n - 1).$$

2. Carl Friedrich Gauss's name was Carl Friedrich Gauss. You may not find that statement surprising, but in fact it is not universally known. Look in ten books that mention Gauss and see how many of them spell his first name as *Karl*. I would guess that at least three make this error. An error it is, since peoples' names are their own property and Gauss's tombstone has on it Carl with a *C*. As a further exercise, you could write to the authors of the inaccurate books informing them of the error of their ways to see what sort of answers you would get. The exercise would be for amusement only, because "everyone" knows that Germans spell Carl with a *K*, so poor Gauss will continue to be misspelled forever.

3. Anecdotes can wander from person to person. For example, anecdote number one about Plato is always attributed to Euclid and I have no idea who told Clifton Fadiman that Plato got there first. For another, what Newton did in anecdote number three was just what Thales did some two thousand years earlier, except the woman in that story was looking for a bundle and Thales told her to look for it in a ditch. Such wandering anecdotes must express something larger than the idiosyncracies of individuals, some semi-eternal semi-universal semi-truths. What ideas do the two anecdotes cited express? Do you know any other anecdotes that have been attached to more than one person? What universals do they express?

4. Here is a problem that I have never been able to solve. The Diderot anecdote has been reprinted and retold numberless times, and the problem is, *why*? The anecdote informs us that

(a) Diderot knew no mathematics, and

(b) Diderot consented to consider a mathematical proof of the existence of God.

Are those two statement not irreconcilable? If you were asked if you would like to comment on a proof in Chinese and you knew not a word of Chinese, would you say, "Sure"? Assuming that the anecdote has some basis in reality, what *really* happened? It is an entire mystery to me.

THE RELATION OF MATHEMATICS TO PHYSICS

by Richard Feynman

Everyone knows that mathematics has all sorts of applications to physics. Just ask anyone about them, and you will get answers like "$F = ma$," "Inclined planes," "Pendulums," and "Throw a rock straight up in the air with an initial velocity of 80 feet per second and you can find out how high it goes." Yes, but mathematics goes deeper than that, much deeper. In this selection, Richard Feynman, a physicist of the first rank (he won the Nobel Prize in 1965), gives his thoughts on the place of mathematics in physics.

Some of them seem slightly curious to a mathematical non-physicist. For example, he is disturbed that Newton's law of universal gravitation does such a good job of predicting how planets move around the sun and how apples fall to earth when it is merely a *mathematical* statement. It has no physics in it! Could it be that physicists are uneasy when they do not have *things* before them? It could be, and it could explain some of the differences between physicists and mathematicians. (They do differ, you know: for one thing, physicists are more serious than mathematicians—they have more gravity.) Mathematicians do not need *things*.

On the other hand, his conclusion that to understand nature you must know mathematics is very comforting to mathematicians, who have suspected it all along.

In thinking out the applications of mathematics and physics, it is perfectly natural that the mathematics will be useful when large numbers are involved in complex situations. In biology, for example, the action of a virus on a bacterium is unmathematical. If you watch it under a microscope, a jiggling little virus finds some spot on the odd shaped bacterium—they are all different shapes—and maybe it pushes its DNA in and maybe it does not. Yet if we do the experiment with millions and millions of bacteria and viruses, then we can learn a great deal about the viruses by taking averages. We can use mathematics in the averaging, to see whether the viruses develop in the bacteria, what new strains and what percentage; and so we can study the genetics, the mutations and so forth.

To take another more trivial example, imagine an enormous board, a checkerboard to play checkers or draughts. The actual operation of any one step is not mathematical—or it is very simple in its mathematics. But you could imagine that on an enormous board, with lots and lots of pieces, some analysis of the best moves, or the good moves or bad moves, might be made by a deep kind of reasoning which would involve somebody having gone off first and thought about it in great depth. That then becomes mathematics, involving abstract reasoning. Another example is switching in computers. If you have one switch, which is either on or off there is nothing very mathematical about that, although mathematicians like to start there with their mathemat-

ics. But with all the interconnections and wires, to figure out what a very large system will do requires mathematics.

I would like to say immediately that mathematics has a tremendous application in physics in the discussion of the detailed phenomena in complicated situations, granting the fundamental rules of the game. That is something which I would spend most of my time discussing if I were talking only about the relation of mathematics and physics. But since this is part of a series of lectures on the character of physical law I do not have time to discuss what happens in complicated situations, but will go immediately to another question, which is the character of the fundamental laws.

If we go back to our checker game, the fundamental laws are the rules by which the checkers move. Mathematics may be applied in the complex situation to figure out what in given circumstances is a good move to make. But very little mathematics is needed for the simple fundamental character of the basic laws. They can be simply stated in English for checkers.

The strange thing about physics is that for the fundamental laws we still need mathematics. I will give two examples, one in which we really do not, and one in which we do. First, there is a law in physics called Faraday's law, which says that in electrolysis the amount of material which is deposited is proportional to the current and to the time that the current is acting. That means that the amount of material deposited is proportional to the charge which goes through the system. It sounds very mathematical, but what is actually happening is that the electrons going through the wire each carry one charge. To take a particular example, maybe to deposit one atom requires one electron to come, so the number of atoms that are deposited is necessarily equal to the number of electrons that flow, and thus proportional to the charge that goes through the wire. So that mathematically-appearing law has as its basis nothing very deep, requiring no real knowledge of mathematics. That one electron is needed for each atom in order for it to deposit itself is mathematics, I suppose, but it is not the kind of mathematics that I am talking about here.

On the other hand, take Newton's law for gravitation, which has the aspects I discussed last time. I gave you the equation:

$$F \;=\; G\,\frac{mm'}{r^2}$$

just to impress you with the speed with which mathematical symbols can convey information. I said that the force was proportional to the product of the masses of two objects, and inversely as the square of the distance between them, and also that bodies react to forces by changing their speeds, or changing their motions, in the direction of the force by amounts proportional to the force and inversely proportional to their masses. Those are words all right, and I did not necessarily have to write the equation. Nevertheless it is kind of mathematical, and we wonder how this can be a fundamental law. What does the planet do? Does it look at the sun, see how far away it is, and decide to calculate on its internal adding machine the inverse of the square of the distance, which tells it how much to move? This is certainly no explanation of the machinery of gravitation! You might want to look further, and various people have tried to look further. Newton was originally asked about his theory—"But it doesn't mean anything—it doesn't tell us anything". He said, "It tells you *how* it moves. That should be enough. I have told you how it moves, not why." But people often are unsatisfied without a mechanism, and I would like to describe one theory which has been invented, among others, of the type you might want. This theory suggests that this effect is the result of large numbers of actions, which would explain why it is mathematical.

Suppose that in the world everywhere there are a lot of particles, flying through us at very high speed. They come equally in all directions—just shooting by—and once in a while they hit us in a bombardment. We, and the sun, are practically transparent for them, practically but not completely, and some of them hit. Look, then,

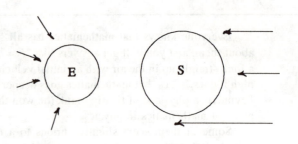

Figure 1

at what would happen (Figure 1): S is the sun, and E the earth. If the sun were not there, particles would be bombarding the earth from all sides, giving little impulses by the rattle, bang, bang of the few that hit. This will not shake the earth in any particular direction, because there are as many coming from one side as from the other, from top as from bottom. However, when the sun is there the particles which are coming from that direction are partly absorbed by the sun, because some of them hit the sun and do not go through. Therefore the number coming from the sun's direction towards the earth is less than the number coming from the other sides, because they meet an obstacle, the sun. It is easy to see that the farther the sun is away, of all the possible directions in which particles can come, a smaller proportion of the particles are being taken out. The sun will appear smaller—in fact inversely as the square of the distance. Therefore there will be an impulse on the earth towards the sun that varies inversely as the square of the distance. And this will be a result of large numbers of very simple operations, just hits, one after the other, from all directions. Therefore the strangeness of the mathematical relation will be very much reduced, because the fundamental operation is much simpler than calculating the inverse of the square of the distance. This design, with the particles bouncing, does the calculation.

The only trouble with this scheme is that it does not work, for other reasons. Every theory that you make up has to be analyzed against all possible consequences, to see if it predicts anything else. And this does predict something else. If the earth is moving, more particles will hit it from in front than from behind. (If you are running in the rain, more rain hits you in the front of the face than in the back of the head, because you are running into the rain.) So, if the earth is moving it is running into the particles coming towards it and away from the ones that are chasing it from behind. So more particles will hit it from the front than from the back, and there will be a force opposing any motion. This force would slow the earth up in its orbit, and it certainly would not have lasted the three or four billion years (at least) that it has been going around the sun. So that is the end of that theory. "Well," you say, "it was a good one, and it got rid of the mathematics for a while. Maybe I could invent a better one." Maybe you can, because nobody knows the ultimate. But up to today, from the time of Newton, no one has invented another theoretical description of the mathematical machinery behind this law which does not either say the same thing over again, or make the mathematics harder, or predict some wrong phenomena. So there is no model of the theory of gravitation today, other than the mathematical form.

If this were the only law of this character it would be interesting and rather annoying. But what turns out to be true is that the more we investigate, the more laws we find, and the deeper we penetrate nature, the more this disease persists. Every one of our laws is a purely mathematical statement in rather complex and abstruse mathematics. Newton's statement of the law of gravitation is relatively simple mathematics. It gets more and more abstruse and more and more difficult as we go on. Why? I have not the slightest idea. It is only my purpose here to tell you about this fact. The burden of the lecture is just to emphasize the fact that it is impossible to explain honestly the beauties of the laws of nature in a way that people can feel, without their having some deep understanding of mathematics. I am sorry, but this seems to be the case. You might say, "All right, then if there is no explanation of the law, at least tell me what the law is. Why not tell me in words instead of in symbols? Mathematics is just a language, and I want to be able to translate the language". In fact I can, with patience, and I think I partly did. I could go a little further and explain in more detail that the equation means that if the distance is twice as far the force is one fourth as much, and so on. I could convert all the symbols into words. In other words I could be kind to the laymen as they all sit hopefully waiting for me to explain something. Different people get different reputations for their skill at explaining to the layman in layman's language these difficult and abstruse subjects. The layman then searches for book after book in the hope that he will avoid the complexities which ultimately set in, even with the best expositor of this type. He finds as he reads a generally increasing confusion, one complicated statement after another, one difficult-to-understand thing after another, all apparently disconnected from one another. It becomes obscure, and he hopes

that maybe in some other book there is some explanation. ... The author almost made it—maybe another fellow will make it right.

But I do not think it is possible, because mathematics is *not* just another language. Mathematics is a language plus reasoning; it is like a language plus logic. Mathematics is a tool for reasoning. It is in fact a big collection of the results of some person's careful thought and reasoning. By mathematics it is possible to connect one statement to another. For instance, I can say that the force is directed towards the sun. I can also tell you, as I did, that the planet moves so that if I draw a line from the sun to the planet, and draw another line at some definite period, like three weeks, later, then the area that is swung out by the planet is exactly the same as it will be in the next three weeks, and the next three weeks, and so on as it goes around the sun. I can explain both of those statements carefully, but I cannot explain why they are both the same. The apparent enormous complexities of nature, with all its funny laws and rules, each of which has been carefully explained to you, are really very closely interwoven. However, if you do not appreciate the mathematics, you cannot see, among the great variety of facts, that logic permits you to go from one to the other. ...

When the problems in physics become difficult we may often look to the mathematicians, who may already have studied such things and have prepared a line of reasoning for us to follow. On the other hand they may not have, in which case we have to invent our own line of reasoning, which we then pass back to the mathematicians. Everybody who reasons carefully about anything is making a contribution to the knowledge of what happens when you think about something, and if you abstract it away and send it to the Department of Mathematics they put it in books as a branch of mathematics. Mathematics, then, is a way of going from one set of statements to another. It is evidently useful in physics, because we have these different ways in which we can speak of things, and mathematics permits us to develop consequences, to analyze the situations, and to change the laws in different ways to connect the various statements. In fact the total amount that a physicist knows is very little. He has only to remember the rules to get him from one place to another and he is all right, because all the various statements about equal times, the force being in the direction of the radius, and so on, are all interconnected by reasoning.

Now an interesting question comes up. Is there a place to begin to deduce the whole works? Is there some particular pattern or order in nature by which we can understand that one set of statements is more fundamental and one set of statements more consequential? There are two kinds of ways of looking at mathematics, which for the purpose of this lecture I will call the Babylonian tradition and the Greek tradition. In Babylonian schools in mathematics the student would learn something by doing a large number of examples until he caught on to the general rule. Also he would know a large amount of geometry, a lot of the properties of circles, the theorem of Pythagoras, formulae for the areas of cubes and triangles; in addition, some degree of argument was available to go from one thing to another. Tables of numerical quantities were available so that they could solve elaborate equations. Everything was prepared for calculating things out. But Euclid discovered that there was a way in which all of the theorems of geometry could be ordered from a set of axioms that were particularly simple. The Babylonian attitude—or what I call Babylonian mathematics—is that you know all of the various theorems and many of the connections in between, but you have never fully realized that it could all come up from a bunch of axioms. The most modern mathematics concentrates on axioms and demonstrations within a very definite framework of conventions of what is acceptable and what is not acceptable as axioms. Modern geometry takes something like Euclid's axioms, modified to be more perfect, and then shows the deduction of the system. For instance, it would not be expected that a theorem like Pythagoras's (that the sum of the areas of squares put on two sides of a right-angled triangle is equal to the area of the square on the hypotenuse) should be an axiom. On the other hand, from another point of view of geometry, that of Descartes, the Pythagorean theorem is an axiom. So the first thing we have to accept is that even in mathematics you can start in different places. If all these various theorems are interconnected by reasoning there is no real way to say "These are the most fundamental axioms", because if you were told something different instead you could also run the reasoning the other way. It is like a bridge with lots of members, and it is over-connected; if pieces have dropped out you can reconnect it another way. The mathematical tradition of today is to start with some particular ideas which are chosen by some kind of convention to be axioms, and then to build up the structure from there. What I have called the Babylonian idea is to say, "I happen to know this, and I happen to know that, and maybe I know that; and I work everything out from there. Tomorrow I may forget that this is true, but remember that something else is true, so I can reconstruct it all again. I am never quite sure of where I am supposed to begin or where I am supposed to end. I just remember enough all the time so that as the memory fades and some of the pieces fall out I can

put the thing back together again every day".

The method of always starting from the axioms is not very efficient in obtaining theorems. In working something out in geometry you are not very efficient if each time you have to start back at the axioms. If you have to remember a few things in geometry you can always get somewhere else, but it is much more efficient to do it the other way. To decide which are the best axioms is not necessarily the most efficient way of getting around in the territory. In physics we need the Babylonian method, and not the Euclidian or Greek method.

...

I should like to say a few things on the relation of mathematics and physics which are a little more general. Mathematicians are only dealing with the structure of reasoning, and they do not really care what they are talking about. They do not even need to know what they are talking about, or, as they themselves say, whether what they say is true. I will explain that. You state the axioms, such-and-such is so, and such-and-such is so. What then? The logic can be carried out without knowing what the such-and-such words mean. If the statements about the axioms are carefully formulated and complete enough, it is not necessary for the man who is doing the reasoning to have any knowledge of the meaning of the words in order to deduce new conclusions in the same language. If I use the word triangle in one of the axioms there will be a statement about triangles in the conclusion, whereas the man who is doing the reasoning may not know what a triangle is. But I can read his reasoning back and say, "Triangle, that is just a three-sided what-have-you, which is so-and-so," and then I know his new facts. In other words, mathematicians prepare abstract reasoning ready to be used if you have a set of axioms about the real world. But the physicist has meaning to all his phrases. That is a very important thing that a lot of people who come to physics by way of mathematics do not appreciate. Physics is not mathematics, and mathematics is not physics. One helps the other. But in physics you have to have an understanding of the connection of words with the real world. It is necessary at the end to translate what you have figured out into English, into the world, into the blocks of copper and glass that you are going to do the experiments with. Only in that way can you find out whether the consequences are true. This is a problem which is not a problem of mathematics at all. Of course it is obvious that the mathematical reasonings which have been developed are of great power and use for physicists. On the other hand, sometimes the physicists' reasoning is useful for mathematicians.

Mathematicians like to make their reasoning as general as possible. If I say to them, "I want to talk about ordinary three dimensional space", they say "If you have a space of n dimensions, then here are the theorems". "But I only want the case 3", "Well, substitute $n = 3$."! So it turns out that many of the complicated theorems they have are much simpler when adapted to a special case. The physicist is always interested in the special case; he is never interested in the general case. He is talking about something; he is not talking abstractly about anything. He wants to discuss the gravity law in three dimensions; he never wants the arbitrary force case in n dimensions. So a certain amount of reducing is necessary, because the mathematicians have prepared these things for a wide range of problems. This is very useful, and later on it always turns out that the poor physicist has to come back and say, "Excuse me, when you wanted to tell me about four dimensions ..."

When you know what it is you are talking about, that some symbols represent forces, others masses, inertia, and so on, then you can use a lot of common-sense, seat-of-the-pants feeling about the world. You have seen various things, and you know more or less how the phenomenon is going to behave. But the poor mathematician translates it into equations, and as the symbols do not mean anything to him he has no guide but precise mathematical rigor and care in the argument. The physicist, who knows more or less how the answer is going to come out, can sort of guess part way, and so go along rather rapidly. The mathematical rigor of great precision is not very useful in physics. But one should not criticize the mathematicians on this score. It is not necessary that just because something would be useful to physics they have to do it that way. They are doing their own job. If you want something else, then you work it out for yourself.

The next question is whether, when trying to guess a new law, we should use the seat-of-the-pants feeling and philosophical principles—"I don't like the minimum principle", or "I do like the minimum principle", "I don't like action at a distance", or "I do like action at a distance". To what extent do models help? It is interesting that very often models do help, and most physics teachers try to teach how to use models and to get a good physical feel for how things are going to work out. But it always turns out that the greatest discoveries abstract away from the model and the model never does any good. Maxwell's discovery of electrodynamics was first made with a lot of imaginary wheels and idlers in space. But when you get rid of all the idlers and things in space the thing is O.K. Dirac discovered the correct laws for relativity quantum

mechanics simply by guessing the equation. The method of guessing the equation seems to be a pretty effective way of guessing new laws. This shows again that mathematics is a deep way of expressing nature, and any attempt to express nature in philosophical principles or in seat-of-the-pants mechanical feelings, is not an efficient way.

It always bothers me that, according to the laws as we understand them today, it takes a computing machine an infinite number of logical operations to figure out what goes on in no matter how tiny a region of space, and no matter how tiny a region of time. How can all that be going on in that tiny space? Why should it take an infinite amount of logic to figure out what one tiny piece of space/time is going to do? So I have often made the hypothesis that ultimately physics will not require a mathematical statement, that in the end the machinery will be revealed, and the laws will turn out to be simple, like the checker board with all its apparent complexities. But this speculation is of the same nature as those other people make—"I like it", "I don't like it",—and it is not good to be too prejudiced about these things.

To summarize, I would use the words of Jeans, who said that "the Great Architect seems to be a mathematician". To those who do not know mathematics it is difficult to get across a real feeling as to the beauty, the deepest beauty, of nature. C. P. Snow talked about two cultures. I really think that those two cultures separate people who have and people who have not had this experience of understanding mathematics well enough to appreciate nature once.

It is too bad that it has to be mathematics, and that mathematics is hard for some people. It is reputed—I do not know if it is true—that when one of the kings was trying to learn geometry from Euclid he complained that it was difficult. And Euclid said, "There is no royal road to geometry". And there is no royal road. Physicists cannot make a conversion to any other language. If you want to learn about nature, to appreciate nature, it is necessary to understand the language that she speaks in. She offers her information only in one form; we are not so unhumble as to demand that she change before we pay any attention. All the intellectual arguments that you can make will not communicate to deaf ears what the experience of music really is. In the same way all the intellectual arguments in the world will not convey an understanding of nature to those of "the other culture". Philosophers may try to teach you by telling you qualitatively about nature. I am trying to describe her. But it is not getting across because it is impossible. Perhaps it is because their horizons are limited in this way that some people are able to imagine that the center of the universe is man.

QUESTIONS AND EXERCISES

1. "It is not necessary for the man who is doing the reasoning to have any knowledge of the meaning of the words in order to deduce new conclusions in the same language."

(a) Yes, as Bertrand Russell observed, in mathematics we do not have not know what it is that we are talking about. For example, from

All glerbs are globs, and all glorps are glerbs

show that

All non-globs are non-glorps.

You did not know what you were talking about, did you?

(b) In addition, as Bertrand Russell also observed, in mathematics we do not have to know that what we say is true. For example, suppose that a is a non-zero number with the property that $a^3 = 0$. Show that $(a + 1)^6 - 1$ is divisible by 21. There aren't any non-zero numbers whose cube is zero, but the conclusion is nevertheless correct.

2. In physics, do you have to know what it is that you are talking about? And do you have to know if what you are saying is true?

3. Rank the following sciences in order from most Babylonian to most Greek: geology, medicine, psychology, chemistry, sociology.

4. "But it always turns out that ... the model never does any good." Why do you think that is?

5. "Perhaps it is because their horizons are limited ... that some people are able to imagine that the center of the universe is man." A strong statement. On the other hand, we have Alexander Pope:

Know then thyself, presume not God to scan;
The proper study of mankind is man.

Do the two statements conflict? Can they be reconciled?

MATHEMATICS AS A CREATIVE ART

by P. R. Halmos

What is mathematics? Different people would give different answers. A student in elementary school would probably say that it was about adding, subtracting, multiplying and dividing. Oh, yes—and about fractions and decimals too. A student in high school would probably say that it is about learning rules and formulas to solve equations. Oh, yes—and learning rules and formulas in geometry too. I'm afraid that all too many students of calculus would also say that mathematics is about rules and formulas and impossible word problems and getting the right answer by the right method. Then, since most people lose contact with mathematics after high school, or after calculus, the average citizen keeps a limited view of mathematics for a lifetime. That is too bad, because those answers are not complete and we should not carry around in our heads any more delusions or distorted views of reality than we have to.

The following selection gives an answer by a professional mathematician. It is not the answer that would be given by every professional mathematician, but it is probably far closer to the truth than the answers that people in general would give. It is well worth the attention of every student of calculus, especially those who will not become professional mathematicians. Most students of calculus will forget how to find equations of tangent lines, but they should remember that mathematics is far closer to an art than it is to the business of equation-solving.

Do you know any mathematicians—and, if you do, do you know anything about what they do with their time? Most people don't. When I get into conversation with the man next to me in a plane, and he tells me that he is something respectable like a doctor, lawyer, merchant, or dean, I am tempted to say that I am in roofing and siding. If I tell him that I am a mathematician, his most likely reply will be that he himself could never balance his check book, and it must be fun to be a whiz at math. If my neighbor is an astronomer, a biologist, a chemist, or any other kind of natural or social scientist, I am, if anything, worse off—this man *thinks* he knows what a mathematician is, and he is probably wrong. He thinks that I spend my time (or should) converting different orders of magnitude, comparing binomial coefficients and powers of 2, or solving equations involving rates of reactions.

C. P. Snow points to and deplores the existence of two cultures; he worries about the physicist whose idea of modern literature is Dickens, and he chides the poet who cannot state the second law of thermodynamics. Mathematicians, in converse with well-meaning, intelligent, and educated laymen (do you mind if I refer to all non-mathematicians as laymen?) are much worse off than physicists in converse with poets. It saddens me that educated people don't even know that my subject exists. There is something that they call mathematics, but they neither know how the professionals use that word, nor can they conceive why anybody should do it.

It is, to be sure, possible that an intelligent and otherwise educated person doesn't know that egyptology exists, or hematology, but all you have to tell him is that it does, and he will immediately understand in a rough general way why it should and he will have some empathy with the scholar of the subject who finds it interesting.

Usually when a mathematician lectures, he is a missionary. Whether he is talking over a cup of coffee with a collaborator, lecturing to a graduate class of specialists, teaching a reluctant group of freshman engineers, or addressing a general audience of laymen—he is still preaching and seeking to make converts. He will state theorems and he will discuss proofs and he will hope that when he is done his audience will know more mathematics than they did before. My aim today is different—I am not here to proselyte but to enlighten—I seek not converts but friends. I do not want to teach you what mathematics is, but only *that* it is.

I call my subject mathematics—that's what all my colleagues call it, all over the world—and there, quite possibly, is the beginning of confusion. The word covers two disciplines—many more, in reality, but two, at least two, in the same sense in which Snow speaks of two cultures. In order to have some words with which to refer to the ideas I want to discuss, I offer two temporary and ad hoc neologisms. Mathematics, as the word is customarily used, consists of at least two distinct subjects, and I propose to call them *mathology* and *mathophysics*. Roughly speaking, mathology is what is usually called pure mathematics, and mathophysics is called applied mathematics, but the qualifiers are not emotionally strong enough to disguise that they qualify the same noun. If the concatenation of syllables I chose here reminds you of other words, no great harm will be done; the rhymes alluded to are not completely accidental. I originally planned to entitle this lecture something like "Mathematics is an art," or "Mathematics is not a science," or "Mathematics is useless," but the more I thought about it the more I realized that I mean that "Mathology is an art," "Mathology is not a science," and "Mathology is useless." When I am through, I hope you will recognize that most of you have known about mathophysics before, only you were probably calling it mathematics; I hope that all of you will recognize the distinction between mathology and mathophysics; and I hope that some of you will be ready to embrace, or at least applaud, or at the very least, recognize mathology as a respectable human endeavor.

In the course of the lecture I'll have to use many analogies (literature, chess, painting), each imperfect by itself, but I hope that in their totality they will serve to delineate what I want delineated. Sometimes in the interest of economy of time, and sometimes doubtless unintentionally, I'll exaggerate; when I'm done, I'll be glad to rescind anything that was inaccurate or that gave offense in any other way.

What Mathematicians Do

As a first step toward telling you what mathematicians do, let me tell you some of the things they do not do. To begin with, mathematicians have very little to do with numbers. You can no more expect a mathematician to be able to add a column of figures rapidly and correctly than you can expect a painter to draw a straight line or a surgeon to carve a turkey—popular legend attributes such skills to these professions, but popular legend is wrong. There is, to be sure, a part of mathematics called number theory, but even that doesn't deal with numbers in the legendary sense—a number theorist and an adding machine would find very little to talk about. A machine might enjoy proving that $1^3 + 5^3 + 3^3 = 153$ and it might even go on to discover that there are only five positive integers with the property that the equation indicates (1, 370, 371, 407), but most mathematicians couldn't care less; many mathematicians enjoy and respect the theorem that every positive integer is the sum of not more than four squares, whereas the infinity involved in the word "every" would frighten and

paralyze any ordinary office machine, and, in any case, that's probably not the sort of thing that the person who relegates mathematicians to numbers had in mind.

Not even those romantic objects of latter day science fiction, the giant brains, the computing machines that run our lives these days—not even they are of interest to the mathematician as such. Some mathematicians are interested in the logical problems involved in the reduction of difficult questions to the sort of moronic baby talk that machines understand: the logical design of computing machines is definitely mathematics. Their construction is not, that's engineering, and their product, be it a payroll, a batch of sorted mail, or a supersonic plane, is of no mathematical interest or value.

Mathematics is not numbers or machines; it is also not the determination of the heights of mountains by trigonometry, or compound interest by algebra, or moments of inertia by calculus. Not today it isn't. At one point in history each of those things, and others like them, might have been an important and non-trivial research problem, but once the problem is solved, its repetitive application has as much to do with mathematics as the work of a Western Union messenger boy has to do with Marconi's genius.

There are at least two other things that mathematics isn't; one of them is something it never was, and the other is something it once included and by now has sloughed off. The first is physics. Some laymen confuse mathematics and theoretical physics and speak, for instance, of Einstein as a great mathematician. There is no doubt that Einstein was a great man, but he was no more a great mathematician than he was a great violinist. He used mathematics to find out facts about the universe, and that he successfully used certain parts of differential geometry for that purpose adds a certain piquancy to the appeal of differential geometry. Withal, relativity theory and differential geometry are not the same thing. Einstein, Schrodinger, Heisenberg, Fermi, Wigner, Feynman—great men all, but not mathematicians; some of them, in fact, strongly antimathematical, preach against mathematics, and would regard it as an insult to be called a mathematician.

What once was mathematics remains mathematics always, but it can become so thoroughly worked out, so completely understood, and, in the light of millennia of contributions, with hindsight, so trivial, that mathematicians never again need to or want to spend time on it. The celebrated Greek problems (trisect the angle, square the circle, duplicate the cube) are of this kind, and the irrepressible mathematical amateur to the contrary notwithstanding, mathematicians are no longer trying to solve them. Please understand, it isn't that they have given up. Perhaps you have heard that, according to mathematicians, it is impossible to square a circle, or trisect an angle, and perhaps you have heard or read that, therefore, mathematicians are a pusillanimous chicken-hearted lot, who give up easily, and use their ex-cathedra pronouncements to justify their ignorance. The conclusion may be true, and you may believe it if you like, but the proof is inadequate.

The Start of Mathematics

No one knows when and where mathematics got started, or how, but it seems reasonable to guess that it emerged from the same primitive physical observations (counting, measuring) with which we all begin our own mathematical insight (ontogeny recapitulates phylogeny). It was probably so in the beginning, and it is true still, that many mathematical ideas originate not from pure thought but from material necessity; many, but probably not all. Almost as soon as a human being finds it necessary to count his sheep (or sooner?) he begins to wonder about numbers and shapes and motions and arrangements—curiosity about such things seems to be as necessary to the human spirit as curiosity about earth, water, fire, and air, and curiosity—sheer pure intellectual curiosity—about stars and about life. Numbers and shapes and motions and arrangements, and also thoughts and their order, and concepts such as "property" and "relation"—all such things are the raw material of mathematics. The technical but basic mathematical concept of "group" is the best humanity can do to understand the intuitive concept of "symmetry" and the people who study topological spaces, and ergodic paths, and oriented graphs are making precise our crude and vague feelings about shapes, and motions, and arrangements.

Why do mathematicians study such things, and why should they? What, in other words, motivates the individual mathematician, and why does society encourage his efforts, at least to the extent of providing him with the training and subsequently the livelihood that, in turn, give him the time he needs to think? There are two answers to each of the two questions: because mathematics is practical and because mathematics is an art. The already existing mathematics has more and more new applications each day, and the rapid growth of desired applications suggests more and more new practical mathematics. At the same time, as the quantity of mathematics grows and the number of people who think about it keeps doubling over and over again, more new concepts need explication, more new logical interrelations cry out for study, and understanding, and simplification, and more and more the tree of mathematics bears elaborate and gaudy flowers that are, to many

beholders, worth more than the roots from which it all comes and the causes that brought it all into existence.

Mathematics is very much alive today. There are more than a thousand journals that publish mathematical articles; about 15,000 to 20,000 mathematical articles are printed every year. The mathematical achievements of the last 100 years are greater in quantity and in quality than those of all previous history. Difficult mathematical problems, which stumped Hilbert, Cantor, or Poincaré, are being solved, explained, and generalized by beardless (and bearded) youths in Berkeley and in Odessa.

Mathematicians sometimes classify themselves and each other as either problem-solvers or theory-creators. The problem-solvers answer yes-or-no questions and discuss the vital special cases and concrete examples that are the flesh and blood of mathematics; the theory-creators fit the results into a framework, illuminate it all, and point it in a definite direction—they provide the skeleton and the soul of mathematics. One and the same human being can be both a problem-solver and a theory-creator, but, usually, he is mainly one or the other. The problem-solvers make geometric constructions, the theory-creators discuss the foundations of Euclidean geometry; the problem-solvers find out what makes switching diagrams tick, the theory-creators prove representation theorems for Boolean algebras. In both kinds of mathematics and in all fields of mathematics the progress in one generation is breathtaking. No one can call himself a mathematician nowadays who doesn't have at least a vague idea of homological algebra, differential topology, and functional analysis, and every mathematician is probably somewhat of an expert on at least one of these subjects—and yet when I studied mathematics in the 1930's none of those phrases had been invented, and the subjects they describe existed in seminal forms only.

Mathematics is abstract thought, mathematics is pure logic, mathematics is creative art. All these statements are wrong, but they are all a little right, and they are all nearer the mark than "mathematics is numbers" or "mathematics is geometric shapes." For the professional pure mathematician, mathematics is the logical dovetailing of a carefully selected sparse set of assumptions with their surprising conclusions via a conceptually elegant proof. Simplicity, intricacy, and above all, logical analysis are the hallmark of mathematics.

The mathematician is interested in extreme cases—in this respect he is like the industrial experimenter who breaks lightbulbs, tears shirts, and bounces cars on ruts. How widely does a reasoning apply, he wants to know, and what happens when it doesn't? What happens when you weaken one of the assumptions, or under what conditions can you strengthen one of the conclusions? It is the perpetual asking of such questions that makes for broader understanding, better technique, and greater elasticity for future problems.

Mathematics—this may surprise you or shock you some—is never deductive in its creation. The mathematician at work makes vague guesses, visualizes broad generalizations, and jumps to unwarranted conclusions. He arranges and rearranges his ideas, and he becomes convinced of their truth long before he can write down a logical proof. The conviction is not likely to come early—it usually comes after many attempts, many failures, many discouragements, many false starts. It often happens that months of work result in the proof that the method of attack they were based on cannot possibly work, and the process of guessing, visualizing, and conclusion-jumping begins again. A reformulation is needed and—and this too may surprise you—more experimental work is needed. To be sure, by "experimental work" I do not mean test tubes and cyclotrons. I mean thought-experiments. When a mathematician wants to prove a theorem about an infinite-dimensional Hilbert space, he examines its finite-dimensional analogue, he looks in detail at the 2- and 3-dimensional cases, he often tries out a particular numerical case, and he hope that he will gain thereby an insight that pure definition-juggling has not yielded. The deductive stage, writing the result down, and writing down its rigorous proof are relatively trivial once the real insight arrives; it is more like the draftsman's work, not the architect's.

Mathematics is a Language

Why does mathematics occupy such an isolated position in the intellectual firmament? Why is it good form, for intellectuals, to shudder and announce that they can't bear it, or, at the very least, to giggle and announce that they never could understand it? One reason, perhaps, is that mathematics is a language. Mathematics is a precise and subtle language designed to express certain kinds of ideas more briefly, more accurately, and more usefully than ordinary language. I do not mean here that mathematicians, like members of all other professional cliques, use jargon. They do, at times, and they don't most often, but that's a personal phenomenon, not the professional one I am describing. What I do mean by saying that mathematics is a language is sketchily and inadequately illustrated by the difference between the following two sentences. (1) If each of two numbers is multiplied by itself, the difference of the two results is the same as the product of the

sum of the two given numbers by their difference. (2) $x^2 - y^2 = (x + y)(x - y)$ (Note: the longer formulation is not only awkward, it is also incomplete.)

One thing that sometimes upsets and repels the layman is the terminology that mathematicians employ. Mathematical words are intended merely as labels, sometimes suggestive, possibly facetious, but always precisely defined; their everyday connotations must be steadfastly ignored. Just as nobody nowadays infers from the name Fitzgerald that its bearer is the illegitimate son of Gerald, a number that is called irrational must not be thought unreasonable; just as a dramatic poem called *The Divine Comedy* is not necessarily funny, a number called imaginary has the same kind of mathematical existence as any other. (Rational, for numbers, refers not to the Latin *ratio*, in the sense of reason, but to the English "ratio," in the sense of quotient.)

Mathematics is a language. None of us feels insulted when a sinologist uses Chinese phrases, and we are resigned to living without Chinese, or else spending years learning it. Our attitude to mathematics should be the same. It's a language, and it takes years to learn to speak it well. We all speak it a little, just because some of it is in the air all the time, but we speak it with an accent, and frequently inaccurately; most of us speak it, say, about as well as one who can only say *"Oui, monsieur"* and *"S'il vous plâit"* speaks French. The mathematician sees nothing wrong with this as long as he's not upbraided by the rest of the intellectual community for keeping secrets. It took him a long time to learn his language, and he doesn't look down on the friend who, never having studied it, doesn't speak it. It is however sometimes difficult to keep one's temper with the cocktail party acquaintance who demands that he be taught the language between drinks and who regards failure or refusal to do so as sure signs of stupidity or snobbishness.

Some Analogies

A little feeling for the nature of mathematics and mathematical thinking can be got by the comparison with chess. The analogy, like all analogies, is imperfect, but it is illuminating just the same. The rules for chess are as arbitrary as the axioms of mathematics sometimes seem to be. The game of chess is as abstract as mathematics. (That chess is played with solid pieces, made of wood, or plastic, or glass, is not an intrinsic feature of the game. It can just as well be played with pencil and paper, as mathematics is, or blindfolded, as mathematics can.) Chess also has its elaborate technical language, and chess is completely deterministic.

There is also some analogy between mathematics and music. The mathologist feels the need to justify pure mathematics exactly as little as the musician feels the need to justify music. Do practical men, the men who meet payrolls, demand only practical music—soothing jazz to make an assembly line worker turn nuts quicker, or stirring marches to make a soldier kill with more enthusiasm? No, surely none of us believes in that kind of justification; music, and mathematics, are of human value because human beings feel they are.

The analogy with music can be stretched a little further. Before a performer's artistic contribution is judged, it is taken for granted that he hits the right notes, but merely hitting the right notes doesn't make him a musician. We don't get the point of painting if we compliment the nude Maya on being a good likeness, and we don't get the point of a historian's work if all we can say is that he didn't tell lies. Mere accuracy in performance, resemblance in appearance, and truth in storytelling doesn't make good music, painting, history: in the same way, mere logical correctness doesn't make good mathematics.

Goodness, high quality, are judged on grounds more important than validity, but less describable. A good piece of mathematics is connected with much other mathematics, it is new without being silly (think of a "new" western movie in which the names and the costumes are changed, but the plot isn't), and it is deep in an ineffable but inescapable sense—the sense in which Johann Sebastian is deep and Carl Philip Emmanuel is not. The criterion for quality is beauty, intricacy, neatness, elegance, satisfaction, appropriateness—all subjective, but all somehow mysteriously shared by all.

Mathematics resembles literature also, differently from the way it resembles music. The writing and reading of literature are related to the writing and reading of newspapers, advertisements, and road signs the way mathematics is related to practical arithmetic. We all need to read and write and figure for daily life: but literature is more than reading and writing, and mathematics is more than figuring. The literature analogy can be used to help understand the role of teachers and the role of the pure-applied dualism.

Many whose interests are in language, in the structure, in the history, and in the aesthetics of it, earn their bread and butter by teaching the rudiments of language to its future practical users. Similarly many, perhaps most, whose interests are in the mathematics of today, earn their bread and butter by teaching arithmetic, trigonometry, or calculus. This is sound economics: society abstractly and impersonally is willing to subsidize pure language and pure mathematics, but not very far. Let the would-be purist pull his weight by teaching

the next generation the applied aspects of his craft; then he is permitted to spend a fraction of his time doing what he prefers. From the point of view of what a good teacher must be, this is good. A teacher must know more than the bare minimum he must teach; he must know more in order to avoid more and more mistakes, to avoid the perpetuation of misunderstanding, to avoid catastrophic educational inefficiency. To keep him alive, to keep him from drying up, his interest in syntax, his burrowing in etymology, or his dabbling in poetry play a necessary role.

The pure-applied dualism exists in literature too. The source of literature is human life, but literature is not the life it comes from, and writing with a grim purpose is not literature. Sure there are borderline cases: is Upton Sinclair's *Jungle* literature or propaganda? (For that matter, is Chiquita Banana an advertising jingle or charming light opera?) But the fuzzy boundary doesn't alter the fact that in literature (as in mathematics) the pure and the applied are different in intent, in method, and in criterion of success.

Perhaps the closest analogy is between mathematics and painting. The origin of painting is physical reality, and so is the origin of mathematics—but the painter is not a camera and the mathematician is not an engineer. The painter of "Uncle Sam Wants You" got his reward from patriotism, from increased enlistments, from winning the war—which is probably different from the reward Rembrandt got from a finished work. How close to reality painting (and mathematics) should be is a delicate matter of judgment. Asking a painter to "tell a concrete story" is like asking a mathematician to "solve a real problem." Modern painting and modern mathematics are far out—too far in the judgment of some. Perhaps the ideal is to have a spice of reality always present, but not to crowd it the way descriptive geometry, say, does in mathematics, and medical illustration, say, does in painting.

Talk to a painter (I did) and talk to a mathematician, and you'll be amazed at how similarly they react. Almost every aspect of the life and of the art of a mathematician has its counterpart in painting, and vice versa. Every time a mathematician hears "I could never make my checkbook balance" a painter hears "I could never draw a straight line"—and the comments are equally relevant and equally interesting. The invention of perspective gave the painter a useful technique, as did the invention of 0 to the mathematician. Old art is as good as new; old mathematics is as good as new. Tastes change, to be sure, in both subjects, but a twentieth century painter has sympathy for cave paintings and a twentieth century mathematician for the fraction juggling of the Babylonians. A painting must be painted and then looked at; a theorem must be printed and then read. The painter who thinks good pictures, and the mathematician who dreams beautiful theorems are dilettantes; an unseen work of art is incomplete. In painting and in mathematics there are some objective standards of good—the painter speaks of structure, line, shape, and texture, where the mathematician speaks of truth, validity, novelty, generality—but they are relatively the easiest to satisfy. Both painters and mathematicians debate among themselves whether these objective standards should even be told to the young—the beginner may misunderstand and overemphasize them and at the same time lose sight of the more important subjective standards of goodness. Painting and mathematics have a history, a tradition, a growth. Students, in both subjects, tend to flock to the newest but, except the very best, miss the point; they lack the vitality of what they imitate, because, among other reasons, they lack the experience based on the traditions of the subject.

I've been talking *about* mathematics, but not *in* it, and, consequently, what I've been saying is not capable of proof in the mathematical sense of the word. I hope just the same, that I've shown you that there is a subject called mathematics (mathology?), and that that subject is a creative art. It is a creative art because mathematicians create beautiful new concepts; it is a creative art because mathematicians live, act, and think like artists; and it is a creative art because mathematicians regard it so. I feel strongly about that, and I am grateful for this opportunity to tell you about it. Thank you for listening.

QUESTIONS AND EXERCISES

1. "It saddens me that educated people don't even know that my subject exists." No doubt, but so what? There are people who spend their days valuing the reserves of group annuities, and educated people don't know anything about *that* either. Should educated people care? Or does the author just have a big ego and wants everybody to know what *he* does?

2. $1^3 + 5^3 + 3^3 = 153$ all right. Why did the author give no examples with squares? Clearly, because there are none. *Prove* that this is so. That is, show that $a^2 + b^2 = 10a + b$ has no solutions if a and b are digits and a is not zero.

3. If you solved that problem, did you enjoy it? If so, how, and why? If not, why not? What sorts of pleasure can come from mathematics?

4. "Mathematics is the logical dovetailing of a carefully selected sparse set of assumptions with their surprising conclusions via a conceptually elegant proof." Now there is a definition! It is certainly not what your ordinary person would respond if asked to define mathematics. How could you take the four elements of the definition—"logical dovetailing," "set of assumptions," "surprising conclusions," and "elegant proof"—and make them plain to an educated person?

5. Can the distinction made between *mathology* and *mathophysics* also be made in other areas of human endeavor? If so, what are some examples; if not, why not?

BEAUTY IN MATHEMATICS

by H. E. Huntley

Beauty, of course, is in the eye of the beholder, and many people can behold mathematics until the cows come home and never see anything beautiful. In fact, to a large fraction of the population, beauty and mathematics are simply unrelated and connecting them seems as strange as would be talking about the glories of dishwashing or the majestic splendor of solid waste disposal. Nevertheless, there is abundant testimony to the beauties contained in mathematics, so we must therefore conclude that it exists, even if we cannot appreciate it ourselves. Some people can sometimes get quite carried away:

—I saw a set bubbling and whirling, then take purpose and structure to itself and become a semigroup, generate a unity element and become a group, generate a second unity element, mount itself and become a field, ringed by rings. Near it, a mature field, shot through with ideals, threw off a splitting field in a passion of growth, and become complex. ...

—I saw the point sets, with their cliques and clubs, infinite numbers of sycophants clustering round a Bolzano-Weierstrass aristocrat—the great compact medieval coverings of infinity with denumerable shires—the conflicts as closed sets created open sets, and the other way round. ...

—I saw the proud old cyclic groups, father and son and grandson, generating the generations, rebel and blacksheep and hero, following each other endlessly. Close by were the permutation groups, frolicking in a way that seemed like the way you sometimes repeat a sentence endlessly, stressing a different word each time.

There was much I saw that I did not understand, for mathematics is a deep. But that world of abstractions flamed with a beauty and meaning that chilled the works and worlds of men, so I wept in futility.

(From "The Mathenauts" by Norman Kagan, reprinted in *The Year's Best SF*, edited by Judith Merrill, Delacorte Press, New York, 1965)

Mathematical beauty must exist, for if it did not, what could have produced an effect like that?

One of the difficulties in talking about mathematical beauty, or for that matter about any kind of beauty, is that you cannot. Beauty is ineffable; since it is experienced and not reasoned about, words are fairly helpless at communicating it. Can you describe in words exactly what it feels like to sneeze? No, you cannot: the best way to communicate that to someone else is to tickle that person's nose until a sneeze erupts. Like sneezing, beauty is best explained by giving examples. Words fail. People have had and continue to have mystical experiences that are brought on by many things, some of them mathematical, but they can never describe what they experienced with words. The author of "The Mathenauts" had what I am sure was at least a semi-mystical mathematical experience and he reported it as well as words could.

Writers nevertheless attempt to use words to make their readers see beauty, but they can fail for a reason other than the inadequacy of words. Sneezes can be induced, but no amount of tickling can stimulate wherever it is that the mathematical esthetic sense resides. It is as possible to be insensitive to mathematical beauty as it is to be unable to appreciate, say, the operas of Wagner. No amount of writing about the wonders of Wagner will convince some people that they are anything but tedium unalloyed, and writing about the beauties of mathematics will be equally unconvincing, or even incomprehensible, to people who lack the capacity to see them.

So, the following selection may strike you as being less than perfect at conveying a sense of mathematical beauty, even if you have a susceptibility, latent or not, to it. That is no fault of its author, who is doing the best he can. It is not easy to express the inexpressible.

There is ambiguity in this chapter heading. It may mean the pleasure derived from the mental activity which the study of mathematics generates; or it may mean the aesthetic feeling evoked by (e. g.) a mathematical theorem, which, on account of this feeling, is regarded as a thing of beauty. In what follows we are concerned with the latter meaning. An illustration may make the distinction clear.

The study of a plane curve such as the parabola involves a succession of discoveries of hitherto unsuspected truths; each of these discoveries gives rise, in a greater or less degree, to an experience of beauty: the creation of harmony out of dissonance. The study is a pleasurable activity. Renan said, "There is a scientific taste just as there is a literary or artistic one," and we are thinking of beauty that has an objective ingredient.

An Example

Let us consider this curve from several points of view.

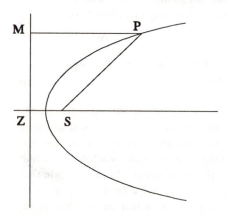

Figure 1

i. In the first place, the parabola (Figure 1) is a curve which is beautiful in itself. In the absence of any mathematical sophistication, merely to contemplate it is a pleasurable sensuous experience, though it might tax the wisdom of the psychologist to explain in what the pleasure lies. Of course, the curve is symmetrical about an axis, but so are many capital letters of the alphabet which can lay no claim to beauty. One might perhaps say that there is the tang of infinity about the curve as it journeys off into uncharted space, which is in contrast to the parochial quality of the region in the neighbor-

hood of the focus S. But how much is that worth? The aesthetic appeal is not to be doubted, but its source is hidden. That is part of the total artistic appreciation which is inborn; all the rest is acquired.

ii. Secondly, the parabola is a *locus* of great simplicity. It is the path traced out in a plane by a point moving in accordance with a simple law briefly stated thus: the point P is equidistant from a fixed point S (the focus) and a fixed line ZM (the directrix). If one tried to run to earth the aesthetic appeal of this, it would be found, in part, in the simplicity of the idea, the neatness of the method of generating a lovely curve. Moreover, the pleasure is enhanced with further education, for this allows a comparison between the parabola ($PS/PM = 1$) and the other conic sections, the ellipse ($PS/PM < 1$) and the hyperbola ($PS/PM > 1$).

iii. Since the marriage by Descartes (1596-1650) of geometry to algebra, it has been possible to represent the parabola in shorthand: $y^2 = 4ax$; this provides a powerful tool for revealing the properties of the curve. There is no appeal to the eye in this, but there is a great deal of aesthetic satisfaction arising from the application of coordinate geometry to the parabola.

iv. The viewpoint from which the parabola is seen in its most beautiful aspect is as a special section of a right circular cone. The most general conic section is an ellipse (AA', Figure 2, which has two extreme forms—the circle and the parabola. In Figure 2 $P'OP$, $Q'OQ$ represent generators of a right circular cone, AA' being a section of the cone by a plane which makes an angle with the axis of the cone greater than half the vertical angle AOA'. When this angle is equal to one half of AOA', the major axis AA' of the ellipse is parallel to the generator of the cone and is therefore of infinite length (Figure 3). This extreme form of the ellipse is the parabola. The spheres inscribed to touch the cone in circles and the plane of the ellipse (Figure 2) touch this plane at the foci S, S'.. Moreover, the planes containing the circles of contact between the spheres and the cone intersect the plane of the ellipse in two lines which are the directrices. These exciting results are applicable, with modification, to the parabola. To grasp these truths and their manifold implications is to glimpse beauty in mathematics.

v. The path of many a comet is a parabola with the sun at its focus. Each drop of spray from a water fountain describes a path which is a near-parabola. It

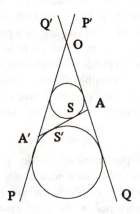

Figure 2

is, in fact, a very elongated ellipse having the earth's center as one of its foci.

This example should serve to show the distinction

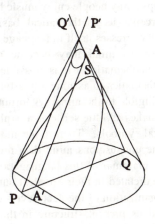

Figure 3

between beauty in mathematics and pleasure in seeking it out. There is joy in rock climbing but it is not to be confused with the pleasure derived from viewing the scenery.

The Aesthetically Ungifted

I have spared little time hitherto to argue in these

pages with those who doubt the reality of beauty in mathematics. They are, I believe, relatively few. But it may be necessary to say a word to those who think that aesthetic appreciation is rare, and these are many. Professor Hogben thinks that "the aesthetic appeal of mathematics may be very real for a chosen few." For these "few," he says, "mathematics exercises a coldly impersonal attraction." On an occasion when I was speaking to a Student Christian Movement of sixth formers I happened to remark, incidentally, that the theorem of Pythagoras was "a thing of beauty." The explosion of derisive laughter that greeted this innocent remark was shattering. The reason for the outburst was, in my view, very simple. Every one knew that what I had said was true, but to admit in involved "wearing one's heart on one's sleeve" and this "isn't done" by sixth formers. One rarely hears the adjective "beautiful" from the lips of an adolescent: his private feelings are not for public display.

The universal popularity of board games with a mathematical basis is an argument against the view that mathematics is for the few. *Go* in Japan and chess in Russia are examples. Chess games and problems are found in many of the world's periodicals. It is relevant to our thesis to note that end-games are described as "beautiful," chess situations as "diverting," the checkmate as "neat," the solution of the problem as "elegant." Is there, then, beauty in chess but not in "mathematics"?

Supporting evidence for the widespread appeal of mathematics is found in the popularity of puzzles; in such fascinating columns as "Mathematical Games" published monthly for many years past in the *Scientific American* under the brilliant editorship of Martin Gardner; and in the dozens of books on "popular mathematics" which have sold as paperbacks by the million.

Ideas in Poetry and Mathematics

If we are to discover the source of satisfaction that arises from the contemplation of a mathematical thesis, we shall do well to consider the more general question of the aesthetic pleasure associated with the creation and appreciation of great art as it is found in poetry or literature or music. Few (if any) have contributed more to the illumination of this question than the psychologist C. G. Jung in the development of his theory of the collective unconscious.

Let us consider as an example the first stanza of Gray's "Elegy Written in a Country Courtyard," reputed to be among the most popular poems in the English language:

The curfew tolls the knell of parting day,
The lowing herd winds slowly o'er the lea.
The ploughman homeward plods his weary way,
And leaves the world to darkness, and to me.

If a teacher of English literature were asked to account for the pleasure aroused by reading these lines, he would probably refer to the rhymes and rhythms, the tempo, the long, slow vowels, and the alliterative echoes. But would he say much about the content, the ideas, the imagery? Housman stated roundly that ideas in poetry are unimportant:

> I cannot satisfy myself that there are any such things as poetical ideas. ... Poetry is not the thing said, but the way of saying it.

If this is true, mathematical beauty must differ radically from that of poetry, for the working material of the mathematician is nothing but ideas. But is it true? It is certain that C. G. Jung would attach primary importance to the *ideas* conveyed by this poem. He quotes Gerhart Hauptmann, "Poetry means the distant echo of the primitive world behind our veil of words," and proceeds to amplify this by reference to his theory of the collective unconscious, which he distinguishes from the personal unconscious of the poet:

> The collective unconscious is in no sense an obscure corner of the mind, but the all-controlling deposit of ancestral experience from untold millions of years, the echo of prehistoric world events to which each century adds an infinitesimally small amount of variation and differentiation.

If we recall Gray's primordial images of the ploughman, of the lowing herd and its winding way, of the progress of the parting day leaving the world to darkness, we may see the relevance of Jung's ideas to his verse. Jung writes:

> The man who speaks with primordial images speaks with a thousand tongues; he entrances and overpowers, while at the same time he raises the idea he is trying to express above the occasional and the transitory into the sphere of the ever-existing. ...

That is the secret of effective art.

The creative process, in so far as we are able to follow it at all, consists in the unconscious animation of the archetype, and in a development and shaping of the image until the work is completed. The shaping of a primordial image is, as it were, a translation into the language of the present which makes it possible for every man to find again the deepest springs of life which would otherwise be closed to him.

Like the appreciation of music, pleasure in the pursuit of mathematics as a mental discipline springs from those deep layers of the human psyche which, having been developed in the early epochs of human evolution, lie buried beneath mental strata of later development. The accepted view that there is a definite connection between musical appreciation and mathematical taste is based not only on the observation that many gifted mathematicians have had a warm appreciation of music (a few, like Einstein, having been skilled instrumental performers), but also on the similarity between the deep-seated structure of musical form and that of mathematical ideas. As Hardy said, "There are probably more people really interested in mathematics than in music." Perhaps many people enjoy music because they intuitively perceive its mathematical basis. As the conscious mind expresses itself in language and gesture, so the unconscious mind may become articulate in music and in mathematics. It is possible for neither music nor mathematics to assume any arbitrary form if it is to be intelligible to the mind. A fortuitous succession of notes makes as little sense as a haphazard chain of mathematical symbols. There is no more articulate language of the unconscious mind than music, but the syntax and the grammar of this language are not capricious; they are dictated in their broadest outlines by the texture and organization of the deep levels of the mind, which assumed its present structure in those aeons of evolutionary time that led up to the coming of *Homo sapiens*. So with mathematics. While we know something of prehistoric man's physical environment, and we can speculate concerning the mental stresses which in the course of vast periods of time evolved a mentality of definite pattern functioning in a particular fashion, we know little or nothing of the reasons why this mentality should find satisfaction in certain types of mathematics rather than in a thousand other possible, unimagined types. The fact remains that the forms of music and the shapes of mathematics which appeal to our minds are directed by a basic mental structure which was itself an

Evolution of Aesthetic Feeling

Now we have reached the nucleus of the argument. The ultimate source of aesthetic sensibility to the various manifestations of beauty in mathematics is to be sought for in the unconscious mind or (frequently) in the collective unconscious by virtue of which man is heir of all the ages. The mental processes evoked in all men for a million years past by their physical environment have deposited a soil in which the roots of the psyche are deeply and securely implanted. Experiences of all the generations of a man's ancestors, repeated millions of times and recorded as memory structures in the brain, are scored ever more deeply as they are transmitted from generation to generation through the centuries, often showing a brief vitality in unaccountable dreams. The mind of a newly born baby is no *tabula rasa*: before he has had the opportunity to develop a conscious mind he is equipped with these inherited memory structures. For example, one of his earliest mental activities impels him to seek his mother's breast.

It is to the emotionally charged experiences of a thousand generations of our ancestors that we must look in order to discover the sources of aesthetic pleasure in art, in poetry, in music, in mathematics, and in other artistic forms. It is not impossible to guess what some of these experiences must be which, either because their repetition is so frequent or because they evoke strong mental excitement, have left their inaudible traces on our mental structure; these traces are a fixed part of our human inheritance and the ground of our aesthetic appreciation.

Ingredients of Beauty

Beauty in mathematics, as in music, is not elemental; it is a compound of several ingredients, which are not found in isolation—ingredients which have this in common: they stir buried memories which rise to awaken feelings in the conscious levels of the mind. Let us consider some of these. They are of a very general character, inherited by every member of the human race.

The alteration of tension and relief is a universal emotion. In reading any form of serious mathematics, we experience alternately perplexity and illumination. Out of chaos comes order. Out of the many, one. This, reaching the deepest levels of feeling, gently stimulates the aesthetic sensibilities. The effect is found in music when the alteration in (e. g.) a hymn tune of the dominant and the tonic—tension and relaxation—contributes to the beauty of the tune. Another example is the familiar one of discord into harmony.

In mathematics a student who finds himself bewildered by the wide variety of the series representations of such functions as e^x, log x, cos x, etc., is delighted to find a theorem like Taylor's which covers them all. "A beautiful generalization" may well be his reaction.

The realization of expectation is a mental pleasure of more ancient standing than the human race itself. An example from music is the familiar sequence: dominant $\rightarrow$ tonic. An example from mathematics is found in the early paragraphs of this chapter, where the inscribed spheres depicted in Figure 2, which touch the ellipse at its two foci, raise the question whether the same result would follow if the ellipse is projected into a parabola. The satisfaction that derives from seeing this expectation fulfilled is a spice which is relished in the flavor of the mathematical beauty of the conic sections.

Surprise at the unexpected, conversely, is an emotion which we have in common with our animal ancestry. When a striking mathematical conclusion which has not been anticipated suddenly presents itself, old established emotions are stirred. An example might be the discovery of the Fibonacci series hidden in the Pascal triangle. Other examples are found on the following pages.

The perception of unsuspected relationships is another pleasurable experience old enough to have been built into our mental structure. One can imagine, for example, the excitement roused in the minds of primitive men when they first realized that there was a connection between the heights of the tides and the phases of the moon.

An example from mathematics might be the relation between the equation of a conic:

$$x^2 + y^2 + 2gx + 2fy + c = 0$$

and its tangent at (x_1, y_1):

$$xx_1 + yy_1 + g(x + x_1) + f(y + y_1) + c = 0.$$

There does not appear, at first glance, to be any connection between the coefficients of the binomial expansion $(x + 1)^n$ as displayed in Pascal's triangle and the coefficients found in a formula for tan $n\theta$. But consider the integers in the following:

$$(x + 1)^5 = x^5 + 5x^4 + 10x^3 + 10x^2 + 5x + 1$$

and compare with

$$\tan 5\theta = \frac{5\tan\theta - 10\tan^3\theta + \tan^5\theta}{1 - 10\tan^2\theta + 5\tan^4\theta}.$$

It turns out on further investigation that trigonometrical

functions can be expressed algebraically without reference to right-angle triangles. This unification and generalization is a source of gratifying surprise. A feeling of increased mathematical power, too, comes by way of the remarkable formula

$$e^{i\theta} = \cos\theta + i\sin\theta \qquad (i = \sqrt{-1})$$

from which are derived

$$\cos\theta = \frac{e^{i\theta} + e^{-i\theta}}{2}$$

$$\sin\theta = \frac{e^{i\theta} - e^{-i\theta}}{2i}.$$

This increase in our resources means, for example, that a problem in sines and cosines, such as

$$\int \sin^8\theta \, d\theta$$

can be changed into a more manageable one in exponentials.

Mathematical beauty is found in *patterns*. The enjoyment of patterns is older than folk dancing. Hardy wrote:

> A mathematician, like a painter or a poet, is a maker of patterns. If his patterns are more permanent than theirs, it is because they are made with *ideas*. ... The mathematician's patterns, like the painter's or the poet's, must be *beautiful*; the ideas, like the colours or the words, must fit together in a harmonious way. ...

"Brevity is the soul of wit." It may be the soul of beauty too. An example from poetry could be the brevity of the metre of Francis Thompson's "To a snowflake":

> Fashioned so purely
> Fragilely, surely
> From what paradisal
> Imagineless metal
> Too costly for cost?

An example from mathematics might be Fermat's famous theorem, in which range and generality are condensed into a couple of lines.

> Given x, y, z integers, the equation $x^n + y^n = z^n$ has no integral solutions if n is an integer greater than 2.

Goldbach's postulate may be quoted. It has been proved for all numbers less than 10,000:

> Every even integer is the sum of two primes.

An elementary example might be the proof of Pythagoras' theorem given by the Indian mathematician Bhaskara (born A. D. 1114). He simply draws four

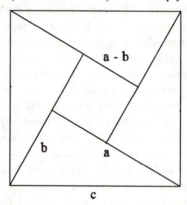

Figure 4

equal right-angle triangles as in Figure 4. The area of each triangle is $ab/2$, so c^2 (the area of the square) is equal to $4ab/2 + (a - b)^2 = a^2 + b^2$. The reader can easily verify the construction.

"Unity in variety" was Coleridge's definition of beauty. It is frequently exemplified in music. As an example of the artistic quality of a mathematical theorem, consider the discovery by Johann Bernoulli (1667-1748) of the beautiful curve called the Brachistochrone.

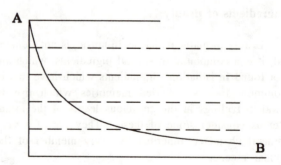

Figure 5

A particle slides down a smooth curve from A to B. What curve makes the time of descent a minimum? Would it be a straight line, an arc of a circle (as Galileo supposed), or some other curve?

Bernoulli compared the path to that of a ray of light

traversing a stratified layer of decreasing optical density (broken lines, (Figure 5)); this is also a "least time" problem. He obtained the equation of the Brachisto-chrone:

$$y\left(1 - \left(\frac{dy^2}{dx}\right)\right) = \text{constant}.$$

This is a cycloid, the curve described by a point on the circumference of a circle that rolls along a straight line.

The linking together of a problem in mechanics with a phenomenon in optics and relating the identical solution of both to a lovely curve—the cycloid, derived from pure geometry—has an artistic appeal that can scarcely be missed. As Polya remarks in this connection, "there is a real work of art before us." It is unity in variety.

There is *sensuous pleasure* to be derived from geometry. One of the gentle satisfactions enjoyed by all our ancestors, which must have left its mark on the unconscious mind, is the smooth sweep of the eye along the many quiet curves found in Nature. The smoothness of their contours is associated with the ease and comfort of the eye's muscular effort. Jagged and jerky lines have been shown by psychologists to produce an opposite mental effect. The curves that the human gaze has followed for a million years include the sea horizon, the skyline of the rolling downs, the rainbow, the meteor track, the parabola of the waterfall, the slingst-one and the arrow, the arcs traced in the sky by the sun and the crescent moon, the flight of a bird, and many others.

Such purely sensuous pleasure is an ingredient of the aesthetic joy found in the geometry of the circle, the ellipse and other conic sections, as well as of the cycloid, the catenary, the graphs of trigonometrical functions, the cardioid, the logarithmic spiral $\log \rho = a\theta$ the limaçon and many other lovely shapes.

A melody may mirror such grace. It is rare to find the jagged contour of widely separated notes in a melody. The melodious phrase may ascend and descend gently, register minor and major climaxes, pirouette like a ballerina, before subsiding smoothly to its point of departure. Such primeval aesthetic satisfactions, mathe-matical and musical, are rooted in the racial unconscious of humanity.

A sense of wonder, even of awe, in the presence of the infinite, is one of the basic human emotions. Through all the aeons of time when man has stood beneath the cold light of stars and gazed into the unbounded depths of space; and especially since man first understood, a century ago, that an age-long stretch of evolutionary history lies behind him, infinity has

been for him an emotionally charged concept. Music has power to arouse this emotion. So has mathematics. A divergent series of any sort induces this sense of infinity even as a convergent series leads to the related idea of the infinitesimal. Both feelings are roused by the spectacle of the curve of the hyperbola streaking off to infinite distance, simultaneously reducing its separa-tion from its asymptote without ever reaching it. These are aspects of the aesthetic experiences of mathematics which easily pass unnoticed as such.

With this we may associate the baffled sense of mystery produced by certain mathematical theorems, the beauty of which is accompanied by an initial feeling of inadequacy to explain such remarkable results.

An example is Pascal's "Mystic Hexagram" (Figure 6):

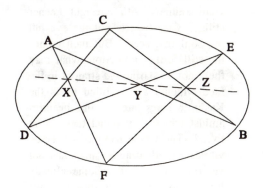

Figure 6

If a hexagon is inscribed in a conic, then the intersections of the three pairs of opposite sides are collinear.

A beautiful theorem! Pascal (1623-1662) proved it when he was only sixteen years old and gave the figure its name.

Brianchon proved a theorem as follows:

If a hexagon is circumscribed about a conic, then the joins of the three pairs of opposite vertices are concurrent.

Many theorems of this type are found in treatises on projective geometry. The trained minds of mathemati-cians have delighted in their beauty for centuries past.

"The trained minds": the enjoyment of beauty in mathematics is for the most part an acquired taste. The eye has to be educated to *see*. How much of beauty the eye misses for lack of training! "Having eyes, they see not." Even the most highly trained mathematician must remain unmoved by much of the splendor because it is hidden from his keenest sight. "I can't see much in your scenery here," said an American tourist to a guide in Wordsworth's country. "Don't you wish you could, sir?" was the apt retort. Did anyone ever "see" more than Wordsworth? We may well doubt it. What we may never doubt is that there is more to be seen. The point has been well made by Sir Francis Younghusband. Moved by the beauty of Kashmir scenery he wrote:

> There came to me this thought, which doubtless has occurred to many another beside myself—why the scene should so influence me and yet makes no impression on the men about me. Here were men with far keener eyesight than my own, and around me were animals with eyesight keener still. ... Clearly it is not the eye but the soul that sees. But then comes the still further reflection: what may there not be staring *me* straight in the face which I am as blind to as the Kashmir stags are to the beauties amidst which they spend their entire lives? The whole panorama may be vibrating with beauties man has not yet the soul to see. Some already living, no doubt, see beauties we ordinary men cannot appreciate. It is only a century ago that mountains were looked upon as hideous. And in the long centuries to come may we not develop a soul for beauties unthought of now?

QUESTIONS AND EXERCISES

1. Show that the author's equation for the tangent line to a conic is correct.

2. Draw a picture to illustrate Brinachon's Theorem.

3. "An example might be the discovery of the Fibonacci series hidden in the Pascal triangle." Here is part of the Pascal triangle:

1						
1	1					
1	2	1				
1	3	3	1			
1	4	6	4	1		
1	5	10	10	5	1	
1	6	15	20	15	6	1 ;

by going up the diagonals you get the Fibonacci numbers:

$$1, 1, 1 + 1 = 2, 1 + 2 = 3, 1 + 3 + 1 = 5,$$
$$1 + 4 + 3 = 8, 1 + 5 + 6 + 1 = 13.$$

Get other numbers by going over and up by knight's moves:

$$1, 1, 1, 1 + 1 = 2, 1 + 3 = 4,$$
$$1 + 6 = 7, 1 + 10 + 1 = 12$$

and see if you can get a formula for them, or a relation like that for the Fibonacci numbers,

$$f_{n+1} = f_n + f_{n-1}, \quad n = 2, 3, \dots .$$

4. "This example should serve to show the distinction between beauty in mathematics and pleasure in searching it out." On a scale of 0 to 10, rate the beauty of the relation between the equation of the conic and the equation of its tangent and on the same scale rate the pleasure that you had in searching it out. If your two numbers are different, why are they different? If your two numbers are both zeros, what is the explanation for that?

5. Calculate $\int \cos^6\theta \, d\theta$

(a) By using $\cos^2\theta = (1 + \cos 2\theta)/2$ three times.

(b) By substituting $\cos\theta = \frac{1}{2}(e^{i\theta} + e^{-i\theta})$.

Was the second method more manageable?

6. Try the construction of Figure 6 inscribing the hexagram in a hyperbola instead of an ellipse.

7. "Those who doubt the reality of beauty in mathematics are, I believe, relatively few." Do you think that Huntley's belief is correct?

8. "A supreme purpose of beauty ... is to serve as a stimulus to creative activity." Do you think that is true? Does beauty have a purpose and, if so, why?

9. "Experiences of all the generations of a man's ancestors, repeated millions of times and recorded as memory structures in the brain, are scored ever more deeply as they are transmitted from generation to generation through the centuries." It seems as if Huntley is arguing for the inheritance of acquired characteristics. Is he saying anything different from asserting that giraffes have long necks because giraffes

have stretched their necks reaching for food millions of times, from generation to generation? If that is what he is saying, how can he maintain that discredited notion? If that is not what he is saying, what *is* he saying?

10. "How much of beauty the eye misses for lack of training!" Can you think of something in which you now find beauty that you once did not? If you can, how did the change from not seeing to seeing take place? How can the eye and mind be trained to see beauty?

NINE MORE PROBLEMS

by Martin Gardner

"Mathematics is fun!" Has anyone ever tried to tell you that? Maybe not, since many people think that it is a statement so obviously false that no one not certifiably insane would say it. Fun and mathematics in the same sentence! Clearly, there can be no connection between the two. A less extreme version is "Mathematics can be fun." If you have heard that, it was probably said by a teacher of mathematics, and you may have dismissed it as just part of the oddness to be expected from teachers, and especially from teachers of mathematics. Peculiar creatures you may think they are, with peculiar ideas: good in their place, explaining how problems should be done, but not to be taken too seriously elsewhere. Be that as it may, when they say that mathematics can be fun they are being neither peculiar nor crazy, but are speaking the sober truth. There is a part of mathematics called recreational mathematics, mathematics for recreation's sake, on which some people like to spend time, for fun. The purpose of this section is to describe, a little, what recreational mathematics is. To know that such a thing exists expands your world, and it is always a good thing to have your world expanded. There is also the possibility that you will come to agree that mathematics, or some parts of it anyway, can in fact be fun. You will thus have discovered a new source of pleasure, and sources of pleasure are always worth having. On the other hand, you may conclude that there is no fun for you in recreational mathematics. There is no shame in that, since no one can enjoy everything that can be enjoyed.

If you find the connection of mathematics with fun new and strange, the reason is that *learning* mathematics is not necessarily fun, and you have probably been spending most of your mathematical time trying to learn it. The process of learning mathematics, as of learning any skill, is likely to be hard work, and hard work is often disagreeable. Of course, there are satisfactions to be gotten from hard work, but those mostly come after it is over. The piano student practices and practices, not because it is fun, but to be able, later, to play the piano. The runner trains and trains, not because it is fun, but to be able, later, to run in races. The student of mathematics does problems and problems, not because it is fun, but to be able, later ... to do what? Most students of mathematics never do anything but practice, and train, and do assigned problems. The reward of being able to use a skill, developed over months and years, never comes. Very few jobs involve actually using all of the mathematics that you know. When people leave school, most of them have solved their last equation, differentiated their last function, and sketched their last curve. How unfair it is: you have been pushed and pushed and you have been filled as full as you can be filled with mathematics that you are never going to have the fun of using. It is as if a piano student after years of practice was told that from now on pianos can no longer be touched, only cellos, or as if a runner was told going outside was no longer permitted. No wonder that most people do not think that mathematics if fun, or can be fun. All they ever do is train for a day that never comes.

But mathematical life does not have to be that bleak, and it is possible for some students of mathematics to get fun out of their efforts. Recreational mathematics is mathematics for the fun of it with no worry about any use outside of amusement. The taste for it is not universal. In fact, the proportion of people with mathematical training who find enjoyment in mathematical play is very small. Whose fault this is, if it is anyone's fault, is not clear, nor is it clear if anything should be done about it, or if anything could be done about it. It is too bad that so many students find mathematics so distasteful, so unfun, that they avoid it whenever possible, but it may be that this is part of the nature of people, or of mathematics. On the other hand, it may be because no one ever told them that any pleasure whatsoever could possibly come from mathematics. It may be because they never found out about recreational mathematics.

One example of recreational mathematics is the four 4s game. What you have to do is take four 4s, exactly four 4s and no other numbers at all, together with the symbols for arithmetic operations and combine them to make expressions equal to 1, 2, 3, ... up as far as you can go. For example, here is one way to represent 1:

$$1 = \frac{4}{4} \cdot \frac{4}{4}.$$

There are other ways as well:

$$\frac{44}{44}, \quad or \quad 4 - 4 + \frac{4}{4}.$$

After 1 comes 2:

$$2 = \frac{4}{4} + \frac{4}{4} = 4 - \sqrt{4} - \sqrt{4} + \sqrt{4}.$$

It's all right to use square roots signs, as above, but of course an expression is nicer if it has no square roots in it, though sometimes they are unavoidable. When you get up to larger integers, you may find the factorial symbol helpful, so that 24 = 4! can be written with one 4, and 26 can be written with four 4s:

$$26 = 4! + \frac{4 + 4}{4}.$$

The decimal point is also legal, which means that 10 = 4/.4 can be efficiently expressed. Exponentials and square roots are allowed, so $6 = \sqrt{4}^{4/4} + 4$.

Doesn't that look like fun? Can't you hardly wait to try to get 3, 4, ... , and that well-known killer, 31? Your answers may be "No—what's the point?" or "I can wait—for a hundred years if I have to" and they are quite legitimate. If the four 4s game is not to your taste, well, there is no arguing with taste. On the other hand, if it is to your taste and you go on to get

$$3 = \frac{4 + 4 + 4}{4}, \quad 4 = (\sqrt{4} + \sqrt{4}) \cdot \frac{4}{4},$$

and 5, and 6, and so on, then you are experiencing recreational mathematics. Writing numbers with four 4s is mathematics, since it is dealing with numbers and how they combine, and it is recreational since it has no use other than passing the time in a pleasant way. Some people claim that the exercise given to the brain by doing recreational mathematics does it good, at least more good than would be done by spending the same amount of time watching soap operas on television, but scientific evidence is lacking. Whether the four 4s game strengthens the mind or not does not matter, since recreation does not need to be justified on the grounds that it is good for you. Fun is its own reward.

Another recreational mathematics problem that you may have met before is the old one of the person going to the river with two jars, one that holds exactly five pints and the other exactly three pints, with instructions to bring back exactly four pints of water: how can that be done? The process of solving the problem is part of what makes it recreational. There are no rules. There is no procedure outlined in some textbook that, if followed exactly, will give the solution. The solution must be *discovered* by the solver.

How to discover the solution? The problem is one in recreational mathematics, so it is solved recreationally. That is to say, playfully. When you play, there are no hard-and-fast rules, there are no procedures you must follow, there is no example in a textbook to be mimicked. You play around. You pour water in and out of the jars until you see how to get the four pints, or until you get tired of trying and quit. Another feature of play, and of recreational mathematics, is that no one makes you do it, and no one cares how long you do it or how well you do. That is one of the things that makes play (and recreational mathematics) so unlike work (and homework problems). Another thing is the reward, the "Aha!" that you say, mentally or out loud, when the solution arrives. The "Aha!" experience is the thrill of discovery, and it is seldom encountered while doing textbook exercises. Even though what you have discovered has been discovered before, the joy of discovery is what makes the time spent worthwhile. You do not get a thrill from calculating twenty derivatives using the quotient rule. That is practice, not play. That is for developing technique, not for using it. I will not spoil the jars problem by giving the solution.

Recreational mathematics goes back as far as mathematics does. Babylonian clay tablets written three thousand years ago contain, just as textbooks do today, mathematical problems that could have no practical significance. They were probably meant to be assigned as homework problems, but whoever made them up was indulging in recreational mathematics. Here is problem 79 in the Rhind papyrus, written more than three thousand years ago:

> An estate consisted of seven houses; each house had seven cats; each cat ate seven mice; each mouse ate seven heads of wheat; each head of wheat could yield seven measures of grain. Houses, cats, mice, heads of wheat, and measures of grain, how many of these in all were in the estate?

In 1202, Leonardo of Pisa published his *Liber Abaci*, The Book of the Abacus, meant to explain all that was known about arithmetic and algebra, and one of the things that it contained was:

> Seven old women went to Rome; each woman has seven mules; each mule carried seven sacks; each sack contained seven loaves; and with each loaf were seven knives; each knife was put up in seven sheaths.

Familiar. The problem survives today, though now it has a twist:

> As I was going to St. Ives,
> I met a man with seven wives;
> Every wife had seven sacks;
> Every sack had seven cats;
> Every cat had seven kits.
> Kits, cats, sacks and wives,
> How many were going to St. Ives?

The answer used to be 19,607, but now it is 1.

The first book devoted solely to recreational mathematics appeared in France in 1612: *Problèmes plaisants et délectables qui se font par les nombres*, Pleasant and Delightful Problems Based on Numbers, by Claude-Gaspar Bachet. Here is the first item in the book. You ask someone to pick a number, any number. Your instructions to the person then are to double it, add five, multiply the result by 5, add ten, multiply by ten, and announce the result. You will then be able to identify the number your victim started with and you can try to claim that you were able to because of your powers of mind-reading. For example, if the number selected was 7, the calculations would go

double it	14
add five	19
multiply by five	95
add ten	105
multiply by ten	1050

and you would get the 7 that the process started with by subtracting 350 from the 1050 to get 700 and then disregarding the zeros. No mind-reading is necessary, nor is it hard to see how the trick works. If you start with n, then the calculations would go

double it	$2n$
add five	$2n + 5$
multiply by five	$10n + 25$
add ten	$10n + 35$
multiply by ten	$100n + 350$

and if you subtract 350 from the last number you have $100n$. Variations on this continue to amaze people today. Try this variation of Bachet's problem out and see if it amazes: ask someone to take the number of his or her birth month, multiply it by five, add four, multiply by ten, add eight, add the number of the birth day, multiply by two, and add four. For example, if someone was born on August 22, the sequence of numbers would be 8, 40, 44, 440, 448, 470, 940, and 944. From the last number you can retrieve the birth month and day by subtracting 100: the first digit is the number of the birth month and dividing the last two digits by 2 give the birth day. The reason this works is that the last number is $100m + 100 + 2d$, where m and d are the numbers of the birth month and day.

Books on recreational mathematics have been coming out ever since 1612. One by Ozanam was first published in 1694 and was reissued in editions in 1741, 1750, 1770, 1790, 1803, 1814, and 1840. In this century, Ball's *Mathematical Recreations and Essays* has gone through more than ten editions. One reason for the long life of recreational mathematics books is that original ideas in recreational mathematics are rare, so that the contents of the books do not become out of date. For hundreds of years people have been going to rivers with 5-pint and 3-pint jars with instructions to bring back four pints exactly. The books contain a wealth of fascinating material—fascinating to those who find it fascinating, of course, and boring to those who can see no fun in it—and anything with the name of Martin Gardner on it is likely to be more fascinating than the average. *Games* magazine, available at newsstands, has in each issue some things that would be classified as recreational mathematics. There is even a *Journal of Recreational Mathematics*, but it does not appear on newsstands and its contents tend to be advanced and scholarly and best left to veterans of years of recreational mathematics experience. Books on recreational mathematics are filed in libraries under QA 95 or, for libraries using the Dewey Decimal system, under 793.74 or perhaps 510.75.

There are hardly any recreational calculus problems. The reason for that is that calculus is not fun. It is *serious*. Calculus books do not contain jokes, nor are they lighthearted. The purpose of calculus is to solve important problems in geometry, physics, and elsewhere, and there is to be no kidding around. It is too bad that we cannot use our skill in calculus for enjoyment, but that is the way of the world. Not every field has recreational branches. There is no recreational plumbing, for example, or recreational banking. The problems of recreational mathematics are almost all drawn from arithmetic, logic, geometry, and algebra. It is just as well that there are none from calculus, since most people know nothing of the subject and even those who have been trained in it tend to forget it quickly when they leave their last calculus course and find that they never have any occasion to use it. There is plenty of fun available outside of calculus.

There follow seven of a collection of nine recreational mathematics problems. They are not easy ones, which is why the solutions are included also. Reading the answers can also be fun.

---

1. Crossing the Desert

An unlimited supply of gasoline is available at one edge of a desert 800 miles wide, but there is no source on the desert itself. A truck can carry enough gasoline to go 500 miles (this will be called one "load"), and it can build up its own refueling stations at any spot along the way. These caches may be any size, and it is assumed that there is no evaporation loss.

What is the minimum amount (in loads) of gasoline the truck will require in order to cross the desert? Is there a limit to the width of a desert the truck can cross?

2. The Two Children

Mr. Smith has two children. At least one of them is a boy. What is the probability that both children are boys?

Mr. Jones has two children. The older child is a girl. What is the probability that both children are girls?

3. Lord Dunsany's Chess Problem

Admirers of the Irish writer Lord Dunsany do not need to be told that he was fond of chess. (Surely his story "The Three Sailors Gambit" is the funniest chess fantasy ever written.) Not generally known is the fact that he liked to invent bizarre chess problems which, like his fiction, combine humor and fantasy.

The [following problem] was contributed by Dunsany to *The Week-End Problems Book*, compiled by Hubert Phillips. Its solution calls more for logical thought than skill at chess, although one does have to know the rules of the game. White is to play and mate in four moves. [The position looks like that at the start of the game, with two changes: Black's king and queen are interchanged, and White has no pawns.] The position is one that could occur in actual play.

4. Professor on the Escalator

When Professor Stanislaw Slapenarski, the Polish mathematician, walked very slowly down the down-moving escalator, he reached the bottom after taking 50 steps. As an experiment, he then ran up the same escalator, one step at a time, reaching the top after taking 125 steps.

Assuming that the professor went up five times as fast as he went down (that is, took five steps to every one step before), and that he made each trip at a constant speed, how many steps would be visible if the escalator stopped running?

...

6. Dividing the cake

There is a simple procedure by which two people can divide a cake so that each is satisfied he has at least half: One cuts and the other chooses. Devise a general procedure so that *n* persons can cut a cake into *n* portions in such a way that everyone is satisfied that he has at least 1/*n* of the cake.

...

8. The Absent-minded Teller

An absent-minded bank teller switched the dollars and cents when he cashed a check for Mr. Brown, giving him dollars instead of cents, and cents instead of dollars. After buying a five-cent newspaper, Brown discovered that he had left exactly twice as much as his original check. What was the amount of the check?

9. Water and Wine

A familiar chestnut concerns two beakers, one containing water, the other wine. A certain amount of water is transferred to the wine, then the same amount of the mixture is transferred back to the water. Is there now more water in the wine than there is wine in the water? The answer is that the two quantities are the same.

Raymond Smullyan writes to raise the further question: Assume that at the outset one beaker holds ten ounces of water and the other holds 10 ounces of wine. By transferring three ounces back and forth any number of times, stirring after each transfer, is it possible to reach a point at which the percentage of wine in each mixture is the same?

Answers

1. The following analysis of the desert-crossing problem appeared in a recent issue of *Eureka*, a publication of mathematics students at the University of Cambridge. Five hundred miles will be called a "unit"; gasoline sufficient to take the truck 500 miles will be called a "load"; and a "trip" is a journey of the truck in either direction from one stopping point to the next.

Two loads will carry the truck a maximum distance of 1 and 1/3 units. This is done in four trips by first setting up a cache at a spot 1/3 unit from the start. The truck begins with a full load, goes to the cache, leaves 1/3 load, returns, picks up another full load, arrives at the cache and picks up the cache's 1/3 load. It now has a full load, sufficient to take it the remaining distance to one unit.

Three loads will carry the truck 1 and 1/3 plus 1/5 units in a total of nine trips. The first cache is 1/5 unit from the start. Three trips put 6/5 loads in the cache. The truck returns, picks up the remaining full load and arrives at the first cache with 4/5 load in its tank. This, together with the fuel in the cache, makes two full loads, sufficient to carry the truck the remaining 1 and 1/3 units, as explained in the preceding paragraph.

We are asked for the minimum amount of fuel required to take the truck 800 miles. Three loads will take it 766 and 2/3 miles (1 and 1/3 plus 1/5 units), so we need a third cache at a distance of 33 and 1/3 miles

(1/15 unit) from the start. In five trips the truck can build up this cache so that when the truck reaches the cache at the end of the seventh trip, the combined fuel of truck and cache will be three loads. As we have seen, this is sufficient to take the truck the remaining distance of 766 and 2/3 miles. Seven trips are made between starting point and first cache, using 7/15 load of gasoline. The three loads of fuel that remain are just sufficient for the rest of the way, so the total amount of gasoline consumed will be 3 and 7/15, or a little more than 3.46 loads. Sixteen trips are required.

Proceeding along similar lines, four loads will take the truck a distance of 1 and 1/3 plus 1/5 plus 1/7 units, with three caches located at the boundaries of these distances. The sum of this infinite series diverges as the number of loads increases; therefore the truck can cross a desert of any width. If the desert is 1,000 miles across, seven caches, 64 trips and 7.673 loads of gasoline are required.

Hundreds of letters were received on this problem, giving general solutions and interesting sidelights. Cecil G. Phipps, professor of mathematics at the University of Florida, summed matters up succinctly as follows:

"The general solution is given by the formula:

$$d = m\left(1 + \frac{1}{3} + \frac{1}{5} + \frac{1}{7} + ...\right),$$

where d is the distance to be traversed and m is the number of miles per load of gasoline. The number of depots to be established is one less than the number of terms in the series needed to exceed d. One load of gasoline is used in the travel of each pair of stations. Since the series is divergent, any distance can be reached by this method although the amount of gasoline increases exponentially.

"If the truck is to return eventually to its home station, the formula becomes:

$$d = m\left(\frac{1}{2} + \frac{1}{4} + \frac{1}{6} + \frac{1}{8} + ...\right).$$

This series is also divergent and has properties similar to those for the one-way trip." ...

2. If Smith has two children, at least one of which is a boy, we have three equally probable cases:

> Boy-boy
> Boy-girl
> Girl-boy.

In only one case are both children boys, so the probability that both are boys is 1/3.

Jones's situation is different. We are told that his older child is a girl. This limits us to only two equally probable cases:

> Girl-girl
> Girl-boy.

Therefore the probability that both children are girls is 1/2.

[This is how I answered the problem in my column. After reading protests from many readers, and giving the matter considerable further thought, I realized that the problem could not be answered without additional data. For a later discussion of the problem, see Chapter 19.]

3. The key to Lord Dunsany's chess problem is the fact that the black queen is not on a black square as she must be at the start of a game. This means that the black king and queen have moved, and this could have happened only if some black pawns have moved. Pawns cannot move backward, so we are forced to conclude that the black pawns reached their present positions from the other side of the board! With this in mind, it is easy to discover that the white knight on the right has an easy mate in four moves.

White's first move is to jump his knight at the lower right corner of the board to the square just above his king. If black moves the upper left knight to the rook's file, white mates in two more moves. Black can, however, delay the mate one move by first moving his knight to the bishop's file instead of the rook's. White jumps his knight forward and right to the bishop's file, threatening mate on the next move. Black moves the knight forward to block the mate. White takes the knight with the queen, then mates with his knight on the fourth move.

4. Let n be the number of steps visible when the escalator is not moving, and let a unit of time be the time it takes Professor Slapenarski to walk down one step. If he walks down the down-moving escalator in 50 steps, then $n - 50$ steps have gone out of sight in 50 units of time. It takes him 125 steps to run up the same escalator, taking five steps to every one step before. In this trip, $125 - n$ steps have gone out of sight in 125/5, or 25, units of time. Since the escalator can be presumed to run at constant speed, we have the following linear equation that readily yields a value for n of 100 steps:

$$\frac{n - 50}{50} = \frac{125 - n}{25}.$$

...

6. Several procedures have been devised by which n persons can divide a cake in n pieces so that each is satisfied that he has at least $1/n$ of the cake. The

following system has the merit of leaving no excess bits of cake.

Suppose there are five persons: A, B, C, D, E. A cuts off what he regards as 1/5 of the cake and what he is content to keep as his share. B now has the privilege, if he thinks A's slice is more than 1/5, of reducing it to what he thinks is 1/5 by cutting off a portion. Of course if he thinks it is 1/5 or less, he does not touch it. C, D, and E in turn now have the same privilege. The last person to touch the slice keeps it as his share. Anyone who thinks that this person got less than 1/5 is naturally pleased because it means, in his eyes, that more than 4/5 remains. The remainder of the cake, including any cut-off pieces, is now divided among the remaining four persons in the same manner, then among three. The final division is made by one person cutting and the other choosing. The procedure is clearly applicable to any number of persons.

...

8. To determine the value of Brown's check, let x stand for the dollars and y for the cents. The problem can now be expressed by the following equation:

$$100y + x - 5 = 2(100x + y).$$

This reduces to $98y - 199x = 5$, a Diophantine equation with an infinite number of integral solutions. A solution by the standard method of continued fractions gives the lowest values in positive integers: $x = 31$ and $y = 63$, making Brown's check $31.63. This is a unique answer to the problem because the next lowest values are: $x = 129$, $y = 262$, which fail to meet the requirement that y be less than 100.

There is a much simpler approach to the problem and many readers wrote to tell me about it. As before, let x stand for the dollars on the check, y for the cents. After buying his newspaper, Brown has left $2x + 2y$. The change that he has left, from the x cents given him by the cashier, will be $x - 5$.

We know that y is less than 100, but we don't yet know whether it is less than 50 cents. If it is less than 50 cents, we can write the following equations:

$$2x = y$$
$$2y = x - 5.$$

If y is 50 cents or more, then Brown will be left with an amount of cents ($2y$) that is a dollar or more. We therefore have to modify the above equations by taking 100 from $2y$ and adding 1 to $2x$. The equations become:

$$2x + 1 = y$$
$$2y - 100 = x - 5.$$

Each set of simultaneous equations is readily solved. The first set gives x a minus value, which is ruled out. The second set gives the correct values.

9. Regardless of how much wine is in one beaker and how much water is in the other, and regardless of how much liquid is transferred back and forth at each step (provided it is not all of the liquid in one beaker), it is impossible to reach a point at which the percentage of wine in each mixture is the same. This can be shown by a simple inductive argument. If beaker A contains a higher concentration of wine than beaker B, then a transfer from A to B will leave A with the higher concentration. Similarly a transfer from B to A—from a weaker to a stronger mixture—is sure to leave B weaker. Since every transfer is one of these two cases, it follows that beaker A must always contain a mixture with a higher percentage of wine than B. The only way to equalize the concentrations is by pouring all of one beaker into the other.

QUESTIONS AND EXERCISES

1. What is "fun"? Why do people want to "have" it?

2. Some sour intellectual once observed that when people say, "Wow! That was really fun!" they are always referring to some activity that used their minds not at all. Is the observation accurate, or only partly so? What is the explanation for whatever amount of truth it has in it?

3. Why would anyone want to make up a recreational mathematics problem? (There is no money in it, nor is there fame, since writers on recreational mathematics have always stolen from each ever without giving credit, from ancient times until today.)

4. "Doing recreational mathematics problems is as sterile an exercise as doing a crossword puzzle, or a word search. When you finish one of those, all you do is throw it away. Recreational mathematics problems are just the same." True, false, or in between? If true, is it to the discredit of recreational mathematics?

5. While recreational mathematics problems no longer appear in newspapers and magazines as frequently as they used to, the number of recreational mathematics books published has increased over the past few decades. What is the reason for that? What do you think lies in the future for recreational mathematics: growth, or decline?

6. Can recreational mathematics problems be used as an aid in mathematics education? Give at least one reason why, and at least one why not. If so, *should* they be? If not, why not?